THE CARBON FIX

Given the growing urgency to develop global responses to a changing climate, *The Carbon Fix* examines the social and equity dimensions of putting the world's forests—and, necessarily, the rural people who manage and depend on them—at the center of climate policy efforts such as REDD+, intended to slow global warming. The book assesses the implications of international policy approaches that focus on *forests as carbon* and especially, *forest carbon offsets*, for rights, justice, and climate governance.

Contributions from leading anthropologists and geographers analyze a growing trend towards market principles and financialization of nature in environmental governance, placing it into conceptual, critical, and historical context. The book then challenges perceptions of forest carbon initiatives through in-depth, field-based case studies assessing projects, policies, and procedures at various scales, from informed consent to international carbon auditing. While providing a mixed assessment of the potential for forest carbon initiatives to balance carbon with social goals, the authors present compelling evidence for the complexities of the carbon offset enterprise, fraught with competing interests and interpretations at multiple scales, and having unanticipated and often deleterious effects on the resources and rights of the world's poorest peoples—especially indigenous and rural peoples.

The Carbon Fix provides nuanced insights into political, economic, and ethical issues associated with climate change policy. Its case approach and fresh perspective are critical to environmental professionals, development planners, and project managers; and to students in upper-level undergraduate and graduate courses in environmental anthropology and geography, environmental and policy studies, international development, and indigenous studies.

Stephanie Paladino is an environmental anthropologist with the Center for Applied Social Research, University of Oklahoma, USA. Her research focuses on how environmental governance strengthens equity and sustainability, most recently in the areas of carbon forestry offset markets, oil spill response, ecologically protected areas, and the Rio Grande basin.

Shirley J. Fiske is Research Professor at the University of Maryland, USA. She is an environmental anthropologist with career experience in the executive and legislative branches of the US government in ocean, climate change, and natural resources management policy and governance.

THE CARBON FIX

Forest Carbon, Social Justice, and Environmental Governance

Edited by Stephanie Paladino
and Shirley J. Fiske

NEW YORK AND LONDON

First published 2017
by Routledge
711 Third Avenue, New York, NY 10017

and by Routledge
2 Park Square, Milton Park, Abingdon, Oxon OX14 4RN

Routledge is an imprint of the Taylor & Francis Group, an informa business

© 2017 Taylor and Francis

The right of the editors to be identified as the authors of the editorial material, and of the contributors for their individual chapters, has been asserted in accordance with sections 77 and 78 of the Copyright, Designs and Patents Act 1988.

All rights reserved. No part of this book may be reprinted or reproduced or utilised in any form or by any electronic, mechanical, or other means, now known or hereafter invented, including photocopying and recording, or in any information storage or retrieval system, without permission in writing from the publishers.

Trademark notice: Product or corporate names may be trademarks or registered trademarks, and are used only for identification and explanation without intent to infringe.

British Library Cataloguing in Publication Data
A catalogue record for this book is available from the British Library

Library of Congress Cataloging in Publication Data
Names: Paladino, Stephanie, editor. | Fiske, Shirley J., editor.
Title: The carbon fix : forest carbon, social justice, and environmental governance / edited by Stephanie Paladino and Shirley J. Fiske.
Description: Abingdon, Oxon ; New York, NY : Routledge, 2017. | Includes bibliographical references and index.
Identifiers: LCCN 2016020591 | ISBN 9781611323320 (hardback : alk. paper) | ISBN 9781611323337 (pbk. : alk. paper)
Subjects: LCSH: Carbon sequestration. | Climate change mitigation. | Climatic changes—Social aspects. | Climatic changes—Political aspects.
Classification: LCC SD387.C37 C36735 2017 | DDC 632/.1—dc23
LC record available at https://lccn.loc.gov/2016020591

ISBN: 978-1-61132-332-0 (hbk)
ISBN: 978-1-61132-333-7 (pbk)
ISBN: 978-1-315-47401-4 (ebk)

Typeset in Bembo
by Apex CoVantage, LLC

CONTENTS

List of Figures *viii*
List of Tables *ix*
Acknowledgements *x*
Contributors *xiii*
Foreword: The Carbon Offsetting Dilemma *xix*
Esteve Corbera

List of Acronyms *xxii*

    Introduction: Carbon Offset Markets and Social
    Equity: Trading in Forests to Save the Planet   1
    Shirley J. Fiske and Stephanie Paladino

SECTION I
Framing the Carbon Regime in the Context of Global Trends   23

1   A Genealogy of Exchangeable Nature   25
    James Igoe

2   Profits and Promises: Can Carbon Trading Save Forests
    and Aid Development?   37
    Kathleen McAfee

3 Forest Carbon Sinks Prior to REDD: A Brief History
   of Their Role in the Clean Development Mechanism   60
   María Gutiérrez

4 Justice and Equity in Carbon Offset Governance: Debates
   and Dilemmas   74
   Mary Finley-Brook

SECTION II
Accounting and Accountability   89

5 The Limitations of International Auditing: The Case
   of the Norway-Guyana REDD+ Agreement   91
   Janette Bulkan

6 Corporate Carbon Footprinting as Techno-political Practice   107
   Ingmar Lippert

7 Regulating Fairness in the Design of California's
   Cap-and-Trade Market   119
   Patrick Bigger

SECTION III
National and Subnational Framings   135

8 Carbon, Carbon Everywhere: How Climate Change
   Is Transforming Conservation in Costa Rica   137
   Robert Fletcher

9 Customary Landowners, Logging Companies and
   Conservationists in a Decentralized State: The Case
   of REDD+ and PES in Papua New Guinea   151
   David Lipset and Bridget Henning

10 Interrogating Public Debates over Jurisdictional REDD+
   in California's Global Warming Solutions Act: Implications
   for Social Equity   171
   Libby Blanchard and Bhaskar Vira

11 Doing REDD+ Work in Vietnam: Will the New Carbon
   Focus Bring Equity to Forest Management?   184
   Pamela McElwee

SECTION IV
REDD, Rights, and Equity — 201

12 Renegotiating REDD: Beyond Social Safeguards
    to Social Contracts — 203
    Michael Brown

13 A Win-Win Scenario? The Prospects for Indigenous Peoples
    in Carbon Sequestration: REDD Projects in Brazil — 220
    Janet Chernela and Laura Zanotti

14 Equity Concerns during REDD+ Planning and Early
    Implementation: A Case from Malawi — 238
    Heather M. Yocum

15 Lessons from Community Forestry for REDD+ Social
    Soundness — 254
    Janis B. Alcorn

SECTION V
Alternative Configurations of Community and Governance — 271

16 Empowering Forest Dependent Communities:
    The Role of REDD+ and PES Projects — 273
    Mark Poffenberger

17 Climate Mitigation Based in Adaptation: El Salvador's
    Restoration of Mangrove Ecosystems, 2011–2013 — 286
    Fiona Wilmot

18 A Critical Reflection on Social Equity in Ugandan
    Carbon Forestry — 302
    Adrian Nel

Index — *321*

FIGURES

9.1	Map of Papua New Guinea indicating locations of case studies further discussed: (1) Kamula Doso Forest Area, (2) April-Salomei Forest Area, (3) Murik Lakes, and (4) Wanang Conservation Area	152
9.2	Two brothers out for a turn in their canoe at low tide, 2012	160
9.3	Wanang villagers use customary exchange to resolve a conflict with BRC, 2012	162
11.1	Map of Vietnam with provinces discussed in chapter highlighted	186
11.2	Poster explaining "Trees Are the Lungs of the Earth" outside a forest ranger station in Lam Dong Province, Vietnam	193
16.1	Map of Khasi Hills project area	275
17.1	Project area, Lower Lempa River and Jiquilisco Bay, El Salvador	289
18.1	Uganda project map	306

TABLES

0.1	Mileposts toward carbon governance	6
4.1	Potential for contradictions in REDD results	81
10.1	Actors and organizations' positions in the REDD+ debate in California	176
18.1	Project types	305
18.2	Project summaries	307

ACKNOWLEDGEMENTS

The Carbon Fix came about in response to the enormous political, scientific, and financial effort being mounted at the global scale, beginning in the mid-2000s, to transform forests into internationally tradable greenhouse gas offsets as an incentive to slow global warming.

Each of the book's co-editors, both anthropologists, had been working in different arenas of climate change-related policy or practice. Starting around 2000, Shirley Fiske was working in Washington, DC, for a US Senator, researching and advising on domestic carbon sequestration and cap-and-trade legislation under consideration by the US Congress. At that time, cap and trade, carbon sequestration, and carbon sinks to mitigate climate change were still relatively new ideas in the US policy arena. While the European Union committed to obligatory carbon control measures, the U.S. continued to skirt compliance approaches and relied on the existence of a voluntary carbon market. Shirley discovered there was little in the way of systematic and documented information on which to judge the social and economic effects of these policy measures.

During this same time frame, Stephanie Paladino was working with El Colegio de la Frontera Sur on an assessment of a carbon agroforestry project in Chiapas, Mexico, that catered mainly to Maya smallholders in Chiapas. The project, Scolel Te', pre-dated the emergence of REDD-style policies by at least a decade, and although it differed significantly from them in its emphasis on farmer-controlled tree plantings, integrated into existing agricultural and social landscapes, the enormous institutional learning of its participants was being relied on at the time to help create newly forming state and national carbon forestry policies. Stephanie found herself amidst compelling claims both in pro and in contra of the social, economic, political, and justice aspects of forest carbon offset markets, but little prior field research on which to evaluate them.

A propitious encounter at the annual American Anthropological Association (AAA) meetings led to the decision by Paladino and Fiske to meld these two threads of experience and perspective to address the knowledge gap. Seeking in-depth, grounded anthropological analysis that could inform critical policy decisions, they convened a special panel, "Carbon Capture and Environmental Services Projects: Who and What Do They Serve?" at the 2009 Society for Applied Anthropology (SfAA) annual meetings. The presentations given by anthropologists and geographers, including academics and NGO practitioners, drew on cutting edge research and experience with carbon forestry projects, and made crystal clear just how complicated the forest carbon offset enterprise is and how delicate and risky it is in justice terms. That discussion, as well as exchanges with countless professionals and participants in the field, reinforced our sense of the glaring lack of systematic, nuanced, socially-oriented analysis, and led to the conception and production of this book.

We acknowledge and thank the participants of the original session for validating the critical need for deeper analysis and enriching our knowledge of how forest carbon offsets work for and against social equity. Wendy-Lin Bartels, Stephan Schwartzman, Maria Gutierrez, Constance Neely, and Elizabeth Shapiro were participants, along with Fiske and Paladino, in the seminal 2009 SfAA panel that inspired this collection. Maron Greenleaf, Shaozeng Zhang, and Pamela McElwee updated the discussion in an AAA roundtable with Paladino and Fiske in 2013. Gabriel Thoumi and John Lewis each provided important perspectives from the private sector on carbon forestry, social and other risks, and finance. Carol Colfer, Janis Alcorn, Maria Gutierrez, and Sarah Strauss provided important feedback on the Introduction to the book. Jeanne Simonelli used her prodigious, editorial slash-and-burn skills to help us carve the Introduction into manageable form. Anonymous reviewers did the yeoman task of providing independent reviews for the contributed chapters. Jennifer Collier, Jack Meinhardt, and Mitch Allen, editors and publisher, respectively, at Left Coast Press, which accepted the book before the company became part of Routledge, provided consistently positive and patient support to our enterprise. Through the many turns during the production of the book, and as we came down to the wire, we are grateful for the insightful participation of Esteve Corbera Elizalde, who offered to write a Foreword for the book, and for the constructive views and encouragement of Jesse Ribot. In the final stage of the book we are most grateful for the outstanding help of Christy Miller Hesed, who was a quick study in a topic new to her, and took on the enormous task of helping us create and produce an index.

Fiske would like to acknowledge the bi-partisan nature and collegial assistance of committee and personal office staff in developing critical grounding in carbon policy for the Congress and the public. Particularly valuable was her association over the years with staff of the Senate Committee on Energy and Natural Resources—Dr. Bob Simon, Jonathan Black, and David Brooks—as we developed legislation and held hearings to examine the shape, potential, and effects of proposed cap and trade legislation.

Finally, conception of the book owes a large debt to the individuals and institutions in Mexico that allowed Paladino to observe carbon forestry in action and see up-close its complexity in relation to environmental governance and justice questions. Very special thanks and acknowledgements are due Dr. Lorena Soto Pinto, of ECOSUR, Chiapas, Mexico, for the opportunity to work with her and a larger ECOSUR team on assessing the experience of Scolel Te' with a grant from CONACyT; and to Sotero Quechulpa Montalvo, Elsa Esquivel Bazán, and a larger team of AMBIO personnel and Scolel Te' farmers, for the opportunity to learn from their intimate knowledge of smallholder-oriented, carbon agroforestry challenges and successes.

CONTRIBUTORS

Janis B. Alcorn is Acting Director of The Tenure Facility and the Senior Director of Programs at Rights and Resources Initiative (RRI) in Washington, DC. Previously, she was Deputy Director Social and Environmental Soundness at a USAID project: Forest Carbon, Markets and Communities (FCMC). She is also an Adjunct Professor at the University of Manitoba's Natural Resources Institute. She received her PhD in Botany and Anthropology from the University of Texas at Austin. She is past president of the American Anthropological Association's (AAA) Anthropology and Environment section. She has over 30 years' experience applying her expertise in natural resources governance, ecological resilience, human rights, and civil society engagement in Asia, the South Pacific, Africa, and Latin America. She has published five books and over 144 articles.

Patrick Bigger is Senior Research Associate at the Pentland Centre for Sustainability in Business, Lancaster University. He holds a BA from the University of Arizona and an MA and PhD from the University of Kentucky, all in geography. His doctoral research, supported by the US National Science Foundation, focused on the creation of California's cap-and-trade program as part of the first economy-wide greenhouse gas reduction policy in the United States. His current work examines the constitution and growth of the European climate bonds market, developing a comparative case for analyzing environmental-financial products across regions and types of asset.

Libby Blanchard is a PhD candidate and Gates Scholar at the University of Cambridge, where she studies political ecology, political economy, and climate change policy in the Department of Geography. Prior to her research at Cambridge, she worked in the specialty coffee industry for 6 years, where she raised over $4 million and directed projects for livelihood improvement and conservation programs with coffee producers in 11 countries. These projects were featured in the Clinton Global Initiative, *National Geographic*'s *Wild Chronicles* television series, and the Stanford Social Innovation Review. She holds a BA from Occidental College and an MPhil from Cambridge.

Michael Brown is Director in the Environment & Natural Resources Practice at Chemonics International (mbrown@chemonics.com). His book *Redeeming REDD: Policies, Incentives and Social Feasibility* was published in 2013. He has focused on tropical forest conservation, dryland resource management, governance, and capacity building issues since the 1970s in Africa, Asia, the Caribbean and south Pacific.

Janette Bulkan is Assistant Professor for Indigenous Forestry in the Faculty of Forestry, University of British Columbia, Canada. She is a social anthropologist with degrees from the University of Manitoba (BA), the University of Texas at Austin (MA), and Yale's School of Forestry and Environmental Studies (PhD). She works collaboratively with indigenous and traditional/local communities in the Amazon regions of Guyana, Suriname, and Peru on community forestry, forest certification, tenure systems, control of illegal logging, anti-corruption, and REDD+. She is a member of the Forest Stewardship Council's (FSC) Policy and Standards Committee, which reviews drafts of all new and revised FSC policies, standards, procedures, and other normative FSC documents.

Janet Chernela is Professor of Anthropology at the University of Maryland and former Research Professor of the National Institute of Amazonian Research (INPA) in Brazil. She has worked on matters of traditional environmental knowledge, conservation policy, and indigenous peoples in the Amazon basin of Brazil for 40 years and is author of the 1993 book, *A Sense of Space: The Wanano Indians of the Brazilian Amazon*. Most recently she has been concerned with language and knowledge flow in local-international partnerships. She is a former president of the Society for the Anthropology of Lowland South America (SALSA).

Mary Finley-Brook is Associate Professor of Geography, International Studies, and Environmental Studies at the University of Richmond. She has conducted fieldwork assessing Clean Development Mechanism (CDM) projects in Costa Rica, the Dominican Republic, Nicaragua, and Panama and addressing REDD readiness efforts in Panama and Peru. She has published about climate and energy justice in journals such as *Water Alternatives* and the *Annals of the American Association of Geographers*.

Shirley J. Fiske is an environmental anthropologist with career experience in the executive and legislative branches of the US government in ocean, climate, and natural resources management policy and governance. Most recently she worked as a senior legislative advisor on carbon sequestration and climate change cap-and-trade legislation in the US Senate. She is currently a Research Professor at the University of Maryland and recent chair of a national task force on climate change for the American Anthropological Association (AAA) (http://www.aaanet.org). "Awarded the 2016 Solon T. Kimball Award by the AAA, for contributions to public and applied anthropology."

Robert Fletcher is Associate Professor in the Sociology of Development and Change Group at Wageningen University in the Netherlands. He is the author of *Romancing the Wild: Cultural Dimensions of Ecotourism* (2014) and co-editor of *Nature^{TM} Inc.: Environmental Conservation in the Neoliberal Age* (2014). His research interests include environmental governance, development, globalization, conservation, ecotourism, and climate change.

María Gutiérrez received her PhD in anthropology from the Graduate Center, City University of New York, with a dissertation on the creation of a market for carbon sinks under the Kyoto Protocol. Since 2000 she has worked as a consultant to various international organizations, in particular the UNFCCC secretariat, on matters related to climate change and forestry, land use and adaptation.

Bridget Henning is a postdoctoral research associate at the Illinois Natural History Survey of the University of Illinois. She holds a PhD from the University of Minnesota in Conservation Science. She has conducted research on cultural perspectives of biodiversity conservation, ecosystem services, and conservation planning.

James (Jim) Igoe is Associate Professor in the Anthropology Department at the University of Virginia. His work, broadly construed, concerns the history of nature in expanding world systems. Specifically he has addressed conflicts between national parks and indigenous communities in East Africa and North America, the emergence of neoliberal conservation at the turn of the millennium, and the role of mass-produced images in mediating people's perceptions of, and relationships to, the environment. His current book project, *Spectacle of Nature, Spirit of Capitalism*, sketches the power of nature as abstract image in contemporary consumer capitalism.

Ingmar Lippert is Assistant Professor at IT University of Copenhagen. His research focuses on the everyday lived experiences and practices of green workers, and their successes and struggles to green capitalism. This research is oriented to nuanced and critical ethnographic engagement with the realities of ecological modernization. His work includes studies of carbon accounting, recycling networks, and environmental knowledge management. He has studied and worked on the border of Environmental Sociology and Science and Technology Studies at National University of Singapore, at Lancaster University, the Institute for Advanced Studies of Science, Technology and Society in Graz, and Augsburg University.

David Lipset is Professor of Anthropology at the University of Minnesota. He has conducted long-term fieldwork in Papua New Guinea since 1981. He is the author of two books, *Gregory Bateson: Legacy of a Scientist* (1982) and *Mangrove Man: Dialogics of Culture in the Sepik Estuary* (1997), as well as articles on a variety of topics relating to culture, change, and masculinity in Melanesia. He is the co-editor (with Paul Roscoe) of *Echoes of the Tambaran: Masculinity, History and the Subject in the Work of Donald F. Tuzin* (2011); co-editor (with Richard Handler) of *Vehicles: Cars, Canoes and Other Metaphors of Moral Imagination* (2014); and co-editor (with Eric K. Silverman) of *Mortuary Dialogues: Death Rites and the Reproduction of Moral Community in Pacific Modernities* (2016).

Kathleen McAfee received a PhD in Geography from the University of California, Berkeley, after a career in community and international development. She is a professor at San Francisco State University, teaching about international political economy, conventional and alternative development strategies, global food systems, biotechnology, and environmental policy. She has been a consultant to UN agencies and a participant-observer in international social movements for agroecology and environmental justice. Building on her work on commodification of genetic resources, her recent research concerns "selling nature to save it" through market-based responses to unsustainable growth and climate change: carbon trading, payments for ecosystem services, and REDD+. Her current work also explores limits and alternatives to growth-obsessed capitalism, such as *buen vivir* in Latin America and degrowth in Europe.

Pamela McElwee is Associate Professor of Human Ecology in the School of Environmental and Biological Sciences at Rutgers, the State University of New Jersey. She has conducted fieldwork on biodiversity, forestry, and climate issues in Vietnam since 1996, and holds a PhD in Forestry & Environmental Studies and Anthropology from Yale University.

Adrian Nel was recently appointed as a Senior Lecturer in the Department of Geography at the University of Kwazulu-Natal. Prior to this he held a Visiting Scholar position with the Institute for Development Studies at the University of Sussex, and a research association with the Institute for Development Studies at the National University of Science and Technology (NUST) in his hometown of Bulawayo, Zimbabwe. He continues to research and teach about political ecology themes in eastern and southern Africa.

Stephanie Paladino is an environmental anthropologist working with the Center for Applied Social Research, University of Oklahoma on river basin and climate change studies, and co-editor of *Culture, Agriculture, Food and Environment*. Her research focuses on how environmental governance interacts with equity, rights, and sustainability, in areas related to natural resources, extraction, agriculture, conservation, and disaster. She has conducted in-depth studies of carbon agroforestry in southern Mexico, and is an editor of a forthcoming volume, ExtrACTION: Impacts, Engagements, and Alternative Futures.

Mark Poffenberger is Founder and Executive Director of Community Forestry International and a Fellow at the East-West Center, University of Hawaii. He has written widely on community forestry in South and Southeast Asia, including *Patterns of Change in the Nepal Himalaya* (1980); *Keepers of the Forest* (1990); *Village Voices: Forest Choices* (1995); and *Cambodia's Contested Forest Domain* (2013), as well as numerous articles and monographs. He has guided the design of community REDD+ projects in Cambodia, Indonesia, and India to explore the potential of payment for environmental services (PES) in sustainable forest management.

Bhaskar Vira is Reader in Political Economy in the Department of Geography, University of Cambridge, and Director of the University of Cambridge Conservation Research Institute. His research interests center on the changing political economy of development, especially in India; and on political ecology, focusing on forests, wildlife and land use change, and the social and political context for biodiversity conservation. His work on incentives for natural resource use and management deals with trade-offs and discourses relating to the concept of ecosystem services, and how this overlaps with poverty and human well-being, as well as values for biodiversity conservation.

Fiona Wilmot is a human-environment geographer with degrees in Archaeology and Anthropology (MA, University of Cambridge, 1983) and Geography (PhD, Texas A&M University, 2014). Most recently she has taught regional and human geography at Blinn College in Texas, although previous career experience included time at the Smithsonian Tropical Research Institute in Panama and with the Florida Keys National Marine Sanctuary. Her work encompasses mangrove-dependent communities, restoration ecology, social equity in the carbon economy, and climate adaptation in Central America. She has served on the board of the Mangrove Action Project for most of the past decade and is starting a new life as an independent scholar in Mérida, Mexico, working on an environmental history of mangroves.

Heather M. Yocum is an environmental anthropologist and political ecologist. She has conducted research on climate change, REDD+, PES, and land and forest management on public and private land in southern Africa and the United States.

She is currently a postdoctoral research associate with the University of Colorado and the National Oceanic and Atmospheric Association's (NOAA's) Cooperative Institute for Research in Environmental Sciences (CIRES).

Laura Zanotti received her PhD in Anthropology from the University of Washington and joined the faculty at Purdue University in 2009. She is an environmental anthropologist and interdisciplinary social scientist whose research program focuses on partnering with communities to determine how local livelihoods and well-being can be sustained for future generations.

FOREWORD

The Carbon Offsetting Dilemma

I am writing this foreword a few days before the opening of the 21st meeting of the Conference of the Parties (COP-21) to the United Nations Framework Convention on Climate Change (UNFCCC). The world's countries have met annually for more than 20 years and have yet to agree on a binding global treaty to reduce the rising greenhouse gas emissions driving anthropogenic climate change. By the time you read this, however, we will know if this lack of will and long-term vision has been addressed and if a global mitigation goal has been set. More importantly, we will know which types of policies, technologies, and funding mechanisms are likely to be settled on to "walk the talk."

The role to be played by (trans)national carbon markets in climate change mitigation, including the well-established European Union Emissions Trading Scheme and others emerging elsewhere, will probably continue to be a cornerstone of international and national climate policies; but how much demand for carbon forestry activities may be generated through these and other emerging markets—or none at all—remains unclear. The price of carbon offsets, including those from forestry activities, has plummeted in the last few years due to an oversupplied market and uncertainty about future demand, both from regulated schemes like the Kyoto Protocol's Clean Development Mechanism (CDM) and from voluntary exchange platforms. This falling demand has meant that many of the early projects are struggling to continue, reminding us about the fragility of using offsetting activities as a means to support both forest management and rural development in the land use sector of the global South.

In the last few years, we have also witnessed the development of the UNFCCC's framework for Reducing Emissions from Deforestation and Forest Degradation, and the sustainable management of forests and enhancement of forest carbon stocks (REDD+), which has led to the design and implementation of many regional and local activities that have been mostly funded by multilateral or bilateral "aid." Whether these activities will be able to cover a share of their implementation costs by trading carbon offsets in regulated or voluntary carbon markets is unknown today, as is the extent to which these costs will be further supported by multilateral and bilateral aid, as mobilized, for example, through the UNFCCC's Green Climate Fund.

In this context of uncertainty about the future of carbon markets and carbon forestry in particular, Stephanie Paladino and Shirley J. Fiske's edited volume is a landmark book that can help us think critically about the present and future of such activities. Its distinguished contributing authors explore the central debates that have emerged around carbon offsetting and forestry activities in the last decade, including:

1. The role of nature commodification narratives in influencing environmental policy design and project implementation at national and local levels, respectively;
2. How "carbon accounting" may obscure the environmental ineffectiveness of carbon markets and the profits attained by powerful and knowledgeable actors;
3. The role of local environmental and social histories, including tenure regimes and political dynamics, in shaping the design and outcomes of carbon offsetting activities;
4. The tensions between formal institutions and procedures at the international level, for example, benefit-sharing frameworks or safeguards, and their actual implementation at national or local levels;
5. The economic, social and environmental impacts—including the distribution of costs and benefits—that result from parachuting new and externally imposed development and conservation blueprints, such as carbon offsetting, into existing landscapes, territories, and communities.

Each of the book's contributions addresses at least one or more of these five issues, and they all delve, directly or indirectly, into one of the central questions of carbon offsetting research:

> Should climate change mitigation be based on actions that transfer the ultimate responsibility of emission reductions away from the responsible parties, especially if they do so at the risk of further impacting already marginalized or poor communities of the global South?

Before your mind sketches an answer to this question, imagine a forest management project developed for *both* timber commercialization *and* carbon trading purposes. Here, villagers with formal rights to the forest feel generally satisfied with the activities and the revenues derived from the project, but say they need more technical and financial support in order for the project to significantly improve their incomes and contribute to livelihoods. Project staff recognize this need, but say that low levels of donor funding constrain their ability to provide a higher level of support. There are also villagers who don't have formal rights to the forest. They have not participated in the design of the project, but some have been able to take on supporting and monitoring roles in the project, and have benefited in that way from a smaller share of timber and carbon revenues than formal rights holders. Project staff argue that benefit-sharing arrangements within the village are an internal matter, and fall to the community to work out. In the project developer's view, that the project causes no evident harm, such as causing increased inequities in access to forest resources, is sufficient to consider the project a success. At the same time, there are national and international social movement organizations advocating that funders should refrain from supporting the project because, on the one hand, its benefits are not equally distributed throughout the community, and on the other, carbon revenues do not fully accrue to the villagers but to the developer as well. Yet project development does have costs that must be met.

This example reflects, of course, only one possible reality in the design and implementation of carbon forestry activities, and I have deliberately posed a mixed picture of positive and negative social outcomes to puzzle you. I have also done so to highlight two more central questions that, in my view, also permeate this edited collection, as well as carbon offsetting research to date:

> Who is legitimately entitled to say "yes" or "no" to carbon offsetting activities?

Is it the scholars and activists contributing to this book? The project developers? The forest-related communities and individuals involved? The project funders, managers, social movements, and readers of this book?

> Whose values should be brought into the discussions about the desirability and the significance of carbon offsetting in a given context or culture?

Together, the three overarching questions I pose here relate to the three pillars of social justice—recognition, procedure and distribution—and they should be simultaneously taken into account when analyzing the present and future of carbon markets, and of carbon forestry activities in particular. The contributions to this collection do a brilliant job in this regard, and highlight either directly or indirectly if, how, and why carbon markets have embraced (or not) these three pillars of justice. Overall, they equip us with the necessary information to resolve, at least in our own minds, a plausible and well-informed response to each of these questions and, particularly, to the editors' neatly sketched ethical dilemma: should we "trade in trees" to save the planet?

Esteve Corbera
Universitat Autònoma de Barcelona,
November 2015

ACRONYMS

ALM	Agricultural land management
ANR	Assisted natural regeneration
A/R	Afforestation and reforestation
CAR	Cap-and-trade program in the state of California, USA
CBC	Community-based conservation
CBNRM	Community-based natural resource management
CCBA	Climate, Community and Biodiversity Alliance
CDM	Clean Development Mechanism
CF	Community forestry
CFI	Community Forestry International
CIFOR	Center for International Forestry Research
COP	Conference of Parties
CSR	Corporate social responsibility
EU ETS	European Union Emissions Trading System
FAO	Food and Agriculture Organization of the United Nations
FPIC	Free, informed, and prior consent
FSC	Forest Stewardship Council
GHGs	Greenhouse gases (carbon dioxide, methane, nitrous oxide, and sulfur hexafluoride)
ICDP	Integrated conservation and development project
IFM	Independent forest monitoring
JCN	Joint Concept Note
JNR	Jurisdictional and Nested REDD+
KP	Kyoto Protocol
LULUCF	Land use, land use change, and forestry
MRV	Monitoring, reporting, and verification

NAPA	National Adaptation Programme of Action
OECD	Organisation for Economic Co-operation and Development
PES	payment for environmental services
REDD	Reducing Emissions from Deforestation and Forest Degradation
REDD+	Extends REDD by including sustainable forest management, conservation of forests, and enhancement of carbon sinks
tCO_2e	Tonne of carbon dioxide equivalent
UNDRIP	United Nations Declaration on the Rights of Indigenous Peoples
UNEP	United Nations Environment Programme
UNESCO	United Nations Educational, Scientific, and Cultural Organization
UNFCCC	United Nations Framework Convention on Climate Change
UN-REDD	United Nations REDD programme
VCM	Voluntary carbon market
VCS	Verified Carbon Standard
WHO	World Health Organization
WMO	World Meteorological Organization

INTRODUCTION

Carbon Offset Markets and Social Equity: Trading in Forests to Save the Planet

Shirley J. Fiske and Stephanie Paladino

Slowing global climate change has become one of the central goals of environmental policy and governance actions during the past two decades. Among the most visible and controversial manifestations of these actions are efforts to create global markets for carbon offset credits in forests, other vegetation, and soils—also called carbon "sinks." This is being done under the guise of the REDD+ (Reducing Emissions from Deforestation and forest Degradation)[1] program established under the United Nations Framework Convention on Climate Change (UNFCCC), as well as through unregulated markets in which organizations and individuals voluntarily invest in carbon sinks to offset carbon emissions ("voluntary markets").

Broadly concerned with the question of changing paradigms of environmental governance, their social implications, and their impacts on rights, justice, and social equity, *The Carbon Fix* examines carbon offsetting in forests as part of a growing trend of "incorporating market logics into environment and conservation policy" (Corson et al. 2013, 1) that has become globally significant in recent decades, even as it has generated substantial concern for its potential to create new sources of political, economic, and ecological disenfranchisement. Offering both promise and risk, at the center of this controversy is an approach to forest governance that has climate change mitigation as its central goal; is inspired by neoliberalist market principles; and uses these principles, as well as non-market financial incentives, to leverage the political will and funding to manage rural lands for slowing global climate change (Brown and Corbera 2003). There are three elements of this approach that are central to the concerns of this book.

The first element is a singular focus on *carbon* as a central goal of environmental management. Over the past few decades, climate change and greenhouse gas (GHG) reduction have become increasingly dominant in driving environmental policy and priorities. During this time, *carbon* (as a stand-in for carbon dioxide and other GHGs) and its management have become a prime directive of environment and conservation policies and initiatives, with international funding priorities frequently shifting to the new REDD+ concept and away from more diversified portfolios of "traditional" conservation activities, thus pushing other environmental management goals, such as biodiversity conservation, into often competing and secondary roles (Peña 2010; Hagerman et al. 2012).

The second element is a reliance on market principles as a key mechanism of governance. Drawing on growing neoliberalist trends in environmental governance (McAfee 1999; Büscher et al. 2014), carbon markets for forest or other sinks are intended to incentivize carbon-conserving land management practices through processes of carbon commodification, marketization, and financialization.[2] A single element of nature—carbon—is abstracted and quantified into units representing 1 tonne of carbon dioxide equivalent (tCO_2e) each, thus making them mutually exchangeable across transnational schemes of carbon emissions accounting (Bumpus 2011). Once transformed into *carbon credits*, these units gain economic value through a form of environmental enclosure (Gutiérrez 2011; Gutiérrez, this volume) and can be used to represent actions taken to reduce GHG emissions anywhere in the world, whether through investment in technological innovations that reduce or avoid emissions at the source, or through the actual removal of carbon dioxide from the atmosphere by trees and vegetation. Carbon credits can then be traded, bought, or sold in a global marketplace to emitting nations or entities desiring to meet their emission reduction goals, or traded for speculative value. Although there are many kinds of carbon offsetting and trading mechanisms, we focus here primarily on the REDD+ model currently evolving within the UNFCCC—and much under debate—that is designed to allow carbon emitters (typically in the global North) to *offset* their carbon emissions by investing in the conservation of carbon sinks (mostly forests in the global South) (see Corbera et al. 2010).

The third element of concern driving the creation of this book is that the *targets* of this intended carbon market governance are the very lands that people depend on—especially indigenous and rural peoples. Carbon sink markets assign a new category of economic value to the forests and lands where rural and indigenous peoples live, draw sustenance, perform sacred practices, and reproduce identity, and *re-value* them in the narrow terms of tonnes of carbon emission reductions produced (Lansing 2012; Leach and Scoones 2015a). This process allows new constellations of public and private actors—with their own sets of interests, and materially, culturally, and politically detached from the forests involved and located anywhere in the world—to have considerable financial and political power in the governance of these lands.

While there is little argument about the merit of keeping and restoring forests worldwide, it is the ways in which the UNFCCC REDD+ model and voluntary forest carbon offset projects are being developed and manifested as forms of environmental governance that are the source of controversy and the impetus behind this book. In its most positive light, REDD+ and carbon forestry could be a triple-win solution: funds and development assistance could be redistributed from global North to South in ways that slow climate change, support forest ecosystems, and strengthen rural populations via carbon income, land tenure reform, or other social and ecological "co-benefits"[3] (Angelsen and McNeill 2012). Initially, forest carbon offsetting was framed as a relatively inexpensive and fast way to effect emissions reduction (Angelsen and McNeill 2012), compared to processes of technological and economic change. Critics, on the other hand, argue that it is another "solution" imposed from above, with a narrow focus on carbon that devalues and subverts all other values of landscapes, and that it will make lands an object of speculation for their carbon storage value, subject to green grabbing, displacement of forest-dependent peoples (Fairhead et al. 2012; Leach and Scoones 2015b), and recentralization of forest governance (Ribot and Larson 2012). Thus it would allow the North off the hook by buying rights to emit rather than instituting real change.[4]

Responses from indigenous and rural peoples and advocacy organizations have been mixed and are evolving. On the one hand, a strong critique has been advanced in events and statements such as the 2009 Belém Letter and the Cochabamba Accord of 2010. At the UNFCCC COP-20 meetings in 2014,[5] mobilizations continued in Lima, Peru, with an estimated 15,000 indigenous peoples, peasants, and climate change activists protesting the effects of climate change and speaking out strongly against REDD+. At the same time, many indigenous groups and organizations have come out with their own proposals for "indigenous REDD+," to be implemented under their own criteria and through alternative governance mechanisms, often working closely with non-profit organizations to seek independent means of financing forest preservation (Tebtebba Foundation 2013).[6]

Consequently, central to the chapters that follow are questions about how the construction and implementation of carbon sink markets affect the relative distribution of communities' rights, roles, risks, and benefits in relation to carbon-targeted rural and forest lands. The book's first section puts the development of forest carbon offsetting, REDD+, and associated concerns about governance and equity into theoretical, cultural, and historical context. The chapters that follow are dedicated to case studies of forest carbon projects from across the globe, where authors examine the consequences for social equity, rights, livelihoods, and full participation in environmental and national governance, going beyond highly polarized debates on whether forest carbon offsetting represents a cost-effective, win-win solution for both North and South, rich and poor, or is yet another intervention that further displaces and marginalizes poor populations. They ask whether policies and practices such as "social safeguards" and "free, prior, and informed consent" can be effective in protecting the interests and rights of indigenous and other forest-dependent peoples, and whether forest communities can make forest carbon trading and financial incentives work for their own objectives and on their own terms. By taking a close look at the details and complexities of governance, accounting, finance, and design, *The Carbon Fix* provides varying assessments of how and whether carbon forestry projects being developed to meet the evolving REDD+ model can meet the desired goals of climate change mitigation, sustainable development, and support to indigenous and rural populations.

It is primarily the experience of REDD+ -oriented projects and policies *in the process of being developed*, and secondarily the experience of forest carbon projects participating in voluntary markets or developed earlier for the CDM, that provide the material that our authors mine here. Many of the forest carbon projects described may not actually be operating in full market conditions, meaning they may have not yet sold carbon credits on an actual market, but have received payments from non-market public or private sector funds for start-up or emission unit production. In fact, within REDD+ negotiations, a mix of market- and non-market incentives for forest conservation is now considered to be essential. Similarly, the processes of institutional reform being demanded at national levels for REDD+ are critical, and in many ways, remain under construction (Corbera, Estada and Brown 2010).

Thus the lessons from many of these studies may not be so much about market-based governance *per se*, as they are about the complexities, conflicts, and trade-offs of moving towards market principles, especially in relation to historically disenfranchised populations.

The rationale for taking a deep look at the diverse and specific realities of carbon forest offsetting is compelling: both market and non-market incentives for forest preservation and enhancement are likely to be included in proposals for future international climate accords. If carbon forestry and agroforestry projects, particularly those tied to international carbon trading, are indeed to become a significant tool for global climate change mitigation, whether within post-Kyoto accords such as the 2015 Paris Agreement, or outside them in voluntary markets, their social and environmental consequences must be better understood. There is too much at stake: the potential gains for slowing climate change and the potential risks to equity and governance are too great. The design of policy and projects, as well as strategies of resistance or change, need the benefit of careful analysis. While the chapters in this volume focus primarily on forest carbon projects, the lessons presented here will have broad applicability to other kinds of carbon-targeted landscapes, as well as to market-based approaches to environmental governance in general.

Antecedents to REDD+

The REDD+ model and voluntary markets in carbon sinks emerged during a period, beginning in the 1970s, of increasing use of neoliberal governance principles by multilateral monetary institutions to liberalize developing countries' economies (Stiglitz 2003); these principles were later extended in multiple and novel ways to environmental governance (Fletcher et al. 2014). Paralleling these policy developments, during the same time period, concerns over the state of the global atmosphere culminated in the first world climate conference (1979) and several treaties to control industrial emissions globally (see Table 0.1). The 1987 Montreal Protocol on ozone-depleting gases and the 1991 Acid Rain Treaty were seen as successful models in reducing harmful gases in the atmosphere (Haas et al. 1993; Gillis 2013) and became models for the fundamental architecture of the Kyoto Protocol, leading ultimately to the use of carbon markets and carbon offsets for environmental governance (see Rayner and Caine 2015; Prins and Rayner 2007).

The Kyoto Protocol, adopted in 1997 within the UNFCCC, provided three market-based "flexibility mechanisms" that participating nations could use to achieve reductions in GHG emissions to agreed-upon levels (an average of 5 percent of 1990 emission levels[7]). Its Joint Implementation mechanism allowed investment in emission reduction projects among developed nations, and its Emission Trading System (ETS) mechanism allowed trade in emissions allowances, again among developed nations. Its Clean Development Mechanism (CDM) was distinctive, however, in that it allowed developed countries to earn emission reduction credits by funding GHG reduction measures in less developed countries.[8] The CDM thus set a precedent for a transnational, market-based framework that allows emitting countries to finance interventions and actions in developing countries in order to *offset* their own emissions. The CDM allowed tree planting in non-forested areas (afforestation and reforestation, or A/R), representing less than 1 percent of its registered projects to date.[9] There have been widespread critiques of the mechanism, focusing on the steep transaction and technical costs to develop projects, and the relative lack of participation and benefits to poorer, rural communities (Boyd et al. 2007; see Finley-Brook, and Nel, this volume).

In parallel to the Kyoto mechanisms, an unregulated voluntary market in carbon emissions reduction and ecosystem services projects has also developed. Project developers and marketers advertise a wide variety of opportunities to invest in carbon sink offsets, including A/R, avoided deforestation, agroforestry, agriculture, and "blue carbon" (marine and coastal sinks, such as mangrove forests), with the volume of transactions in avoided deforestation projects expanding greatly in recent years (Hamrick and Goldstein 2015). Freer to experiment with methodologies, financing arrangements, benefit-sharing, and project governance arrangements, the voluntary carbon market is based on the same neoliberal governance principles of commodifying and marketizing nature that inform the Kyoto Protocol mechanisms, and draws on similar technical and carbon accounting practices. The Kyoto Protocol and voluntary markets exist in dialogue with each other, each responding to methodological and procedural developments in the other, as well as to economic and policy signals generated (Hamrick and Goldstein 2015).

TABLE 0.1 Mileposts toward carbon governance

1972	UN Conference on Human Environment (Stockholm).
1979	The First World Climate Conference (convened by WMO, UNEP, FAO, UNESCO, and WHO).
1987	Montreal Protocol (chlorofluorocarbon and halon cap-and-trade).
1988	Intergovernmental Panel on Climate Change (IPCC) established.
1990	IPCC's first assessment report released (AR1), calling for an international treaty; UN General Assembly begins negotiations on a framework convention; US Global Change Research Act signed into law.
1991	Acid Rain Treaty (US-Canada) uses emissions trading and market to control sulfur dioxide and nitrogen oxides in atmosphere.
1992	UNCED (UN Conference on Environment and Development), Rio de Janeiro, Brazil. UN Framework Convention on Climate Change (UNFCCC); Convention on Biodiversity; and Convention to Combat Desertification adopted.
1994	UNFCCC enters into force.
1995	COP-1 "Berlin Mandate" sets up set negotiation track for Kyoto, including transfer of payments from North to South; IPCC publishes guidelines for national greenhouse gas inventories.[1]
1997	COP-3 Kyoto Protocol adopted.
2001	COP-7 "Marrakesh Accords" set the stage for national ratification of the Kyoto Protocol rules for implementation.
2005	Kyoto Protocol entered into force. EU European Union Emissions Trading System commences.
2005	COP-11 Montreal. Coalition of Rainforest Nations proposes consideration of "reducing emissions from deforestation and degradation" (REDD).
2006	CDM (Clean Development Mechanism) operational.
2007	COP-13 Bali Action Plan agreed to, a negotiation track that includes a commitment to set up and finance REDD+ (the plus sign in REDD refers to conservation, sustainable forest management, and enhancement of forest carbon stocks in developing countries).
2008	Joint Implementation commences.
2009	COP-15 Copenhagen Accord. Agreed to establish a mechanism to mobilize resources from developed countries to finance REDD+ activities.
2010	COP-16 Cancun. REDD+ adopted, along with requirements for national strategies and reporting on how countries are implementing REDD+ safeguards (to avoid negative social and environmental outcomes); World People's Conference on Climate Change and the Rights of Mother Earth ("Cochabamba Accord"), Cochabamba, Bolivia.
2012	COP-18 Doha Amendment to the Kyoto Protocol. Launches a second commitment period, starting on January 1, 2013, until 2020.
2014	UN Secretary General's Climate Summit 2014, New York. Establishes Green Climate Fund across corporate, civil society, and world leaders; IPCC releases AR5, identifies human influence as the "dominant cause" of global warming.
2015	COP-21 Paris Agreement. Agreement reached on new policy structure to reduce threat of climate change, applicable to *all* parties to the KP, through Intended Nationally Determined Contributions (INDCs). Agreement to link international carbon markets and trading broadly across buyers and sellers of emissions units through structures known as internationally transferred mitigation outcomes (ITMOS)—to be developed by 2020.
2016	COP 22 Marrakesh, Morocco.

[1] Review and interpretation of critical milestones toward carbon governance have been provided by Dr. Louis Verchot, Research Director, Forests and Environment, Center for International Forestry Research (CIFOR). He points out that an important but little-known role of the IPCC is to provide guidelines for national greenhouse gas inventories—these are mandates to signatories to the KP. The inventories, first published in 1995, serve as the source of information to policy makers and are the accountability backbone of the UNFCCC. Improvements in the inventory methods over time are critical to the inclusion of forests and forest degradation in the development of REDD+ (Verchot, Louis, 2015, personal communication with editors, November 21.)

Conserving Forests as a Priority: REDD+

The preservation and improvement of standing forests had not been included as an option in the CDM, in large part because of the impermanence of trees and vegetation, as well as difficulties in identifying contributions to reductions or growth in overall emissions ("additionality" and "leakage"). Yet the world's forests were acknowledged as one of the most significant and readily available opportunities for storage of above- and below-ground carbon (IPCC 2007); and more recently, forests were estimated to have removed 30 percent annually of global anthropogenic emissions in the period 1990–2007 (Pan et al. 2011). In addition, developing countries argued that forest conversion to agriculture and development was the greatest source of their own growing carbon emissions, while avoiding deforestation offered their greatest opportunity to participate in carbon offset markets.

In 2005, a coalition of rainforest-containing nations, including Papua New Guinea, Costa Rica, and eight others pressed for investment in standing forests to be included as an option in international climate agreements. In 2007, commitments were made to develop REDD as a climate mitigation option within the UNFCCC, and in 2008, the REDD concept was expanded to include sustainable management of forests, conservation, and the enhancement of degraded forests, with a plus sign added to the acronym to indicate this (REDD+).[10] In 2010, at the annual COP16[11] meetings in Cancun, agreements were reached on a set of principles—including "free, prior, and informed consent," "social safeguards," and reporting on safeguard progress, to be designed and implemented by each nation—that would protect the rights of indigenous peoples and local communities under a REDD+ climate governance mechanism (UNFCCC 2014).

Each nation interested in participating in the REDD+ program is to begin the work of developing the policy and legal frameworks needed to tailor and operationalize REDD+ activities to individual national contexts. REDD+ readiness activities in some form, funded by multilateral and bilateral entities, are being prepared by over 75 tropical forest countries (Akong Minang et al. 2014), even as many details of the mechanism are still being negotiated. These readiness activities are supposed to include, in addition to development of social safeguards, the resolution of outstanding land tenure issues for forest-dependent populations; revision of policies and laws that conflict with social and environmental REDD+ goals, including the explicit protection of indigenous and local communities' rights; and the creation of national carbon baseline, accounting, and administrative capacities. Theoretically, they also involve capacity building for the rural and indigenous peoples most affected by REDD+ implementation.

As the REDD+ concept and program continues to be debated, participating nations continue to work on readiness, and forest carbon offsetting projects continue to be developed on the ground in anticipation of a future REDD+ compliance market. In the absence of formal international commitments to REDD+ and related funding uncertainties, many projects anticipate selling carbon credits in the voluntary markets. Meanwhile, challenges to carbon funding have caused some projects to be abandoned and others to be repackaged as different kinds of conservation/development projects in order to seek other sources of financing (Sills et al. 2014; Nel, this volume).[12]

Governance by Carbon and Markets: Equity and Justice Considerations

We return here to the elements of this emerging environmental governance regime that are at the heart of this book's focus and concern, and signal some of the potential risks to the rights, welfare, livelihoods, and participation in governance of indigenous and forest-dependent peoples in particular.

Markets as a Fix?

A fundamental question is the fitness of using market principles to meet REDD+ and carbon forestry's climate mitigation *and* social/development goals. One school of critique suggests the fallacies of looking to markets to fix the very problems that they have caused (Büscher 2012) as drivers of climate change: carbon-intensive economic growth, and uneven distributions in the use of resources among global North and South, center and periphery, urban and rural. Defining environmental problems and poverty as the results of market failures, we are warned, also implies that the solutions are market fixes, and diverts attention from the fundamental but politically difficult changes needed in underlying structural and power dynamics (Lohman 2014).

Related arguments suggest that market-based governance approaches not only reproduce but also *depend on* the existence of uneven power relationships in order to function, and are therefore antithetical to poverty alleviation and social equity goals (Munden Project 2011; Marino and Ribot 2012; McAfee 2012). From this perspective, carbon forestry in developing countries can only be considered more cost-effective relative to other mitigation options precisely because of the relative poverty of the regions and populations involved, the relatively few alternative economic options, and the underlying income disparities between emitters and offsetters, buyers and producers—otherwise there would be little incentive for private capital investment.

Moreover, as with other commodities, the majority of the carbon rents produced are likely to go to the multiple layers of intermediaries and financial actors, and relatively little to the producers, the people who protect and grow the trees (Munden Project 2011). This may be even more the case with forest carbon, where the commodity being traded is not the trees, nor even the carbon itself, but the carbon certificate for each tonne captured. This carbon derivative is what is valued, bought, sold, and held speculatively for the best returns, while the tree guardian receives a fixed payment based on market conditions at a particular moment.

Moreover, carbon markets are laden with asymmetries of information and power, not just profit, though they are intertwined. Commodification involves conversion of this intangible entity, carbon, into a unit that is measurable, trackable, and globally tradable. This is accomplished through complex technical and administrative procedures that are site-specific and require specialized skills not usually in the toolkit of the rural poor. This tends to decrease local communities' control and put them in continued relations of dependence on local and international elites and funding sources; paying for these procedures also puts a heavy cost burden on the carbon revenues generated, which then diminishes the amount available to producers (Munden Project 2011). However, there are projects and movements designed explicitly to support greater community control over as many aspects of the carbon projects as possible, including technical and administrative aspects (Poffenberger, this volume; Skutsch 2010).

Finally, we are cautioned that to focus justice and equity critiques on markets per se is to risk misjudging the extent to which carbon markets are constructed in particular ways by particular actors, and therefore reflect particular interests. The pioneering payment for environmental service (PES) programs of Mexico and Costa Rica require high levels of structuring by national governments, as well as subsidization by public sector and donor funds (McAfee and Shapiro 2010; Fletcher and Breitling 2012). Similarly, compliance carbon markets, including the Kyoto mechanisms and a possible future REDD+ market, are structured by state negotiations, the selected non-state actors allowed to participate, and the political bargains they strike. These are processes that indigenous, rural, and other vulnerable populations have largely been excluded from in meaningful ways (Bigger, Blanchard and Vira, Bulkan, Gutiérrez, this volume). Whether and on what terms forest-dependent peoples can participate in forest carbon markets is very much affected by how states interpret and act on their varying interests, including in relation to competing industries and sources of national income (Corbera et al. 2010; Lohman 2014).

Carbon versus Social Goals?

Another area of concern is the extent to which the extreme focus on carbon and its commodification competes with, obscures, or excludes other social and environmental goals and relationships surrounding forests. Carbon markets create a commodity—carbon credits—that exist in abstraction from other social or environmental values in the socio-natural landscape. These credits can then be traded, bundled, manipulated, and accumulated as sources of economic value and profit in their own right, anywhere in the world, divorced of any necessary relationship or accountability to the people, places, or landscapes that produce carbon (Sullivan 2013; Fletcher 2014). Much of the indigenous and rural critique of REDD+ is that the enormous institutional and financial resources being dedicated to the carbon in forests do not support the other values and relationships associated with forests that are critical to rural peoples, such as livelihood, biodiversity, and spiritual practices, nor do they adequately address the causes of deforestation in the first place, such as lack of tenure and economic opportunity (Leach and Scoones 2015b, 5).

Poffenberger, Chernela and Zanotti, and Wilmot in this volume, describe cases where public and private sector entities work with local communities to support carbon projects on terms favorable to forest communities; other chapters describe carbon projects that disregard or interfere with local tenure, usufruct, and social and sacred rights within carbon-targeted landscapes.

In addition, it is argued that socially just carbon is neither 'fast' nor 'cheap' (Paladino 2011). Creating processes at both national and project levels that strengthen local peoples' rights—such as developing "social contracts" that recognize rights and responsibilities, clarifying and sanctioning use and tenure rights (Brown 2013, Alcorn, this volume), providing training and capacity building (Chernela and Zanotti, this volume), strategic involvement in decision-making, development and monitoring of social safeguard practices—require substantial financial investment, which the carbon prices generated by markets have thus far not generally supported. The risk is that the economic rationale of carbon markets will drive investments toward projects that maximize the economic efficiency of carbon credit production (e.g., by targeting large areas of forest with one or few owners, such as the state, and therefore minimal transaction costs and political or social risks) over investment in projects that work with many smallholders, and build community capacity and control from the ground up (Tschakert 2007; Jindal et al. 2008). Reviews of PES projects suggest they tend to favor the economic terms of ecosystem service buyers rather than those of ecosystem service providers (Pagiola et al. 2005; McAfee 2012). In response, some projects try to access higher carbon prices through socially responsible certification standards, such as Plan Vivo, CCBS, or Social Carbon; and carbon offsets with sustainability and equity "stories" have become more valuable in the voluntary market (Hamrick and Goldstein 2015). Moreover, staff who work with community-based REDD+ projects suggest that the costs of pro-poor carbon projects need not be exceptionally high, through economies of scale of working collectively with many farmer-participants, and by farmers taking on substantial amounts of the administrative and technical roles (Mark Poffenberger, personal communication, 2015).

Rights to Livelihood, Resources, Land, Carbon, and Participation in Governance

For indigenous and other forest guardian communities, the clarification and defense of different kinds of rights, versus 'interests,' has continued to be thorny: "We are rights holders in these discussions, not just stakeholders."[13]

Protection of local rights to livelihood, land, culturally important resources, and full participation in governance are embodied in several international legal conventions, including the UN Declaration on the Rights of Indigenous Peoples, which UNFCCC nations have agreed should guide REDD+ policies and projects. These rights protections are seen as critical to managing sources of ecological, social, and financial risk to forest carbon projects, but also to delivering the redistributive, social and environmental "co-benefits" that are a part of the REDD+ political bargain. In 2010 and 2011, the UNFCCC nations established common principles for the protection of indigenous and local communities' rights and for reporting on performance on these goals, without specifying how any of this should be accomplished (UN-REDD Programme 2012; McElwee, this volume). Within REDD+, each nation is charged with clarifying tenure and carbon rights issues, and establishing free, prior, and informed consent and "social safeguards", within their countries. Guidelines have continued to evolve on a non-binding basis; they call for comprehensive democratization of all levels of governance, where vulnerable populations participate strategically from project level to policy evaluation (FCMC 2012; UN-REDD Programme 2012), and promote forest community participation in safeguard monitoring (MacFarquhar and Goodman 2015). While agencies funding REDD+ projects often impose their own social and environmental protection standards, within the UNFCCC REDD+ framework there are currently no strong mechanisms to ensure national compliance with the intended social safeguards or social reform.

National progress on rights reform has been slow, complicated by conflicting interests in forest and carbon resources, and challenged by entrenched political and economic patterns that have historically marginalized the very populations whose rights are to be protected, even when nations adopt rights-based guidelines such as those of the UN-REDD program (Cuéllar et al. 2013). Locally based forest management by indigenous and rural communities is recognized as effective at preserving forest ecosystems and, by extension, carbon (Alcorn, this volume; Chhatre and Agrawal 2009). Yet the diverse, complex, overlapping, gendered, and seasonally changing forms of tenure, use, and sacred rights of indigenous and other local people are often not legally recognized in postcolonial states, and frequently contested by elites or other local populations (Sommerville 2011). Even local governance institutions can pose justice issues, as they can be subject to domination by local elites, genders, or age groups. Even where local use or tenure regimes are acknowledged, states make formal claims to much of the forested land in the global South and often lease extraction rights to private businesses. States have historically poor records of protecting local rights and preventing illegal forest uses; have facilitated displacement of local communities for protected areas or extraction leases (Sunderlin et al. 2005; Beymer-Farris and Bassett 2012; Ribot and Larson 2012; McElwee, and Nel, this volume); and may have conflicts of interest in protecting rural populations from "green grabbing" for carbon market purposes.

Determining who has the rights to carbon in the trees (and therefore carbon payments) also continues to be a challenging issue. Rights to carbon revenues and rights to carbon-producing resources (forests) may be linked, but are not coterminous. With their key gatekeeper roles in developing national REDD+ carbon rights and benefit-sharing policy, nations may claim rights to carbon revenues even where customary land tenure is well established and legally recognized (Lipset and Henning, this volume); and the absence of clear national policy can prevent investment in carbon projects controlled directly by local communities (Chernela and Zanotti, and Nel, this volume). A wide variety of carbon revenue-sharing arrangements exist or are contemplated, from states providing forest carbon development rights to private entities for a percentage of the return, to percentages of carbon sales going to collective needs of forest communities, such as local infrastructure.

REDD+ and voluntary market carbon projects have been presented in carbon governance discourse as both ways of strengthening rural livelihoods, through the income streams and "co-benefits" they may generate; and as sources of threat, where people or their livelihood activities are restricted or displaced in order to favor land management for maximum carbon credit production. On the first count, strengthening livelihoods, the record in this volume leaves much to be desired: especially for indigenous and rural populations, livelihoods imply an ability to manage a range of activities for household welfare, and an income stream is but one consideration. Low carbon prices on the international market and the costs of carbon credit production, accounting, and marketing mean that the income stream generated has not been enough to become a source of livelihood in itself, and typically becomes a supplement or bonus to existing activities (Paladino 2011; Sills et al. 2014; Arhin and Atela 2015). The sustainable livelihood objectives of carbon projects are often met through investments in community projects (such as public infrastructure or schools, environmental restoration projects, or in local training), but thus far are usually funded by non-market sources.

On the second count, as sources of threat, the record is mixed. Displacements of people and of livelihoods (including access to resources) are reported, either as a direct result of carbon project development or where carbon financing has become the intended funding source for forest management activities in state-protected areas, as in much of Africa (Leach and Scoones 2015; Nel, this volume). However, there are also projects that integrate carbon activities into peoples' existing environmental practices and livelihoods with little negative impact on them, if also generating little substantial change in livelihood status (Poffenberger; and Wilmot, this volume). Agroforestry is a model that shows particular promise for incorporating carbon forestry into rural landscapes on beneficial terms (Paladino 2011); there are also projects aiming explicitly for livelihood enhancement.[14]

Organization of the Book and Chapter Summaries

Section I: Framing the Carbon Regime in the Context of Global Trends

Section I puts the emergence of the carbon governance regime in context historically, philosophically, and globally, critically exploring the paradigms, value bases, and theoretical assumptions on which the new carbon regime is based.

In Chapter 1, James Igoe challenges the landscape of readers' perceptions about carbon offsetting and carbon markets. He lays bare the "contours and parameters of our cultural models of reality" by exploring the evolution of human-nature imaginaries into what Igoe calls "exchangeable nature." One of the long-standing corollaries of capitalism is that profit-seeking and price will determine the fate of nature; but the author adeptly argues that, with the emergence of exchangeable nature, carbon trading will invert this truism by putting an exchange value on a social-ecological good such as reductions of atmospheric carbon. Igoe raises elemental questions about the limitations in our epistemologies, a particularly critical consideration global carbon governance, as worldviews clash in the negotiations about how the world's climate and forests will be governed.

Next, Kathleen McAfee critiques the paradigm behind the green economy, environmental economics, and carbon markets on evidentiary and logical levels. In Chapter 2, McAfee presents the case that the green economy will not be able to reach its two main goals via carbon markets, namely (1) social equity/poverty alleviation and (2) carbon mitigation and sustainable development. McAfee shows how the assumptions inherent in environmental economics define environmental costs (global warming and the escalation of carbon dioxide) as a form of capital, thereby "bringing nature into 'the market.'" In this way, the market objectifies and commoditizes carbon, isolating it from the global economy, consumption, and other forces that drive carbon emissions. Discussions of the UNFCCC, international financial institutions, and national governments can thereby skirt around issues of equity between North and South or equity impacts on local communities. As McAfee states, "Climate policies that depend on profit making are therefore likely to have highly unequal North-South consequences." She argues conclusively that because carbon markets are *inequitable* and *ineffective* in containing carbon, they will not reach their stated goals of saving forests and aiding development.

María Gutiérrez reminds us that the CDM was the UNFCCC's first official foray into a market for carbon sinks, which preceded and informed REDD and REDD+ in ways that institutionalized North-South biases and inequities. In Chapter 3, Gutiérrez notes that while "sinks in trees" had been lurking as a conservation premise since 1977, it was only in the Kyoto Protocol, where atmospheric issues (climate change and GHGs) were framed in the context of a market, that the concept gained political traction. The new market required the creation of scarcity conditions that would give value to carbon and offset credits, and from the beginning the CDM had built-in inequities, with little to no participation of communities, indigenous peoples, or even of forests. Gutiérrez reminds us we need to look to the roots to understand the outcomes.

In Chapter 4, Mary Finley-Brook takes a head-on, comprehensive aim at justice and equity in carbon offset governance, asking whether it can solve widespread, pressing issues such as income inequality and environmental injustices. She assesses what we have learned from three environmental market governance strategies to date—PES, CDM, and REDD+—and provides valuable parallels from earlier Integrated Conservation and Development Projects (ICDPs). She finds that while we have learned a great deal from earlier efforts, there is a multitude of improvements that need to be made, from the human rights level, to structural and governance changes, to the project level. Finley-Brook addresses whether REDD+ can generate transformative change through capacity building in the REDD readiness phase, and argues that despite transfers of billions of dollars, REDD+ cannot deliver unless there are real, deep-seated institutional reforms and true collaboration in participation.

Section II: Accounting and Accountability

In Section II we delve into the sometimes opaque and byzantine world of accountability, the underlying institutional architecture of market systems used to operationalize carbon offsets so that they are globally accountable and verifiable. In this section we see that carbon markets are far from neutral and can be used to mask conflicting motives and structure carbon trading and markets for the interests of particular actors over others. Here we find important insights into the interest-laden implementation of "objective" carbon accounting and auditing procedures, as well as the design and negotiation of the regulatory and institutional components of carbon markets.

In Chapter 5, Janette Bulkan provides a revealing case from one of the first REDD+ bilateral agreements, between Norway and Guyana, on the limitations of international auditing. In her case study, she highlights the tricky issue of vested national interests (by both nations) and other organizational self-serving interests that can affect the results of carbon audits at national and international levels. She underscores the continued, ongoing inequities upon which the REDD-type structures may be superimposed, and which are likely to exacerbate Amerindians' situations in Guyana and elsewhere.

In Chapter 6 on carbon accounting, Ingmar Lippert shifts the view to the private sector to examine how the "the social" is embedded in carbon emissions and carbon accounting in a corporate setting. The author distinguishes between the formal reality of carbon accounting and the situated reality as it is elaborated in this case example. He underscores the importance of analyzing how unexamined practices shape outcomes in hidden ways, even as they are effectively treated as neutral, objective, value-free, and politics-free.

Turning to the public sector, important policy choices and compromises around equity are made in the process of negotiating and constructing state-level regulatory agreements for carbon markets. In Chapter 7, Patrick Bigger elucidates the emergence and design of California's cap-and-trade legislation and market. He examines how debates about fairness and equity permeate the development of the carbon market statute in the state of California—the only US state to initiate a compliance market for carbon. Even with ostensibly open and transparent public debate, vulnerable populations are, in the end, virtually left out of the equation on environmental justice and social equity.

Section III: National and Subnational Framings

As a number of authors in this volume make clear, colonial legacies, as well as recent political regimes with highly centralized state control, leave important and complex imprints on forest tenure and governance at the nation-state level. The chapters in this section offer insights into how carbon policy and markets are being framed at national and local levels, or subnational political jurisdictions, and offer insights into how and why disarticulations and disjunctures with social justice issues among scales and levels seem to be embedded within carbon governance and REDD+ objectives and expectations (see Beymer-Farris and Bassett 2012).

At the outset of Section III, Robert Fletcher examines how the focus on climate change and carbon in Costa Rica has affected the way environmental issues are defined and addressed in that nation—long at the forefront of successful biodiversity conservation and noted for its pioneering national PES program. In Chapter 8 he argues that the social costs of carbon control are high: climate change has revived interest in building hydroelectric dams, with all their attendant social and environmental consequences; income inequality has increased as carbon payments go predominantly to banks, investment firms, corporations, and agribusinesses rather than small landholders. Fletcher cautions that by internalizing the externalities (both social and economic), neoliberal, market-based governance in turn creates more externalities.

David Lipset and Bridget Henning reflect on the emergence of carbon offset agreements and national policies in Papua New Guinea (PNG), a highly *decen*tralized state with a large tropical forest base and unique legal recognition of customary land tenure—and one of the early proponents for REDD+. Through two contrasting case examples in Chapter 9, the authors highlight how politics and power dynamics operate at the local level, including the many complications and perceptions, as REDD projects develop in national and local settings and link carbon projects to actual places and people. These multiple perspectives and socio-political power bases interact to create unanticipated outcomes for participation—at both levels—in REDD+. The authors conclude that "strides are indeed being made" with national strategic planning, but that the future of REDD+ in PNG is uncertain at best.

In Chapter 10, Libby Blanchard and Bhaskar Vira look to the subnational scale of Jurisdictional and Nested REDD+ (JNR) programs for insight into how social equity is framed in the debate over California's amended cap-and-trade program (post-AB32 passage). The initiative seeks to forge a transnational market for offsets among subnational jurisdictions with the states of Quebec (Canada), Acre (Brazil), and Chiapas (Mexico). Using the methodology of "environmental narrative analysis," they focus on questions of representation and legitimacy in the debate; the extent to which communities that will be most affected by the proposed policies have a voice in the debate; and who has the privilege of having a voice in the debate that will affect peoples' livelihoods in distant forests.

Pamela McElwee asks critical questions about REDD readiness in Chapter 11; for example, "can REDD+ motivate more participatory, livelihood-positive benefits for marginalized forest peoples?" Given the highly centralized state control of 99 percent of the forested areas in Vietnam, McElwee underscores the importance of national forest policies and their impact on local communities and tenure rights. The author finds that real participation (not just consultation) at the local level is rarely achieved through national REDD+ readiness activities in Vietnam. She focuses on three areas where REDD+ processes can make a difference in the lives of forest dwellers in Vietnam: participatory processes in forest management; support for livelihoods; and the development of social safeguards.

Section IV: REDD, Rights and Equity

Section IV brings together analyses of REDD+, integrating the perspective of on-the-ground projects with larger questions of how current configurations of REDD+ shape social equity and sustainability at various scales of practice and the prospects of "fixing" them.

Michael Brown contends that REDD+ is no "further along the pathway to sustainability than biodiversity conservation or development programming" and is not likely to be successful as it is currently designed. In Chapter 12 he argues persuasively that there is no "credible social contract that convincingly binds stakeholders together in REDD," and that such contracts are essential from a rights and social justice perspective—but also because they are efficient in building markets and evaluating effectiveness. Brown concludes with a specific challenge to social sciences *and* to investors and financers of REDD projects: to underwrite the development of new and improved social contracts that will provide viable incentives to carbon producers in forests and rangelands across the globe.

In Chapter 13, Janet Chernela and Laura Zanotti pose the question of whether a win-win solution is possible with REDD+. They examine three projects involving indigenous peoples in the Amazon—the Suruí, the Cinta Larga, and the Xingu Indigenous Lands Ecosystem Services Project—in various states of implementation, with a range of configurations in project design and sponsorship. They provide an extremely valuable inspection of the national context, structures, and nuances that produce different project outcomes. The authors argue that "an inherent trade-off exists between land set aside for carbon trading and land to be used for other purposes," and conclude that win-win solutions for REDD are difficult if not impossible due to trade-offs that are built into to the architecture of REDD+ programs.

Heather M. Yocum reports on two cases in Malawi, both in the voluntary carbon sector, where REDD-related projects are located in and around protected areas. In Chapter 14 she shows that specific historical, social, and ecological contexts of the projects need to be taken into consideration to achieve equitable distributions of risks and benefits. The author makes the point, in parallel with others in this volume, that the entire life cycle of a project must be assessed for impacts on environmental justice and social equity issues *prior to* validation. She highlights the disjuncture between what is considered to be adequate social safeguard protections and participation in project planning, and what is needed on the ground.

Janis B. Alcorn identifies multiple lessons for REDD+ programs from 30 years of global experience with community forestry. We are reminded that the social soundness practices being promoted will have limited impact unless attention is also paid to reforming the key overarching policies that set the rules of the REDD+ program, "including those that legitimize or undermine community tenure rights and governance systems." In Chapter 14 Alcorn underscores the admonition that the most ominous outcome of the REDD+ programs and systems will likely be the increasing temptation of nations to consolidate sovereign rights over forest areas and dispossess forest communities of their lands and resources.

Section V: Alternative Configurations of Community and Governance

In Section V, we look into alternative configurations of state political regimes, customary tenure systems, community agency, and equity outcomes. Contributors highlight a diversity of ways in which carbon projects develop, in concert with nongovernmental organizations, national and ministerial level actions, and in voluntary and compliance markets—and their respective implications for equity issues.

In the first contribution, Mark Poffenberger examines the Khasi Hills project in India, one of the first REDD+ initiatives in Asia to be developed by indigenous tribal governments on communal and clan land, in a national context where both community forestry and tenure are recognized by the state. He makes the point in Chapter 16 that outside intervention is seen by the Khasi Hills communities primarily as a mechanism to fulfill their immediate environmental issues and needs rather than a response to climate change policies per se. The project is selling carbon credits on the voluntary market. Poffenberger identifies

10 critical factors that both challenged and enhanced the Khasi Hills project—factors that are likely to have widespread application across many project types and geographic areas.

In Chapter 17, Fiona Wilmot illustrates how El Salvador's national political rhetoric and philosophy have had determinative effects on the tenor of environmental governance and expression of PES and REDD—in this case, where the left-leaning, pro-social justice agenda provided a supportive basis for local organizing efforts for environmental restoration. A number of enabling events and high-level individuals, coupled with activism and agency of a key community-generated organization (along with state and non-state actors) culminated in the El Llorón mangrove restoration project in the Lower Lempa region.

Adrian Nel uses a comparative approach to critically analyze social equity in nine carbon forest projects in Uganda in Chapter 18. He notes the contemporary change in forest governance from land-based or territory-based management to neoliberal "flow-based" governance (Sikor et al. 2003) and "deterritorialization." Importantly, he notes that conservation policy and practice is turning to market-based interventions "to reconcile the growing conflicts between environmental conservation and rural livelihood needs." He presents an indictment across the nine projects on three equity concerns: asymmetrical benefits, expulsion and marginalization, and false promises. The author adroitly summarizes that carbon "interventions, underpinned by a logic of market environmentalism, seem to be falling short in their equity commitments."

In their recent volume, *The New Carbon Economy*, Newell, Boykoff and Boyd contend that we have entered "an historically unparalleled experiment in marketised environmental governance" (2012, 3). In accord, we focus in this volume on the intersection of the new carbon governance regime and its derivatives—the carbon market and forest carbon offset projects—with people, their livelihoods, forests, and intermediary institutions that deal with them on the ground in case studies from villages, forest settlements, and agro-forestry communities across the globe. At time of writing, negotiations subsequent to the COP-21 Paris Agreement are underway. GHG emission reductions will be broadly based across developed and developing nations through nationally determined commitments (Intended Nationally Determined Contributions, or INDCs). Market-based carbon governance remains a key element in the evolution of the Kyoto Protocol and UNFCCC that is unlikely to disappear in the next decade. In fact, carbon trading, or internationally transferred mitigation outcomes (ITMOS), will have an expanded scope in the development of a new mechanism, the Sustainable Development Mechanism (SDM), planned to succeed the CDM and Joint Implementation. Given the central role of carbon offsets and trading in the climate change policy and political scenarios of the future, we hope that this volume exposes the scales of complexities, contexts, frailties, and strengths of the very notion of carbon governance, with social justice and equity considerations front and center.

Notes

1. The plus sign was added to the acronym in 2008, when the REDD program was extended to include activities in "conservation, sustainable management of forests and enhancement of forest carbon stocks in developing countries."
2. "Financialization," in that the carbon credits or certificates generated by carbon offsets offer "a new frontier for speculative investment and the creation of additional 'value'-accumulating financial instruments" (Sullivan 2013, 199); this in turn attracts the scale of private capital needed to fund global forest conservation (Munden Project 2011).
3. http://www.unredd.net/index.php?option=com_unsubjects&view=unsubject&id=6&Itemid=787.
4. http://no-redd.com/.
5. See the REDD Monitor for a helpful articulation of indigenous resistance, dates and actions. http://www.redd-monitor.org.
6. See, for instance, http://wwf.panda.org/what_we_do/where_we_work/amazon/?235956/Amazon-Indigenous-REDD-launched-at-UNFCCC-COP-20 and http://earthinnovation.org/events/cop-20-side-event-nesting-indigenous-lands-jurisdictional-redd/.
7. http://unfccc.int/kyoto_protocol/items/2830.php.
8. http://unfccc.int/kyoto_protocol/mechanisms/items/1673.php.
9. As of October 2015. See "Distribution of Registered Projects by Scope," http://cdm.unfccc.int/Statistics/Public/CDMinsights/index.html.
10. http://unfccc.int/land_use_and_climate_change/redd/items/4547.php.
11. The Conference of the Parties (COP), which meets annually, is the decision-making body of the UNFCCC.
12. For a fuller narrative of how REDD+ began and has evolved, see Angelsen and McNeill (2012).
13. "The International Alliance of Indigenous and Tribal Peoples of the Tropical Forest on Behalf of International Indigenous Peoples. Forum on Climate Change. Submission to Subsidiary Body for Scientific and Technological Advice for Parties (SBSTA) on Item 11 of FCCC/SBSTA/2008/L.23, Draft Conclusions Proposed by Chair February 2009" (p. 2), http://unfccc.int/resource/docs/2009/smsn/ngo/108.pdf.
14. http://www.bioredd.org/docs/Bioredd+_2014.pdf.

References

Akong Minang, Peter, Meine Van Noordwijk, Lalisa A. Duguma, Dieudonne Alemagi, Trong Hoan Do, Florence Bernard, Putra Agung, Valentina Robiglio, Delia Catacutan, Suyanto Suyanto, Angel Armas, Claudia Silva Aguad, Mireille Feudjio, Gamma Galudra, Retno Maryani, Douglas White, and Atiek Widayati. 2014. REDD+ Readiness Progress Across Countries: Time for Reconsideration. *Climate Policy*. doi:10.1080/14693062.2014.905822.

Angelsen, Arild and Desmond McNeill. 2012. The Evolution of REDD+. In *Analysing REDD+; Challenges and Choices*. A. Angelsen, M. Brockhaus, W. Sunderlin, and Louis Verchot, eds., 31–49. Bogor, Indonesia: CIFOR.

Arhin, Albert and Joanes Atela. 2015. Forest carbon projects and policies in Africa, pp. 43–57. In *Carbon Conflicts and Forest Landscapes in Africa*. Leach, Melissa and Ian Scoones, eds., London & NY: Routledge.

Beymer-Farris, Betsy and Thomas Bassett. 2012. The REDD Menace: Resurgent Protectionism in Tanzania's Mangrove Forests. *Global Environmental Change* 22:332–341.

Boyd, Emily, María Gutiérrez, and Manyu Chang. 2007. Small-Scale Forest Carbon Projects: Adapting CDM to Low-Income Communities. *Global Environmental Change* 17(2):250–259.
Brown, Katrina and Esteve Corbera. 2003. Exploring Equity and Sustainable Development in the New Carbon Economy. *Climate Policy* 3(sup1):S41–S56.
Brown, Michael I. 2013. *Redeeming REDD: Policies, Incentives, and Social Feasibility for Avoided Deforestation*. Abingdon: Routledge.
Bumpus, Adam G. 2011. The Matter of Carbon: Understanding the Materiality of tCO$_2$e in Carbon Offsets. *Antipode* 43(3):612–638.
Büscher, Bram. 2012. Payments for Ecosystem Services as Neoliberal Conservation: (Reinterpreting) Evidence from the Maloti-Drakensberg, South Africa. *Conservation and Society* 10(1):29–41.
Büscher, Bram, Wolfram Dressler, and Robert Fletcher, eds. 2014. *Nature™ Inc.: Environmental Conservation in the Neoliberal Age*. Tucson: University of Arizona Press.
Chhatre, Ashwini and Arun Agrawal. 2009. Trade-Offs and Synergies between Carbon Storage and Livelihood Benefits from Forest Commons. *PNAS* 106(42):17667–17670. http://www.pnas.org_cgi_doi_10.1073_pnas.0905308106.
Corbera, Esteve, Manuel Estrada, and Katrina Brown. 2010. Reducing Greenhouse Gas Emissions from Deforestation and Forest Degradation in Developing Countries: Revisiting the Assumptions. *Climatic Change* 100:355–388. doi:10.1007/s10584-009-9773-1.
Corson, Catherine, Kenneth Iain MacDonald, and Benjamin Neimark. 2013. Introduction: Grabbing "Green": Markets, Environmental Governance and the Materialization of Natural Capital. *Human Geography* 6(1):1–15.
Cuéllar, Nelson, Susan Kandel, Andrew Davis, and Fausto Luna. 2013. *Indigenous Peoples and Governance in REDD+ Readiness in Panama*. San Salvador: PRISMA. http://www.prisma.org.sv, accessed June 14, 2013.
Fairhead, James, Melissa Leach, and Ian Scoones. 2012. Green Grabbing: A New Appropriation of Nature. *Journal of Peasant Studies* 39(2):237–261.
Fletcher, Robert and Jan Breitling. 2012. Market Mechanism or Subsidy in Disguise? Governing Payment for Environmental Services in Costa Rica. *Geoforum* 43(3):402–411.
Fletcher, Robert, Wolfram Dressler, and Bram Büscher. 2014. Nature™ Inc. The New Frontiers of Environmental Conservation. In *Nature™ Inc.; Environmental Conservation in the Neoliberal Age*. Bram Büscher, Wolfram Dressler and Robert Fletcher, eds., 3–21. Tucson: University of Arizona Press.
Forest Carbon, Markets and Communities Program (FCMC). 2012. *REDD+ Social Safeguards and Standards Review*. Burlington, VT: Tetra Tech.
Gillis, Justin. 2013. The Montreal Protocol, a Little Treaty That Could. *New York Times*, December 9. http://www.nytimes.com/2013/12/10/science/the-montreal-protocol-a-little-treaty-that-could.html, accessed August 20, 2015.
Gutiérrez, María. 2011. Making Markets Out of Thin Air: A Case of Capital Involution. *Antipode* 43(3):639–661.
Haas, Peter M., Robert O. Keohane, and Marc A. Levy. 1993. *Institutions for the Earth: Sources of Effective International Environmental Protection*. Cambridge, MA: MIT Press.
Hagerman, Shannon, Rebecca Witter, Catherine Corson, Daniel Suarez, Edward M. Maclin, Maggie Bourque, Lisa Campbell. 2012. On the Coattails of Climate? Opportunities and Threats of a Warming Earth for Biodiversity Conservation. *Global Environmental Change*. doi:10.1016/j.gloenvcha.2012.05.006.
Hamrick, Kelley and Allie Goldstein. 2015. *Ahead of the Curve; State of the Voluntary Carbon Markets 2015*. Washington, DC: Forest Trends' Ecosystem Marketplace.

IPCC. 2007. Summary for Policymakers. In *Climate Change 2007: Mitigation*. Contribution of Working Group III to the Fourth Assessment Report of the Intergovernmental Panel on Climate Change. B. Metz, O. R. Davidson, P. R. Bosch, R. Dave, L. A. Meyer, eds. Cambridge and New York: Cambridge University Press.

Jindal, Rohit, Brent Swallow, and John Kerr. 2008. Forestry-Based Carbon Sequestration Projects in Africa: Potential Benefits and Challenges. *Natural Resources Forum* 32:116–130.

Lansing, David M. 2012. Realizing Carbon's Value: Discourse and Calculation in the Production of Carbon Forestry Offsets in Costa Rica. In *The New Carbon Economy: Constitution, Governance and Contestation*. Peter Newell, Max Boykoff, and Emily Boyd, eds., 135–157. West Sussex: Wiley-Blackwell.

Leach, Melissa and Ian Scoones, eds. 2015a. *Carbon Conflicts and Forest Landscapes in Africa*. London and New York: Routledge.

Leach, Melissa and Ian Scoones. 2015b. Political Ecologies of Carbon in Africa. In *Carbon Conflicts and Forest Landscapes in Africa*, 1–42. London and New York: Routledge.

Lohman, Larry. 2014. Performative Equations and Neoliberal Commodification: The Case of Climate. In *Nature™ Inc.; Environmental Conservation in the Neoliberal Age*. Bram Büscher, Wolfram Dressler, and Robert Fletcher, eds. 158–180. Tucson: University of Arizona Press.

MacFarquhar, C. and L. Goodman. 2015. *Demonstrating 'Respect' for the UNFCCC REDD+ Safeguards: The Importance of Community-Collected Information*. Oxford: Global Canopy Programme.

Marino, Elizabeth and Jesse Ribot. 2012. Special Issue Introduction: Adding Insult to Injury: Climate Change and the Inequities of Climate Intervention. *Global Environmental Change* 22(2):323–328.

McAfee, Kathleen. 1999. Selling Nature to Save It? Biodiversity and Green Developmentalism. *Environment and Planning D: Society and Space* 17(1):133–154.

McAfee, Kathleen. 2012. The Contradictory Logic of Global Ecosystem Services Markets. *Development and Change* 43(1):105–131.

McAfee, Kathleen and Elizabeth Shapiro. 2010. Payment for Ecosystem Services in Mexico: Nature, Neoliberalism, Social Movements, and the State. *Annals of the Association of American Geographers* 100(3):579–599.

Munden Project. 2011. *REDD and Forest Carbon: Market-Based Critique and Recommendations*, March 7. http://www.mundenproject.com/forestcarbonreport2.pdf, accessed August 29, 2013.

Newell, Peter, Max Boykoff, and Emily Boyd, eds. 2012. *The New Carbon Economy: Constitution, Governance and Contestation*. Chichester: Wiley-Blackwell.

Pagiola, Stefano, Agustin Arcenas, and Gunars Platais. 2005. Can Payments for Environmental Services Help Reduce Poverty? An Exploration of the Issues and the Evidence to Date from Latin America. *World Development* 33(2):237–253.

Paladino, Stephanie. 2011. Tracking the Fault Lines of Pro-Poor Carbon Forestry. *Culture, Agriculture, Food and Environment* 33(2):117–132.

Pan, Yude, Richard A. Birdsey, Jingyun Fang, Richard Houghton, Pekka E. Kauppi, Werner A. Kurz, Olliver L. Phillips, Anatoly Shvidenko, Simon L. Lewis, Josep G. Canadell, and Phillippe Ciais. 2011. A Large and Persistent Carbon Sink in the World's Forests. *Science*, 333(6045):988–993. doi:10.1126/science.1201609.

Peña, Pablo. 2010. NTFP and REDD at the Fourth World Conservation Congress: What Is In and What Is Not. *Conservation and Society* 8(4):292–297. doi:10.4103/0972-4923.78143.

Prins, Gwyn, and Steve Rayner. 2007. *The Wrong Trousers: Radically Rethinking Climate Policy*. James Martin Institute for Science and Civilization, University of Oxford and the MacKinder Centre for the Study of Long-Wave Events, London School of Economics and Political Science, Oxford, UK.

Ribot, Jesse and Ann Larson. 2012. Reducing REDD Risks: Affirmative Policy on an Uneven Playing Field. *International Journal of the Commons*, 6(2):233–254. http://www.thecommonsjournal.org.

Sikor, T., G. Auld, A. J. Bebbington, T. A. Benjaminsen, B. S. Gentry, C. Hunsberger, A.–M. Izac, M. E. Margulis, T. Plieninger, H. Schroeder, and C. Upton. 2013. Global Land Governance: From Territory to Flow? *Current Opinion in Environmental Sustainability* 5(5):522–527.

Sills, E. O., S. S. Atmadja, C. de Sassi, A. E. Duchelle, D. L. Kweka, I.A.P. Resosudarmo, and W. D. Sunderlin, eds. 2014. *REDD+ on the Ground: A Case Book of Subnational Initiatives Across the Globe*. Bogor, Indonesia: CIFOR.

Skutsch, M., ed. 2010. *Community Forest Monitoring for the Carbon Market*. London: Earthscan.

Sommerville, Matt. 2011. *Land Tenure and REDD+: Risks to Property Rights and Opportunities for Economic Growth*. Property Rights and Resource Governance Briefing Paper No. 11. USAID Issue Brief, August. http://usaidlandtenure.net/usaidltprproducts/issuebriefs/issue-brief-land-tenure-and-redd-risks-toproperty-rights-and-opportunities-for-economicgrowth, accessed September 2, 2011.

Stiglitz, Joseph E. 2003. *Globalization and Its Discontents*. New York: W. W. Norton.

Sullivan, Sian. 2013. Banking Nature? The Spectacular Financialization of Environmental Conservation. *Antipode* 45(1):198–217.

Sunderlin, W., A. Angelsen, B. Belcher, P. Burgers, R. Nasi, L. Santoso, and S. Wunder. 2005. Livelihoods, Forests, and Conservation in Developing Countries: An Overview. *World Development* 33(9):1381–1402.

Tebtebba Foundation. 2013. *Indigenous Peoples, Forests & REDD Plus: Sustaining & Enhancing Forests through Traditional Resource Management*, Vol. 2. Baguio City, Philippines: Tebtebba Foundation. http://www.tebtebba.org.

Tschakert, Petra. 2007. Environmental Services and Poverty Reduction: Options for Smallholders in the Sahel. *Agricultural Systems* 94:75–86.

UNFCCC. 2014. *Key Decisions Relevant for Reducing Emissions from Deforestation and Forest Degradation in Developing Countries*. http://unfccc.int/land_use_and_climate_change/lulucf/items/6917.php, accessed December 4, 2015 (REDD+).

UN-REDD Programme. 2012. *UN-REDD Programme Social and Environmental Principles and Criteria*. UNREDD/PB8/2012/V/1. http://www.unredd.net.

SECTION I
Framing the Carbon Regime in the Context of Global Trends

1

A GENEALOGY OF EXCHANGEABLE NATURE

James Igoe

Introduction

Carbon offsetting, and related modes of environmental fixing, depend fundamentally on something so commonplace that we rarely give it a second thought: money. In modern culture money is both our blessing and our curse. Money makes a world of qualitatively incomparable objects and services quantitatively comparable. It gives everything an exchange value or a price. Exchange value makes modern standards of living possible, and so money is a blessing. It is in this exact same sense that money is a curse. Far-flung suburbs are connected by superhighways full of cars, and cities are connected by flight paths full of jumbo jets, supplied by supermarkets stocked with products of industrial farms and luxuries from all over the world. Such arrangements depend on exchange value, which is therefore implicated in our current carbon predicament. To quote geographer Neil Smith (1984, 77–78):

> With the development of capitalism at a world scale . . . the relation with nature is before anything else an exchange value relation . . . Capitalist production is accomplished not for the fulfillment of needs in general, but for the fulfillment of one particular need: profit. In search of profit, capital stalks the whole earth. It attaches a price tag to everything that it sees and from then on it is this price tag, which determines the fate of nature.

Smith's first claim is that profit is the driving force of a capitalist society. If it is profitable to bring bananas from Ecuador, lamb from New Zealand, or flowers from Burundi to US supermarkets, then they should be brought. If it is profitable for people to fly continuously, then they should. Smith's next point is that, consequently, price determines the fate of nature. Carbon trading, and a proliferation of similar trading mechanisms, seeks to invert this relationship by channeling exchange value for ecological and social good. The price tag is still imagined as determining the fate of nature, but now that fate will be a positive one.

Though I will critically engage the logic and efficacy of these mechanisms in this chapter, my main intention is to show that they are cultural. In modernity the fact of something being cultural is often greeted as impractical and therefore irrelevant. My position, however, is that nothing could be more practical or relevant. Culture mediates our perceptions of what is real (e.g., what we might call nature), what is possible, (e.g., how nature might be protected and for whom), and what is desirable (e.g., if, and how much, economic growth is a good thing). Our current zeitgeist, in which money is cast as nature's salvation, turns on new models of the possible. By explicitly extending money's exchange value into the realm of nature, this zeitgeist seems to make a clean break with a pervasive cultural notion that nature is priceless and should be protected as such.

One of my central arguments in this chapter, however, is that these seemingly new modes of exchangeable nature are actually rooted in popular desires about nature as a realm of potential transcendence outside the mundane and unpleasant realities of modern life. I will present this argument in two sections: the first deals with explicit and recent formulations of nature for speculation, and the second will trace the roots of these models to an older and culturally resonant nature for contemplation. Together these sections illuminate assumptions, cultural logics, and mutually reinforcing relationships underpinning an exchangeable nature. In so doing I hope to broaden our thinking about what is possible when it comes to human caring for nature.

Recent Historical Emergences of Nature for Speculation

Abstract models of reality are a hallmark of cultural modernity. Geometry, Newtonian physics, and neoclassical economics are all modes of knowledge that turn on such models and deploy them to make modern life possible. Like the modern culture of which they are a part, these models are universalizing. Laws of motion, detachment, and supply and demand seem to apply everywhere, regardless of contextual differences. Of course reality is much more complex and diverse than such a worldview would allow, but this is nevertheless our cultural perception.

At first blush there is nothing particularly exceptional or problematic about this state of affairs in the larger scheme of human culture(s). We humans, like all creatures, apprehend our environment via cognitive models of those environments. What matters for us, and for all creatures, is not so much whether these models accurately represent these realities in all their detail. Indeed they can't, or they wouldn't be useful models. We have to be able to filter for relevant features of our environment, or else we would be paralyzed by its overwhelming complexity. What matters most, therefore, about models of reality is whether or not they work for the purposes for which they are required (Levi-Strauss 1966, 15; Bateson 1977, 315–344).

The problem, as anthropologist Gregory Bateson argued, is that we humans are uniquely capable of making models that are at several removes from the realities they portray. Abstract models can be the basis for new abstract models, which in turn can inform other abstract models, and so on, till original correspondences between model and reality become very difficult to trace (1977, 461; also see Baudrillard 1995, 4). As Bateson noted, this "mistaking the map for the territory" (1977, 455) is particularly worrying relative to the unprecedented ecological destruction humans are unleashing on our planet today. People in modern cultures are connected to environments at multiple scales and locales, meaning that the socio-ecological effects of our activities and relationships are far-reaching and impossible to see clearly in their entirety. We are thus exceedingly dependent on abstract models of reality, with few practical means of verifying them.

This situation has been compounded by recent trends in the global economy that favor abstraction over material reality. Money itself is of course an abstract model of reality, which imagines all goods and services as universally exchangeable. On some level the abstract exchange value of money must relate to the material qualities of what gets exchanged, or what we might call their use value. Since the 1980s, global markets have grown increasingly oriented to exchange values having little discernable connection to use value, such as, for instance, credit, commodity futures, and derivatives, which geographer David Harvey (1991) calls "fictitious capital."

In the housing bubble of the early 2000s, for instance, investors were much more concerned with the future exchange value of houses than with their current use. Indeed many did not invest in houses at all, but in assets based on real estate debt and the interest it would putatively accrue. Eventually, however, it became clear that much of this debt would never be repaid and that these assets were truly "fictitious" (or "toxic" in popular discourse). This revelation figured centrally in the financial meltdown of 2008. Fictitious capital is amenable to rapid economic growth (like a bubble increasing in size), but it is also prone to sudden dissipation (like a bubble bursting).

When the 2008 financial meltdown occurred, I was attending the "world's largest and most important conservation event":[1] the World Conservation Congress in Barcelona, Spain. The tone and aesthetic of the congress "projected growing synergies between growing markets and effective biodiversity conservation" (Igoe, Neves, and Brockington 2010, 487). As part of a group of scholars investigating convergences of capitalism and conservation, I was quick to note that high-profile conference spokespeople were either silent about the collapse or quick to urge attendees not to lose faith in markets at this crucial moment (also see MacDonald 2010).[2]

Early signs of a convergence between global conservation and global capitalism began to appear in the 1990s, but were not widely acknowledged or recognized as such till the early 2000s. One of the earliest commentators, geographer Kathleen McAfee (1999), succinctly described this convergence as "selling nature to save it." Pointing to institutional transformations and policy paradigms that were emergent at the time (e.g., the Convention on Biological Diversity and the "Greening of the World Bank"), McAfee (1999) argues that they were reimagining nature as "a global currency," with ecosystems reimagined as warehouses and service providers (also see Goldman 2005). Anthropologist Mac Chapin (2004), a longtime insider to conservation and development, subsequently alleged that a handful of large NGOs (nongovernmental organizations) were dominating global conservation with conservation models that prescribed large NGOs with large budgets, supported in part by corporate funding and partnerships. In *Nature Unbound* (Brockington, Duffy, and Igoe 2009), we note that the 1990s witnessed an unprecedented growth of protected areas on a global scale, and argue that this growth was driven in large part by the economic boom of this period, and in relation to the kinds of processes and relationships documented by McAfee, Chapin, and others.

The World Conservation Congress of 2008 marked a watershed moment in this history, in which the arrangements I have briefly outlined went from being controversial and potentially shameful to being openly celebrated and promoted. The years that followed have witnessed a scramble for natural capital, which Sian Sullivan (2013a, 200) aptly describes "saving nature to trade it." This has been especially visible in global institutional realignments around TEEB (The Economics of Ecosystems and Biodiversity, http://www.teebweb.org) (MacDonald and Corson 2012), an international initiative created by G8+5 in 2007. In the words of its architect, former Deutsche Bank director Pavan Sukhdev, TEEB is dedicated to reversing "the economic invisibility of nature."[3] TEEB's mission to "make nature's value visible" is enhanced by mechanisms like ARIES (Artificial Intelligence for Ecosystem Services) a web-based program designed to "help users discover, understand, and quantify, environmental assets and factors influencing their values."[4]

These complex arrangements turn on commonsense cultural values, at least within the culture of modernity. People value money tremendously. Therefore if you can show people that something is worth a lot of money, or even better that they can make a lot of money from it, they will be more likely to value and take care of it. These seemingly straightforward propositions turn on a surprising inversion of material and abstract. For according to their logic, nature's material use values seem abstract and inconsequential, while its abstract exchange values seem compellingly material and crucially important. And in the language and logic of this worldview, this exchange value is inherent in a nature that is undervalued to the tune of trillions of dollars, which must be plumbed by teams of ecologists and economists so these exchange values can be brought to light

(Sullivan 2009). These processes depend on elaborate modes of representation including maps, and the quantification of ecosystem services according to elaborate future scenarios.

The relationship of these models to exchange value and use value is trickier and more problematic than their cultural common sense would imply. Somewhere in there is the idea that exchange value will positively influence humans to take better care of nature, but before this exchange value can be realized nature itself must be rendered exchangeable. Theoretically we could accomplish this by documenting nature's myriad use values and putting a price tag on those. This in itself would be an elaborate undertaking, but of course the matter is complicated by the fact that we would like to make nature an underlying asset for exchange values that can be realized by *not* using it (for detail see Büscher 2014). Such speculative ventures depend fundamentally on abstract mechanisms, which seem to have a life and logic of their own. To quote Büscher and Fletcher (2014, 15) with respect to carbon trading:

> Once a particular patch of forest has been certified capable of providing a certain number of carbon credits, these credits are then detached from direct connection with this forest and can be purchased by anyone, anywhere, for the purpose of emission offsets and mitigation. Credits can then be further traded, held as collateral for other investments, packaged with other environmental "products," and so forth. Over time their value becomes abstracted from the [use and exchange] value of the forest parcel from which they were originally derived.

Two elements of this passage are worth highlighting here. First, connections between the exchange value of carbon credits and actual patches of exchange forest become increasingly difficult to substantiate over time. Next, the original value of these credits is not the conserved forests at all, but the environmental harm from which the exchange value of the conserved forests is derived. As environmental harm is abundant these days, it is perhaps no surprise that offsetting and mitigation are prominent among the ecosystem services on offer today, as seen in the rise of biodiversity and species "banks" and related trading mechanisms (Sullivan 2013b, 81–82). Indeed, Fairhead, Leach, and Scoones (2012, 242) suggest that we are living in a global "economy of repair," in which

> it is the repair of damaged nature, and the efforts to price the downside of growth that have brought into being and enhanced the value of . . . offsets of all kinds. The economy of repair has been smuggled in with the rubric of 'sustainability', but its logic is clear, that sustainable use 'here' can be repaired by sustainable practices 'there,' with one nature subordinated to the other. Once this logic of repair is grasped, so a new interplay can be discerned, which is doubly valuing nature: for its use and for its repair.

Sullivan (2013b) meticulously delineates this "logic of repair" in her analysis of offsetting and mitigation services related to nuclear power, from uranium mines inside of parks in Namibia to the proposed storage of radioactive waste inside of parks in England's Lake District. On the basis of this research, she identifies premises and principles of offsetting mechanisms: (1) *residual environmental harm*, which assumes that capitalist development is unavoidable and necessarily causes some environmental harm; (2) *off-site mitigation*, which allows the putatively unavoidable harm to be offset somewhere other than where it occurred; (3) *eco-system metrics*, which quantify eco-system qualities to permit exchangeability; and (4) *additionality*, which affirms that conservation would not have occurred without the offsetting mechanism (Sullivan 2013b, 84–86).

By the logic of these premises and principles it is possible to calculate a net environmental gain, in other words that the Earth is better off as a result of *residual harm* than it otherwise would have been. From a planetary perspective in which life on this planet is bound by an atmosphere and connected by ecological webs, this proposition appears nonsensical. Relative to the conditions and relationships I have described throughout this section, however, it is difficult to gain such a perspective, which in any case seems to be of little immediate relevance or value. Abstract cultural common sense seems much more reassuring: although capitalist development, and thus a certain amount of harm, is inevitable, the power of money can still be used to achieve human prosperity with ecological health. The cultural common sense of this logic is rooted in cultural ideals of a timeless nature, outside modernity, which we can visit to refresh our spirits.

The Deeper Cultural Legacy of Nature for Contemplation

A long-cherished value of modern culture is that the most valuable things are *priceless*, and nature is one of those things. This ideal is exemplified in these words from Sierra Club founder John Muir (1912, 262): "Devotees of raging commercialism seem to have a perfect contempt for nature, and, instead of lifting their eyes to the god of the mountain, lift them to the almighty dollar." Current efforts to make nature exchangeable seem to invert this ideal, and to be sure there is some visible effort in popular culture to convince people that treating nature as a priceless thing is a luxury that we can ill afford (as are many priceless things).[5]

My argument here, however, is the opposite: the creation of exchangeable nature actually depends on the kind of priceless nature that Muir was defending. To begin with, Muir was writing against the wise use movement of the early twentieth century. Today we are operating with abstract models of reality in which *not* using nature will be how we manage to make money from nature. These models posit significant compatibilities between turning one's eyes up to the god of the mountain (i.e., universal nature) and to the almighty dollar (i.e., universal exchange value). Indeed a key point of compatibility is revealed in the metaphor of "turning one's eyes up," which suggests a passive contemplation of an abstract power beyond the mundane knowledge and experience of the persons who are turning up their eyes. Elsewhere I have synthesized key arguments from the rich body of scholarship on historical abstractions of nature in modern culture in relation to more recent formulations of exchangeable nature for speculation (Igoe 2014). The remainder of this section outlines key points of that synthesis.

In the first place, abstract nature revolves around iconic views (in the literal sense of looking), which we denizens of modernity readily recognize (just drive along any scenic byway, stop at any scenic overview, or watch any nature film). These views are best appreciated from a detached distance; we turn our eyes up at them (see especially Williams 1973; Wilson 1991; Neumann 1998; Hughes 2010). Environmental historian William Cronon (1996) argues that seeing (literally) in this way fosters a kind of awe-inspired detachment in relation to a specialized realm of nature that seems to exist beyond the unpleasant conditions of modern culture and capitalist value making. What this view conceals, however, is that modern culture and capitalist value making are the source of this awe-inspiring nature. The creation of conservancies and parks around the world over the past 300 years has accompanied, and to a large extent been financed by, capitalist expansion (Brockington, Duffy, and Igoe 2009). It has also consistently entailed forced removals of people from landscapes designated as nature (Dowie 2009).

This last point is particularly important, since it was by virtue of such forced removals, and enforced separation of newly created nature from contiguous spaces, that parks could appear as "places without people,"[6] and by extension places without history. This feature of parks lends them a universal quality. For early conservationists, in fact, nature protected by parks was "the basis of universal truth, available through direct experience and study" (Tsing 2005, 97). So here we have a quality also definitive of the abstract models that I described in the previous section: a universal concept or rule that seems to apply everywhere regardless of cultural and historical particularities. From this perspective an individual park appears more as "a window onto the universal" than a particular place with a particular history in relation to surrounding places.

Moreover, the possibility of one object (abstract nature) standing for an entire class of other objects (particular nature parks) is a prerequisite for exchangeability. The great Serengeti wildebeest migrations in East Africa, for instance, are often equated with the great bison migrations of nineteenth-century North America, such that the experience of one seems to be exchanged for the experience of the other. The *Lonely Planet Guidebook* features the Serengeti migrations as one of Mother Nature's greatest hits—spectacular events that nature lovers can mix and match by traveling to different parts of the world.[7] Moreover, the possibility of transforming this abstract universal nature into abstract universal exchange value has intensified in relation to tourism's ascendancy as a major sector of the world economy (West and Carrier 2004, 483–484),[8] which is reflected in a global upswing of nature parks (West and Carrier 2004; West, Igoe, and Brockington 2006; Brockington et al. 2008). Abstract modes of exchangeability are thus driving socio-ecological transformations of global significance.

But this is only part of the story, for the exchangeability of traditional modes of nature for contemplation have been made to circulate beyond specific nature parks, and thus even further from the constraints of contextual particularities, via simulated environments and photographic technology. Via simulated environments, exchangeable nature for contemplation is now on exhibit at zoos, aquariums, botanic gardens, museums, and theme parks around the world, and by extension environments like airports, hotels, shopping malls and restaurants. Via coffee table books, glossy magazines, cinema, television, and of course the Internet, people who will never visit nature parks can still experience their views thanks to the power of photographs. So much like currency are these photographs that they can even be deposited in "image banks" (Goldman and Papson 2011, 137), to be withdrawn for the purposes of marketing and advertising, corporate image enhancement, NGO fundraising, and so forth. They can also produce visually compelling renditions of nature that exist nowhere in reality, but are nevertheless familiar, such as the planet Pandora in the 3-D blockbuster of the same name. While this exchangeable nature for contemplation can never be exactly like money, it is a powerful complement to money, and by extension to the kinds of exchangeable nature for speculation that I outlined in the previous section.

For modern consumers, spectacular images of nature appear as compelling visual evidence that their individual purchases, and their lifestyles in general, are connected to positive environmental effects at locations that are usually distant and exotic (from the perspective of the consumer). The push of a virtual button or the swipe of a plastic card appears to initiate a chain of events ending in the protection of a family of Arctic polar bears or an acre of tropical rainforest. Significantly, these arrangements extend the logic of a global economy of repair to individual purchases. Narratives and visual representations accompanying green consumerism typically invoke expert knowledge and vaguely specified interventions that appear to offset or mitigate consumption. On Earth Day 2011, for instance, consumers in New York's Times Square were encouraged to text the word TREE to a specified number, to make a $5 donation to an organization that would use the money to plant five trees in Mexico, Kenya, or the Philippines. A glowing virtual forest "growing" on a giant digital screen, along with the names of those doing the texting, visually embellished these notional exchanges (for further examples see Igoe 2013).

Pervasive images of nature for contemplation are similarly indispensable to mechanisms dedicated to "making nature's value visible," as a first step toward creating exchangeable modes of "nature for speculation." Images of panoramic land and seascapes, endangered animals, and dark-skinned rural people intermix with maps, satellite imagery, flowcharts, and stock trading screens on the websites of ARIES, TEEB, and the Convention on Biological Diversity (for details see Igoe 2014).[9] Photographic images of nature are likewise essential to policy forums, where such frameworks and mechanisms are launched, celebrated, and rolled out. Here these images combine with other spectacular presentations, including circus acrobats and experts talking on giant screens, to project the appearance of consensus and urgency for discovering and trading "natural capital" (Macdonald 2010; MacDonald and Corson 2012).

A widely recognized nature for contemplation thus provides accessible visual reference points for the confusing rationale of exchangeable nature for speculation. While seemingly material and immutable, nature for contemplation is rooted in abstract and seemingly ahistorical views, which lend themselves readily to a surprising diversity of human activities and enterprises.

Conclusion

While at first glance the idea of pricing nature appears as an inversion of deeply held cultural beliefs that nature is priceless and should be protected, I have endeavored to show throughout this chapter that our pervasive cultural vision of nature as a timeless realm beyond modernity is actually a crucial antecedent to contemporary formulations of exchangeable nature. By way of conclusion, therefore, I would like to briefly outline what I see as the main limitations of this culturally pervasive vision of nature.

When we denizens of modernity call something natural, we mean that it exists in a state that some other-than-human entity intended (usually a god or some impersonal process like natural selection). The cultural common sense that humans are innately acquisitive and self-serving animals, for instance, underpins neoclassical economics and is also evident in the idea that nature's exchange values are the key to nature's salvation: that money will motivate people (including corporations) to take better care of nature.

Nature for contemplation turns on visions of a pristine realm, timeless and outside of history. However the authenticity of these visions often, and paradoxically, requires active erasure and concealment of violent histories that, in many contexts, have continued to the present day. Melissa Checker (2009) traces the complex interconnections of the global carbon economy to reveal their human costs and ecological contradictions. Cavanagh and Benjaminsen (2014) follow on to further reveal how spectacles of nature figure in the creation of exchange in carbon trading, while concealing violent dispossessions and the demise of certified forests (cf. Igoe 2010).

It is almost certainly no coincidence that global struggles for indigenous rights and cultural self-determination have lately been turning their attention to carbon trading mechanisms and market environmentalism. These struggles highlight a fundamental divide between cultural modes of relating to nature as an essentially passive and exchangeable object and other modes of relating to nature through animate and sentient relationships in which we are all enmeshed (Kricheff 2012; Sullivan 2014). Such differences even figure in modernist cultural commitments to communities of people that have emerged from intergenerational commitments to place (e.g., slow food and localist/regionalist movements—for further examples see Ehrenfeld 2008, 203–258), which stand in stark contrast to older models of exchangeable nature to be managed by experts in possession of universally applicable right answers. Regular people, looking on as individuals from wherever they happen to be, are now invited to support the efforts of these experts via consumptive choices and charitable donations (see Igoe 2013, 24–25), leaving little scope for collective environmental learning and collective democratic action. "We are critically impoverished as human beings," writes Sian Sullivan (2009, 25), "if the best we can come up with is money as the mediator of our relationships to the non-human world."

A common response to the kind of analysis I have now finished presenting here is often something like, "if solutions like these are so problematic, then what solutions do you propose instead?" As Bram Büscher often points out, however, it took a lot of people 500 years to create the mess that we find ourselves in today. It is unrealistic, therefore, to expect that one person might come up with the solution. What I can contribute as an anthropologist, however, is to illuminate the contours and parameters of our cultural models of reality, to actively remember and remind us that we dealing with maps and not territories. We might call this recognizing the limits of our own epistemologies, or just staying open to other possibilities. Whichever way, we will begin to see that we have discarded a lot of wisdom from our own experiences and cultural values (Ehrenfeld 2008), and that there is much we can learn from the experiences and cultural values of others (Sullivan 2009). Such recognition is crucial to the difficult work that David Ehrenfeld calls "Becoming Good Ancestors," which is undertaken with sincerity but without guarantees.

Notes

1. According to the International Union for the Conservation of Nature, which hosts the Congress, http://www.iucnworldconservationcongress.org/about/ (accessed July 31, 2014).
2. http://green.blogs.nytimes.com/2008/10/08/the-failing-business-of-conservation/?_php=true&_type=blogs&_r=0 (accessed August 1, 2014).
3. Sukhdev's overview of TEEB: http://vimeo.com/58715498 (accessed August 1, 2014).
4. http://www.ariesonline.org/about/intro.html (accessed August 1, 2014).
5. See for instance *Time* magazine's "Ten Ideas That Are Changing the Way We Live," http://content.time.com/time/magazine/article/0,9171,2108014,00.html (accessed August 2, 2014).
6. To borrow a phrase from Dan Brockington from the film of the same title, which documents Maasai evictions from Tanzanian's Serengeti National Park, http://www.anemon.gr/films/film-detail/place-without-people (accessed August 2, 2014).

7. http://www.lonelyplanet.com/travel-tips-and-articles/76694 (accessed August 2, 2014).
8. According to the World Tourism Organization, the global industry now generates over a trillion dollars annually, http://media.unwto.org/press-release/2014-05-13/international-tourism-generates-us-14-trillion-export-earnings (accessed June 24, 2014).
9. http://www.cbd.int/incentives/teeb/, http://www.ariesonline.org/, http://www.teebweb.org/ (all accessed August 3, 2014).

References

Bateson, Gregory. 1977. *Steps to an Ecology of the Mind*. London: Arenson Enterprises.
Baudrillard, Jean. 1995. *Simulation and Simulacra*. Ann Arbor: University of Michigan Press.
Brockington, Dan, Rosaleen Duffy, and Jim Igoe. 2009. *Nature Unbound: Capitalism and the Future of Protected Areas*. London: Routledge.
Büscher, Bram. 2014. Nature on the Move: The Value and Circulation of Liquid Nature and the Emergence of Fictitious Conservation. In *Nature™ Inc.: Environmental Conservation in a Neoliberal Age*, Büscher, Bram, Wolfram Dressler, and Robert Fletcher eds., 183–204. Tucson: University of Arizona Press.
Büscher, Bram and Robert Fletcher. 2014. Accumulation by Conservation. In *New Political Economy*, advance online publication from Taylor and Francis online.
Cavanagh, Connor and Tor Benjaminsen. 2014. Virtual Nature, Violent Accumulation: The Spectacular Failure of Carbon Offsetting at a Ugandan National Park. *Geoforum* 56:55–65.
Chapin, Mac. 2004. A Challenge to Conservationists. *World Watch* 17, 6:17–31.
Checker, Melissa. 2009. Double Jeopardy: Carbon Offsetting and Human Rights Abuses. *Counter Punch*, September 2009. http://www.counterpunch.org/2009/09/09/double-jeopardy-carbon-offsets-and-human-rights-abuses/, accessed August 3, 2014.
Cronon, William. 1996. The Trouble with Wilderness or Getting Back to the Wrong Nature. In *Uncommon Ground: Rethinking the Human Place in Nature*, Cronon, William ed., 69–90. New York: W. W. Norton.
Dowie, Mark. 2009. *Conservation Refugees: The 100-Year Conflict between Global Conservation and Native Peoples*. Cambridge, MA: MIT Press.
Ehrenfeld, David. 2008. *Becoming Good Ancestors: How We Balance Nature, Technology, and Community*. New York: Oxford University Press.
Fairhead, James, Melissa Leach, and Ian Scoones. 2012. Green Grabbing: A New Appropriation of Nature. *Journal of Peasant Studies* 39, 2:237–261.
Goldman, Michael. 2005. *Imperial Nature: The World Bank and Struggles for Social Justice in the Age of Globalization*. New Haven: Yale University Press.
Goldman, Robert and Stephen Papson. 2011. *Landscapes of Capital*. Cambridge: Polity Press.
Harvey, David. 1991. *The Condition of Post-Modernity: An Enquiry into the Origins of Cultural Change*. Malden, MA: Blackwell.
Hughes, David. 2010. *Whiteness in Zimbabwe: Race, Landscape and the Problem of Belonging*. New York: Palgrave Macmillan.
Igoe, James, Katja Neves, and Daniel Brockington. 2010. A Spectacular Eco-Tour around the Historic Bloc: Theorizing the Divergence of Biodiversity Conservation and Capitalist Expansion. *Antipode* 42, 3:486–512.

Igoe, Jim. 2010. The Spectacle of Nature in the Global Economy of Appearances: Anthropological Engagements with the Spectacular Mediations of Transnational Conservation. *Critique of Anthropology* 34, 2:375–397.

Igoe, Jim. 2013. Consume, Connect, Conserve: Consumer Spectacle and the Technical Mediation of Neoliberal Conservation's Aesthetic of Redemption and Repair. *Human Geography* 6, 1:16–28.

Igoe, Jim. 2014. Nature on the Move II: Contemplation Become Speculation. In *Nature*™ *Inc.: Environmental Conservation in a Neoliberal Age*, Büscher, Bram, Wolfram Dressler, and Robert Fletcher, eds., 205–221. Tucson: University of Arizona Press.

Kricheff, Daniel. 2012. Market Environmentalism and the Re-Animation of Nature. *Radical Anthropology* 6:17–25.

Levi-Strauss, Claude. 1966. *The Savage Mind*. Chicago: University of Chicago Press.

MacDonald, Kenneth. 2010. Business, Biodiversity, and the New "Fields" of Conservation: The World Conservation Congress and the Renegotiation of Organizational Order. *Conservation and Society* 8, 4:256–275.

MacDonald, Kenneth and Catherine Corson. 2012. "TEEB Begins Now": A Virtual Moment in the Production of Natural Capital. *Development and Change* 43, 1:159–184.

McAfee, Kathleen. 1999. Selling Nature to Save It? Biodiversity and Green Developmentalism. *Environment and Planning D* 17, 2:133–154.

Muir, John. 1912. *The Yosemite*. New York: Century Books.

Neumann, Roderick. 1998. *Imposing Wilderness: Struggles Over Livelihoods and Nature Preservation in Africa*. Berkeley: University of California Press.

Smith, Neil. 1984. *Uneven Development: Nature, Capital, and the Production of Space*. Athens: University of Georgia Press.

Sullivan, Sian. 2009. Green Capitalism and the Cultural Poverty of Constructing Nature as Service Provider. *Radical Anthropology* 3:18–27.

Sullivan, Sian. 2013a. Banking Nature: The Spectacular Financialization of Environmental Conservation. *Antipode* 45, 1:198–217.

Sullivan, Sian. 2013b. After the Green Rush: Biodiversity Offsets, Uranium Power, and the Calculus of Casualty in Greening Growth. *Human Geography* 6, 1:80–101.

Sullivan, Sian. 2014. Nature on the Move III: (Re)countenancing and Animate Nature. In *Nature*™ *Inc.: Environmental Conservation in a Neoliberal Age*, Büscher, Bram, Wolfram Dressler, and Robert Fletcher eds., 222–245. Tucson: University of Arizona Press.

Tsing, Anna. 2005. *Friction: An Ethnography of Global Connection*. Princeton: Princeton University Press.

West, Paige and James Carrier. 2004. Eco-Tourism and Authenticity: Getting Away from It All? *Current Anthropology* 45, 4:483–498.

West, Paige, Jim Igoe, and Dan Brockington. 2006. Parks and People: The Social Impacts of Protected Areas. *Annual Review of Anthropology* 35:251–277.

Williams, Raymond. 1973. *The Country and the City*. New York: Oxford University Press.

Wilson, Alexander. 1991. *The Culture of Nature: North American Landscapes from Disney to the Exxon Valdez*. New York: Between the Lines.

2
PROFITS AND PROMISES

Can Carbon Trading Save Forests and Aid Development?

Kathleen McAfee

Scientists and policymakers worldwide have come to acknowledge the likelihood of severe, even catastrophic consequences if concentrations of atmospheric carbon dioxide and other greenhouse gasses (GHGs) continue to rise (IPBES/UNEP 2011; TEEB 2010; UNEP 2011; World Bank 2012b).[1] Nevertheless, international policy action to reverse this trajectory has failed almost entirely. Negotiations toward meaningful implementation of the 1994 United Nations Framework Convention on Climate Change (UNFCCC) have yielded meager results, stalemated by disputes over which countries should be held accountable for past and future global warming and what actions they can be required to take.

Carbon markets are currently the most widely endorsed policy response to this climate conundrum. Along with fossil fuel taxation, markets in carbon allowances and offsets are lynchpins of the green economy, the dominant discourse that frames international climate negotiations and, increasingly, the environmental policies of governments. The general idea is to use the mechanisms of capitalism, guided by minimal but judicious regulation, to save globalized capitalism from its most ecologically damaging effects (McAfee 2014). Green economy proponents hope for a multiple-win outcome: slower global warming, investments in low-carbon technology that will stimulate economic growth, and development gains for low-income countries.

The use of so-called market instruments for international environmental aims is not entirely new. Trade in ecosystem services—functions of nature that are useful to humans—has been taking place for more than a decade under the rubric of the Clean Development Mechanism (CDM) of the Kyoto Protocol on climate change, in markets for climate-friendly export commodities such as shade-grown coffee, and in the form of payments for environmental services

(PES) programs in the global South. The CDM and PES are now models for REDD+, the only new plan for global action on climate that has been agreed to, at least in principle, since the adoption of the Kyoto Protocol in 1997. REDD+, or Reduced Emissions from Deforestation and forest Degradation, is a set of international initiatives aimed at maintaining and enlarging forests as sinks for the storage of CO_2, primarily in the tropics. Many REDD+ supporters, although not all of them, see carbon markets as the best way to finance forest conservation. Some are convinced that programs such as REDD+ can also channel benefits to the rural poor, especially indigenous peoples.

Green-economy advocates aim to lift climate policy above politics to a more rational, politically neutral realm where economic reasoning and ecological science hold sway. They hope that market-based climate governance can thus transcend the conflicts of interests and perceptions that have stymied global climate policy agreements. In this chapter I argue the contrary: first, that the reasoning marshaled in support of market-based carbon governance is misleading and contradictory; and second, that the results of markets in environmental services in the global South provide further reason to doubt that carbon markets can generate real climate-mitigation and sustainable-development gains. At issue is not whether or not the rural poor and other landowners ought to be supported by policy and compensated materially for their past, present, and future contributions to ecological sustainability. The concerns, rather, are whether carbon trade can indeed slow global warming and whether market-based criteria for allocating carbon trade revenues can foster development that is sustainable and just.

Green Economy, Carbon Markets, and REDD+: Overview of the Controversies

International versions of a green-economy framework have been elaborated by multilateral agencies including the World Bank, the UN Environment Programme (UNEP), and new multilateral initiatives, notably the Intergovernmental Platform on Biodiversity and Ecosystem Services (IPBES), and The Economics of Ecosystems and Biodiversity (TEEB) program (IPBES 2011; TEEB 2010; UNEP 2011; World Bank 2012b).[2] Green-economy proponents contend that climate change mitigation and ecosystem conservation can best be achieved by means of the scientific measurement, and monetary pricing and—in some versions— globally coordinated, market-based trading of environmental benefits and costs.

> Looking forward, stabilizing temperatures will require a global mitigation effort. At that point carbon will have a price worldwide and will be traded, taxed, or regulated in all countries. Once an efficient carbon price is in place, market forces will direct most consumption and investment decisions toward low-carbon options.
>
> *(World Bank 2010, 271)*

Further, carbon markets appear to solve the problem of how transitions to low-carbon economies in the global South can be financed, given the unwillingness of most governments to provide the funds or enforce regulations for this purpose. In theory, because it would make investments in conservation and low-carbon technologies profitable, market-based green-economy policy could largely finance itself by means of private investments. The World Bank and many climate-policy experts envision carbon markets as a key source of private financing for conservation and for the development and diffusion of low-carbon technologies. However, the monetary valuation nature, and carbon markets in particular, are highly controversial. For reasons I outline in this chapter, this is especially true of policies that involve the buying and selling of greenhouse-gas emissions allowances—carbon credits or offsets—that are derived from climate-mitigating activities in developing countries.

The World Bank and other green-economy advocates envision that revenues from carbon markets of the latter sort will become the major source of future funds for global REDD+. The two main institutional sponsors of REDD+ are the United Nations, through the UNFCCC and UNEP, and the World Bank, through its Forest Carbon Partnership Fund (FCPF) and related agencies. REDD+ builds on models of PES projects in Latin America, Asia, and Africa since the late 1990s. By promising benefits to all parties from trade in carbon offsets—governments and land users in the global South, private investors, and humanity as a whole—market-based REDD+ attempts to skirt the tensions between seemingly contradictory goals: slowing global warming and promoting economic development.

According to green-economy theory, trade in carbon offsets sold by lower income countries and communities will be the least costly means of preserving forest carbon sinks in the near and medium terms. Further, by compensating developing countries for conserving and enhancing forests, REDD+ payments are expected to help resolve, or at least circumvent, the continuing disputes about the relative responsibilities of highly industrialized and formerly colonized countries. Some assert that in addition, REDD+ payments distributed strategically to land owners and the poor can ease the conflicts over land and territorial rights that embroil many regions of the global South.

Country signatories to the UNFCCC have agreed in principle to guidelines for REDD+, although some Latin American governments have expressed strong reservations (McAfee 2016). None of the REDD+ projects sponsored by UN-REDD or the World Bank had reached the implementation stage as of late 2015, and the issue of whether the program ought to be financed by public funds or by private investments remains unresolved. In the meantime, many REDD+ preparatory activities are being financed by grants and loans from governments, UN-REDD, the FCPF, and related World Bank funds, which themselves are supported primarily by donations from a few global North governments, led by Germany and Norway.

The amounts available or pledged by governments for REDD+ are far lower than the amounts that proponents say are needed, but these public sources are seen as catalysts for market-based REDD, to be financed on a much larger scale by for-profit trade in carbon credits. Additional sources of advocacy and support for REDD+ are nongovernmental organizations (NGOs) that focus on conservation and have international scope—such as the Environmental Defense Fund, The Nature Conservancy, Conservation International, and the Worldwide Fund for Nature—as well as conservation NGOs based in developing countries. These organizations have been among the most active developers of preparatory REDD+ projects. Other conservation NGOs, such as Greenpeace and Friends of the Earth International, oppose carbon offsetting and REDD+.

At international environmental events such as the 2012 Rio+20 Earth Summit and negotiating sessions of the Climate Change and Biodiversity Treaties, REDD protesters have massed in the streets and lobbied in the corridors. Among them have been activists from the global North and delegations of peasants and their allies from Brazil, Indonesia, Mexico, India, Mali, and many other countries. Opposition is also emerging at state and national levels and against regional carbon trade pacts, such as those signed by the state of California with Acre, Brazil, and Chiapas, Mexico. At the Rio+20 summit, a vocal minority of governments also argued forcefully against REDD+, the green-economy concept, and what they described as the "commodification of life" that green economy entails (McAfee 2016). Some international indigenous peoples' organizations have cautiously endorsed REDD+; others have denounced carbon offset trading and REDD+ altogether (Lang 2012; Tauli-Corpuz and Baer 2010).

Recent scholarship on carbon markets, PES, and REDD raises doubts about whether market-based climate mitigation can be both market efficient and compatible with social development and poverty alleviation (Pattanayak et al. 2010; Wunder 2013). It also points to significant reasons why REDD+ and similarly conceived, market-based climate policies may not be able to achieve net reductions in GHG emissions (McAfee 2012). The near collapse of carbon markets since 2011 casts additional doubt on their potential (Point Carbon 2014). Other published analysis contends that both PES and REDD+ are already contributing to "green grabbing": the dispossession of indigenous forest dwellers and small-scale farmers for ostensibly environmental purposes (Beymer-Farris and Bassett 2012; Fairhead, Leach, and Scoones 2012; Rocheleau 2015).

Perhaps most importantly, critiques arising from rural-based social movements and their intellectual allies have questioned the foundational categories and assumptions behind carbon markets and market-based environmentalism more broadly. These counter-discourses to green economy challenge the conventional consensus that economic growth is the sine qua non of human progress. They put forward quite different goals, such as the achievement of *buen vivir*: living well and living cooperatively.

The next three sections of this chapter outline the main features of carbon markets, carbon offset trading, and REDD+, and then examine the theoretical assumptions on which these green-economy strategies are based. The sections that follow consider questions about the environmental efficacy of market ecosystem services that have been raised by literature on PES and discuss the tensions between the environmental goals of green economy and the needs and aspirations of communities that live in or make use of landscapes targeted for carbon offset generation through PES and REDD+.

International Carbon Markets to the Rescue?

In carbon markets, responsibility for climate-warming GHG emissions becomes a commodity that can be bought and sold within national or regional jurisdictions, and even between continents, in the form of carbon allowances or carbon offset credits.[3] Many economists and policy experts view this approach as vital to a successful UNFCCC (Stewart, Kingsbury, and Rudyk 2009). Other carbon market proponents have all but given up on the UNFCCC multilateral framework and are instead seeking market-based programs that can be implemented at national or regional levels, or apart from governments altogether, in voluntary carbon markets (VCMs).

In compliance markets, carbon allowances are created by national governments or by alliances of state, provincial, or national governments that establish regulations to limit GHG emissions within their territories. While some of these systems are financed by taxes on fossil fuels, most such programs thus far are cap-and-trade schemes designed to raise the cost of polluting without stifling the profitability of regulated industries. Regulators set legal limits (caps) on how much GHGs may be emitted in the entire region and by particular industrial enterprises or economic sectors. They then give or sell GHG-emissions allowances to the enterprises under regulation. Allowances are permits to continue emitting a specified quantity of CO_2 or other GHGs for a certain time period, after which the number of allowances is meant to be reduced.

In cap-and-trade systems, companies that do not reduce their emissions to the regulatory maximum may remain in compliance with the law by purchasing allowances from another entity, such as a regulated firm that has reduced its emissions enough that it can comply with the law without using all of its own allowances. This is the "trade" part of cap-and-trade. In theory, as the supply of allowances is reduced by law, the rising price of available allowances is expected to serve as an incentive for managers of polluting businesses to invest in technological innovations and greener practices to minimize their need to purchase costly allowances. In this way, advocates contend, cap-and-trade can foster a transition to more sustainable energy production without slowing economic growth.

Some cap-and-trade schemes include the use of offsets. These are credits that an enterprise can add to its allowances to increase the amount of pollution it may legally emit. Offsets are not obtained by reducing GHG emissions at the source: they are generated by investments in climate-mitigating actions at another site within the regulated territory or, in some cases, outside it. For example, under the cap-and-trade component of California's policy on global warming, companies may offset a portion of their GHG emissions by paying for projects in the United States and Quebec that capture GHGs or prevent their production. These may be activities that destroy ozone-depleting chemicals, support urban and rural tree planting or improved forest management, capture methane from pig and cattle manure, collect methane from coal mines, or reduce methane emissions from rice fields (CARB 2012). Rules that would allow California companies to offset part of their emissions by paying for forest conservation and management in Mexico and Brazil were still under consideration as of this writing (see also Blanchard and Vira, this volume).

When traded from one world region to another, offsets allow continued emissions of certain amounts of GHGs, mainly in the industrialized economies, in exchange for investments in greening, mainly in the less industrialized countries of the global South. Such credits may be based on investments in biofuel production, improved management of GHG-emitting landfills, less polluting or more energy-saving manufacturing, or cleaner energy production (e.g., hydropower instead of coal, or agricultural practices that produce fewer emissions of methane, carbon dioxide, or nitrous oxides). The CDM, created by the Kyoto Protocol, was the first internationally sponsored carbon market outside of the European Union, which has had its own cap-and-trade system since 2005. CDM projects, most of which have been based on industrial greening and hydropower investments, have been widely criticized for dubious claims about how much GHG emissions they reduce or prevent, if any (Newell and Phillips 2011).

Thus far, the goals of the various multilateral and government schemes to cap GHGs have been extremely modest in comparison to the amount of GHG reductions that most climate scientists agree are needed to avoid a increase in mean global temperatures of less than 4 degrees centigrade (IPPC 2014). Some have failed to even get off the ground, largely as the result of political pressure from fossil fuel–related industries. Although governments, international agencies and even some NGOs have been trying to prop up carbon markets (Kossoy 2014) by means of subsidies and market interventions to boost the prices of credits, carbon markets have remained too weak to attract investors on the scale anticipated by their sponsors. Consequently, the prices of tradable emissions allowances and offset credits have been far too low to persuade polluters to reduce their GHGs substantially instead of buying the inexpensive credits. Unless a strong, global-scale compliance regime emerges from the ongoing negotiations of the UNFCCC or another international process, this is likely to remain the case.[4]

Carbon markets are nevertheless lucrative for many. Commonly, for-profit consulting firms collect fees for identifying or setting up credit-earning projects and shepherding them through the process of qualifying for offsets under the

CDM or other official regimes. Profits can also be made from carbon trading by finding or creating carbon offset projects and marketing them to VCM investors. Other players who stand to benefit from transnational carbon markets are banks, commodity trading firms, oil and power companies, energy speculators, and the pension, hedge, and private equity funds that invest in carbon credits, as well as firms established specifically to broker the new carbon commodity.[5]

Carbon Offsets and REDD+

Forest carbon offset buyers typically pay for the preservation or expansion of tropical rainforests or other ecosystems that sequester carbon.[6] In the case of REDD+, revenue from the sale of offsets is meant to be used to pay governments that pledge to reduce deforestation and forest degradation by various means, including payments to landholders for making forest-saving land use decisions. However, there is no single, standardized architecture, set of criteria, or financing mechanism for REDD/REDD+. Many players in carbon markets are looking to REDD+ projects as sources of future offset credits.

In the meantime, although the CDM offers some credits for forest investments, most forest-related credits are created and sold through VCMs that are regulated privately, if at all. VCMs are not linked to government legislation or to international treaties. Rather, they are self-policing schemes created by private sector industry groups and companies that want to reduce their carbon footprints or want to create a greener public image. A variety of for-profit firms and nonprofit organizations have been established to manage VCMs. While compliance carbon markets—those linked to government regulations or treaties—have foundered in recent years, VCMs have continued to grow (Forest Trends 2013).

By early 2014, 44 countries were pursuing REDD preparatory processes in conjunction with the World Bank's Forest Carbon Partnership Facility (FCPF). At least 50 countries were partners with UN-REDD, established in 2008 as a joint project of the UN Food and Agriculture Organization, the UN Development Programme, and UNEP and backed by a multi-donor trust fund. Still other projects designated as REDD by their sponsors are independent initiatives by governments, NGOs, and private companies, including firms that participate in VCMs. All this makes REDD a something of a wild frontier, populated by dedicated environmentalists trying to establish effective forest conservation policies and projects alongside those pursuing publicity or profits from climate change mitigation.

In the market-based version of REDD preferred by the World Bank, proceeds from the sale of REDD credits to GHG-emitting enterprises or to other investors in carbon markets are to be channeled to governments that demonstrate improved practice in limiting deforestation (Busch n.d.; World Bank 2012a). Those governments might use REDD funds to maintain protected areas, regulate logging and improve forest management, or slow the conversion of forests or carbon-rich peat lands for agriculture. REDD+ sponsors expect that much of this can be achieved by means of PES. Forest PES schemes typically compensate landowners who promise not to fell forested tracts for pasture or timber, pay farmers who agree to plant trees instead of crops, or compensate indigenous communities and others to desist from cutting wood, grazing animals, or practicing swidden agriculture in forested areas. In these ways, REDD+ builds upon the precedents of the market-oriented PES projects that have been implemented in Latin America, Asia, and Africa since the late 1990s.[7]

In theory, for the sake of market-based efficiency in the use of conservation funds, payment to landholders should be just high enough to persuade the recipients to make the more environmentally beneficial choice between conserving forest or felling trees, but no higher, so that maximum conservation gains will be achieved at minimal cost. Also in theory, the amount of carbon that will be stored or the amount of GHG emissions from land conversions that will be prevented as a result of the payments will be equivalent, in terms of their effects on the atmosphere, to the amount of the GHGs that the ultimate buyers of the offsets will continue to release thanks to their possession of those credits. Both these premises are proving dubious in practice, for reasons outlined later in this chapter.

First, it is useful to consider some theoretical assumptions and methodological commitments that support arguments for market-based management of the climate crisis.

Market-Based Environmental Governance as Anti-politics

A series of assumptions links the logic of green economy to carbon trading and market-based REDD+. The version of green economy advanced by mainstream environmental and development agencies is largely based on environmental economics, which is itself built upon neoclassical economics. Environmental economists also apply tools of institutional economics to analyze market imperfections, complexities, and transaction costs (Ferraro 2008; Swallow et al. 2007). Some who distinguish themselves as ecological economists envision the economy as situated within and limited by the ecosphere (Boulding 1966; Daly and Farley 2003). In any of these approaches, environment and economy are understood as distinguishable, at least conceptually. Environmental economics aims to solve environmental problems by constructing environmental costs and benefits as forms of capital and then bringing nature into "the market." For instance, UNEP's Green Economy manifesto asserts that causes of crises such as looming food and water shortages, as well as "persistent social problems," share a common feature: "the gross misallocation of capital," exacerbated by "existing policies and market incentives" that allow "unchecked social and environmental externalities" (UNEP 2011, 14–15).

In this approach, nature is made subject to monetary valuation and property rights. The exchange values, that is, the prices of natural capital and the monetary costs of environmental damages, are to be determined either by means of expert estimates or discovered through the workings of supply and demand. In theory, once the values of environmental externalities—the unintentional costs or benefits of economic activities—have been calculated, the costs of different choices about the fate of any one landscape or species can be compared. The environmental costs of the preferred choice can then be internalized into the accounting of individuals, enterprises, and states. This monetary pricing of nature's assets and services allows them to be treated as commensurable from place to place and over time. Then, so long as it is clear who owns them, ecosystem services such as the carbon-sequestration functions of forests can be bought and sold, as in they are in PES and carbon offset markets.

In keeping with neoclassical premises, most green-economy thinkers view land users, businesses, and consumers as rational individuals who respond mainly to material incentives. They see competition and market exchange among private actors as more effective than government command-and-control regulatory policies (Chichilnisky and Heal 2000). Therefore, voluntary trade in environmental goods and bads is expected to produce optimal, least-cost outcomes with minimal state involvement (Coase 1960). In this way, the invisible hand of the market can minimize conflicts between different industries and interests and achieve a maximally efficient allocation of land and resources for manufacturing, energy production, and agriculture.[8] In other words, green-economy advocates expect that environmental markets can supersede politics.

That said, green-economy thinkers do not expect that the pricing, ownership, and trading of environmental goods and bads will arise spontaneously, or that this will be sufficient to ensure that climate mitigation goals are reached. Most agree that the new environmental markets must be informed by science and constructed and supported by state interventions. The latter may include taxes on fossil fuel extraction or use, fines or legal caps on GHG emissions and other pollution, and tax reductions or other subsidies that increase the market values of non-renewable resources and ecosystem services and create incentives for their sustainable use.

This market-centric model invites the use of quantitative concepts and methods. As business leaders convening during the Rio+20 summit proclaimed, "we can't manage what we can't measure" (Suarez 2013). People trained in the methodological individualism of neoclassical economics, as are most World Bank economists, are occupationally inclined to imagine the natural world as the sum of interchangeable parts amenable to being counted and modeled. In any case, market-oriented thinking in environmentalism is part of the broader trend of market-centric approaches in policymaking that has prevailed in much of the world since the 1980s, which helps to explain why carbon markets have become a leading response to the global warming crisis.

A related reason why neoclassical economics seems naturally fit for managing environmental crisis has to do with the idea of scarcity. The orthodox economics that frames most green-economy thinking is essentially about the administration of scarcity: markets are the means to manage human wants and desires—which are assumed to be unlimited—for a finite supply of goods and resources (Luks 2010). Moreover, environmentalism itself has long been framed by beliefs about scarcity—resource shortages and population excesses—especially since the 1960s (Barbier 2011; Hardin 1968; cf. Latour 2008). Concepts common in environmental policy conversation in the anglophone global North, such as the tragedy of the commons, lifeboat ethics, carrying capacity, and the $I = PAT$ formula, hold the notion of scarcity at their core.[9] Ecological economics in particular takes scarcity as its starting point: living within the planet's limits is its analytical and programmatic priority.

International environmental negotiations were cast from the start in terms of scarcity: ecological limits and the consequent need for constraints on economic development. Worried that the proposed new environmental treaties would block their development aspirations, global South governments insisted that the first Earth Summit in 1992 address environment *and* development, not only conservation. Twenty-five years later, such concerns persist. If there are supposed to be insufficient material resources and ecological space for economic growth and development as it has been known, then political questions immediately arise: whose development shall be restricted, and who shall pay the costs?

Green-economy thinkers hope to answer these questions by means of economic rationality and fine-tuned ecological science, that is, by substituting hard data and cold calculations for seemingly intractable political struggles and difficult ethical choices. However, policy choices based on such putatively objective criteria have implications that are anything but neutral. If scarcity is assumed (i.e., if trade-offs between conservation and development are required), then leaders and citizens of global South states fear that it is their countries that will be required to sacrifice. This is especially true given that, since the first Earth Summit, neoliberal development policy has failed to reduce the North-South wealth gap in most world regions. For many, faith in the neutrality of market-based solutions appears to be wavering. This was reflected in the suspicion of green-economy language expressed by G77 governments at Rio+20 and in accusations by some Latin American delegations that green economy is a new form of colonialism (UN Commission on Sustainable Development [UNCSD] 2012b).[10]

At national and subnational levels, too, the application of market logic can be politically charged. In PES schemes where market-efficiency standards are used to determine who should receive payments, the results often tend to reward the better off more than the poor, as outlined in the next section. This is a major reason why some governments, such as Mexico's, have resisted the use of exclusively market-based criteria in the implementation of PES, much to the dismay of PES advisors affiliated with the World Bank (McAfee and Shapiro 2010). Critics of carbon marketing as a strategy for conservation and sustainable development offer two main lines of argument: that carbon markets are inequitable, for the reasons explained in the following section; and that carbon trading is ineffective against climate change, as outlined in the section after that.

Will Carbon Markets Lead to Greater Inequality?

Many proponents of carbon markets acknowledge that carbon trading, if not designed specifically to protect or reward the poor, can have inequitable consequences (Angelsen et al. 2012). Some contend that market-based REDD+ or any carbon markets that involve offset trading are inherently and inevitably disadvantageous to those with little wealth or power.[11] Most of the for-profit businesses, brokers, and speculators in carbon markets will buy offset credits only insofar as such investments offer profit advantages over investments in other places or other activities. The profit opportunities that proponents hope will drive North-South carbon trading are derived from the fact that offset credits can be obtained at less expense in economically poorer regions.

To its advocates, the market-based efficiency derived from this wealth difference is the great virtue of transnational trade in carbon credits. But any trading of environmental goods and bads—atmospheric pollution, biodiversity conservation, or other assets and damages—involves shifting the costs of conservation from one location and one group of people to another. In the language of orthodox economics, carbon credits in low-income countries are cheaper because opportunity costs (e.g., the potential income lost if land is used for carbon sinks instead of crops or cattle) are smaller where prices of labor, land, and other factors of production are lower and life expectancies are shorter. In other words, investment in greening in the global South is economically efficient because nature and human lives are cheaper, in effect, and bargaining power is weaker in places where people are poor. Climate policies that depend on profit making are therefore likely to have highly unequal North-South consequences.

Market-based allocation of conservation payments can reinforce inequalities at local and provincial levels, too. Regardless of whether conservation funds come from public grants, taxes, or carbon market investments, market efficiency in the distribution of those funds requires that payments for carbon sequestration or other environmental services be made to those landholders whose decisions whether to conserve trees will be influenced by the payments. To ensure that this is the case, project designers use estimates of opportunity costs to determine where in the world to obtain ecosystem services, from whom, and how much to pay for them.

Resource economists view opportunity cost as a neutral benchmark that can determine the proper amounts and allocation of payments (Janvry and Sadoulet 2006). However, the use of opportunity costs in this way is anything but politically neutral. The opportunity cost concept itself masks the power relations that determine *whose* opportunities are more or less costly and *whose* land use choices shall prevail. It is more labor-intensive and therefore less economically efficient to enroll many smallholders and monitor their compliance with PES or REDD+ project requirements than to pay a smaller number of larger-scale landholders. In the language of institutional economics, there is an inverse relationship between the scale of a project and its transaction costs. It can also be more expensive to enroll less literate people, women who have multiple work obligations, and those who lack formal land tenure credentials. In market-oriented PES or REDD+ projects, measures meant to facilitate the participation of smallholders or the landless may therefore be ruled out on economic efficiency grounds.

Middle-sector landholders, those with more modest opportunities for profit but with the capacity and intention to deforest, are seen as the most appropriate PES recipients (Wunder 2007). Payments to relatively wealthy stakeholders, such as logging or mining operations, large-scale ranching or agricultural plantations, shopping centers, or resorts, would typically be inefficient because the payments would need to be large enough to match the high opportunity costs of abstaining from such profitable activities and therefore would be too high. Yet these activities are major drivers of deforestation and wetlands loss, and these actors often have decisive political power.

Payments for ES targeted to the very poor achieve relatively little in forest conservation gains. With or without payments, such land users usually lack the necessary tenure rights, capital, or access to timber or agricultural markets that would enable them to engage in significant deforestation. More conservation per dollar could be bought elsewhere (Chomitz 2007). Many proto-REDD+ projects have nevertheless targeted such "marginalized" or indigenous communities. There is much debate in PES literature about whether or not such "pro-poor" projects are providing net conservation gains or social benefits (Kronenberg and Hubacek 2013; Luttrell et al. 2013; Pattanayak, Wunder, and Ferraro 2010; Wunder 2013). But, as World Bank economists and others have noted, to the extent that poverty alleviation is a priority, market efficiency is likely to be compromised. One architect of World Bank PES policy has written, PES schemes

> cannot, for example, target their interventions to areas of high poverty, as these may not be the areas that generate the desired services. PES programs also cannot choose to promote particular land use practices solely on the basis of the poor being able to undertake them.
>
> *(Pagiola 2007)*

Farm and forest carbon-sequestration projects, like many conservation schemes, can also have directly negative effects on the rural poor. In many global South regions, forests and swamps are being reconceptualized as carbon sinks and farmlands repurposed as biofuel plantations (Akram-Lodhi 2012; Fairhead et al. 2012; Li 2010; Moore 2010). In the context of increased integration and financialization of the global economy, world-market prices of many food, fiber, and mineral commodities have soared since the late 2000s. Along with the anticipation of profits from carbon market investments, this has led to rising land prices. This has accelerated the processes that critics call land grabbing (the illegal or unjust acquisition of land by the economically powerful) and green grabbing (expulsions of forest dwellers and small-scale farmers for ostensibly environmental goals) (World Bank 2012b).

Projects carried out under the rubric of PES and REDD+ appear to be contributing to this trend by displacing or threatening to displace peasant and indigenous communities (Beymer-Farris and Bassett 2012; Rights and Resources 2014). Even where land users are not evicted, they may face reduced access to sites of cultural significance, passageways, and sources of food, forage, medicines, and shelter materials. In the context of resurging struggles over land and territorial sovereignty in many regions, markets in carbon and biodiversity offsets have inevitable political implications at local and national levels, regardless of their intended scientific neutrality and regardless of whether their sponsors are guided by conservation efficiency or pro-poor priorities.

In addition, no matter how efficient carbon offset markets might become, the buying and selling of offset credits, in itself, does nothing to stop the production and release of GHGs.

Can Carbon Markets Reduce Global Greenhouse Gas Emissions?

As noted earlier, trade in offset credits, in itself, does nothing to stop the production of GHGs. Offsets are a form of permit to continue polluting. It is therefore impossible for international carbon trading alone to reduce the amounts of GHGs in the atmosphere or even to slow the rate of increases of emissions (Böhm and Dabhi 2009; Lohmann 2005).

The argument that carbon offset trading can reduce emissions depends on the expectation that there will be real, enforceable limits on the legally permitted amounts of GHG emissions and that these limits will be reduced regularly and substantially. This would require a strong international regime adhered to by all or nearly all industrialized and industrializing countries: a global cap-and-trade system. As in any such system, *only the level of the cap would count* toward the achievement of net global emissions reductions. Without a low and constantly lowered cap, international carbon trading entails no more than the shifting of activities that produce, absorb, or avert GHG emissions from place to place around the world. There is no global government to enforce this, and efforts to achieve even modest, quantitative, and legally binding targets for country-by-country compliance seem even less achievable today than they were two decades ago when the Kyoto Protocol was hammered out.

Most backers of global carbon markets are well aware of this problem. Some hope that, as global warming affects more of the world, an effective climate accord can yet be achieved (cf. Machin 2013; Stern 2009). Even without such an accord, or until it is achieved, some believe that the quest for offset credits can drive a virtuous cycle of investments in carbon-sink conservation, alternative energy, and low-carbon industrialization in the global South as well as the North (Newell and Paterson 2010). Some are convinced that technological breakthroughs stimulated by these processes will be profitable enough to blunt the opposition to a strong carbon management regime by the petroleum, coal, automobile, and other industries that rely on fossil extraction and combustion, in spite of the political influence and deep economic entrenchment of these interests (Funk 2014; cf. Hamilton 2013).

Whatever may be agreed at the international level, the effectiveness of any policy to slow climate change by reducing GHG emissions or conserving carbon sinks depends on what happens at the local level of the factory, forest, ranch, or farm. The record of climate-oriented PES projects, which are meant to maintain or increase carbon sequestration in forests and, to a lesser extent, in agricultural lands, is therefore directly germane to the question of whether carbon markets can combat global warming.

What follows here is a summary of the main problems in achieving environmental gains through PES that have been widely noted by both advocates and critics of market-oriented PES. Elsewhere I have discussed why it is particularly difficult to achieve both environmental *and* anti-poverty objectives by means of market-based PES or REDD (McAfee 2012).

Problems of Ecosystem Services Markets as a Conservation Strategy

PES projects have been in place since the early 1980s; most have been in Latin America.[12] Ecologists and economists are working to fine-tune methods of measuring carbon flows and criteria for calculating the values of ecosystem services. However sophisticated these methods become, there are ample reasons to doubt the effectiveness of PES as a conservation and climate change mitigation strategy. Problems arise, first, from scientific and technological issues:

- *Ecological complexity* and *scientific uncertainty*: Human knowledge about the relationships among various land uses and species conservation, water supplies, and, especially, the sequestration of carbon in soils, peat lands, and vegetation is limited and much disputed, even among climate and forestry scientists. Consequently it is very difficult to determine or predict how much if any carbon is stored or released, or water conserved or consumed, as the result of activities paid for through PES programs. It is all but impossible to devise methods and formulae for applying such estimates over various time scales and across ecosystems, which are complex, dynamic, and always unique.

Additional challenges to determining the environmental effects of ES payments stem from the various socio-economic and institutional contexts of such projects.

These problems plague market-based offset programs such as PES, as well as the CDM, and are likely to apply to carbon offset credits generated under the rubric of REDD.

- *Leakage* occurs when environmentally destructive activities, such as logging or farming for profit or for subsistence, are shifted from the places targeted for conservation to other sites. It is often impracticable to prevent a destructive activity that is banned under the terms of an ES payments contract from being relocated. It may be impossible to determine whether this escape from accountability is occurring.
- *Non-additionality* refers to cases where payments are made for practices, such as abstaining from felling trees or constructing a coal-fired power plant, which would have occurred even if the PES, CDM, or REDD project did not exist. It is often impossible to know with certainty whether a credit-generating activity would have taken place in the absence of payments for better environmental behavior.
- *Impermanence*: Where payments are the incentive that motivates landholders to conserve, how long will such payments continue, and what happens after they end, or when investment priorities shift, or when climate change transforms ecological relationships? Many PES contracts cover periods as short as three to five years. Some cover 25 years or more, which may be the time it takes for a forest to grow enough to be harvested profitably. In PES schemes where water users help pay for watershed conservation, payments may continue as long as water users are able and willing or compelled to pay for the water they receive. Longer-term funding may be less likely in schemes that depend upon state, multilateral, or NGO subsidies, as do most PES programs. If carbon markets are the source of payments, recipients may be subject to changing investment decisions by buyers looking for cheaper sources of credits.
- *Perverse incentives* may arise when expectations of conservation payments prompt states or landholders to threaten to engage in more deforestation, or more polluting production methods, than they actually intend to carry out. Individual landowners and also governments can claim to qualify for higher payments if they overestimate past deforestation rates in order to demonstrate conservation progress.[13]
- *Rent-seeking and/or other forms of moral hazard* are linked to the conflicting priorities of officials, NGOs, or consultants in charge of monitoring, enforcing, or certifying compliance with PES or REDD project requirements, on the one hand, and ecosystem services buyers or project sponsors, on the other hand. Even when outright corruption is not a factor, more subtle conflicts of interest can create incentives to base project designs and claims of project success on selected, favorable data or on optimistic but unsupported assumptions. For example, agencies or consultants paid to determine whether

ecosystem services have in fact been produced may be tempted to cut costly corners or to certify carbon credits, regardless of spotty evidence, so that they can obtain future contract work.

As a result of such problems, estimates of net environmental losses or gains from PES or REDD projects necessarily rely on best-guess approximations, counterfactual scenarios and assumptions about future human decisions, and questionable claims about the commensurability, fungibility, and economic values of ecosystem functions. These challenges have given rise to a growth industry of careers and consultancies that proffer expertise in carbon calculation. Efforts to devise and promote methods for certifying the ecological legitimacy of credit-earning investments have also become a major undertaking, enrolling thousands of economists, business and legal experts, ecologists, and more rarely, social scientists. Several agencies and alliances that cater to investors in VCMs have established criteria and brand-name labels meant to ensure that projects that offer offset are legitimate enough that the credits they generate are likely to retain their commercial and public relations value (Forest Trends 2014).

Some REDD+ project designers and investors are crafting jurisdictional REDD+ methods meant to strengthen accountability and link REDD+ projects at subnational and national scales (ROW 2013; VCS 2013). In light of continuing inaction at the global level, emerging carbon market plans by national governments, such as China, and states or provinces, such as California in the United States and Acre in Brazil, are drawing increasing attention. But these efforts, too, face daunting technical and political challenges: uncertainty about carbon cycles, conflicts of interests in project implementation and monitoring, and politically laden decisions about whether "leakage" can be detected and how baselines for the measurement of "improved forest management" are determined.

Conclusion

Advocates of carbon markets promise a triple-win solution for business, nature, and humanity. Because carbon trading does not directly challenge existing patterns of consumption and the distribution of wealth and power among nations, and does not threaten the goals of powerful political and economic actors, it appears as if this approach can slow global warming without major lifestyle sacrifices and without confrontation with powerful vested interests.

There are well-grounded reasons, outlined in this chapter, to doubt that carbon markets can achieve significant conservation gains. In addition, the tension between market efficiency and development goals is widely acknowledged, as is the danger that PES and REDD+ will be implemented at the expense of politically marginalized communities. Flagging carbon markets make it less likely that for-profit investment will be a major source of finance for programs such as REDD+. Debates about whether international carbon trading is socially just are ongoing.

Nevertheless, growing numbers of NGOs and government agencies are working to implement PES and REDD+ in the global South.[14] Whether these activities yield net environmental benefits, social benefits, or both depends on the scale at which the question is posed and the circumstances under which such activities are carried out. For example, Shapiro-Garza (2013) reports that Mexico's national PES program has not led to dispossession in the areas studied and has benefited many project participants, although not always in the ways envisioned by market-efficiency maximalists. She attributes the program's partial success to factors specific to the times and places where PES has been carried out: a high degree of transparency during the design and implementation process, relatively robust governance structures and land tenure in participating communities, and the influence of rural social movements.

In light of concerns about violations of human rights, several multilateral organizations and NGOs have turned their attention to making REDD+ safer for the powerless and more beneficial to the poor. They are formulating social safeguards, such as stronger criteria for certifying that people about to be affected by REDD+ have indeed given their free, prior, and informed consent. But even if safeguards to protect peasant and indigenous communities can be agreed and effectively enforced, any version of REDD+ in which funds are raised by means of for-profit carbon trading and allocated according to market-efficiency criteria will not prioritize payments to the poor, for reasons addressed earlier.

Many rural communities, although certainly not all of them, already manage their landscapes in sustainable ways, especially where indigenous peoples maintain significant control over their farming, hunting, fishing, and gathering territories. Several studies of PES projects have shown that forest dwellers and small-scale farmers will sometimes take extra measures for the sake of conservation even when the costs of doing so, in terms of labor and restrictions on gazing and farming, outweigh the payments they receive. The record of community forestry in Mexico illustrates that self-governing indigenous and *ejido* communities, given the chance, can manage forests for both conservation and income goals (Klooster 2003). Small- and medium-scale farms are not only more productive in land use and energy terms and more supportive of biological diversity than are industrial-agriculture plantations, they also sequester more carbon in farm soils (Rosset 2000). The systematic application of agroecological practices can boost food production and increase carbon storage through the use of low-till and intercropping practices and green and animal manures, as the record of agroecology in many parts of Latin America, particularly Cuba, has demonstrated (Machin Sosa et al. 2008). The international federation of peasant organizations, La Via Campesina, proclaims that small farmers help to "cool the planet" (La Via Campesina 2007).

Conceivably, material aid and technical assistance through programs such as REDD+ could support these sustainable farm and forest management practices and help to replicate them for the benefit of the rural poor and local and global environments. But to the extent that communities manage to obtain net social and environmental gains under PES and REDD+, it will be in spite of or instead of the model of REDD+ that is currently promoted by green-economy analysts— that is, REDD+ financed by carbon markets and with payments to land users allocating according to opportunity cost criteria.

REDD+ is engaging the time, skills, and dedication of many thousands of well-intentioned multilateral agency officials, government workers, community leaders, NGO staff, scientists, and students. The danger is that REDD+ may be an immense red herring, distracting resources and attention from the underlying and even the immediate causes of destruction of forested areas and mixed-use landscapes that support rural communities in much of the world.

Consider the case of Chiapas, Mexico.[15] REDD+ supporters and critics debate whether California ought to allow enterprises in the state to offset part of their GHG emissions by paying for REDD+ projects in Chiapas, Acre, Brazil, and other states and provinces in the tropics. Some community spokespeople in Chiapas are supportive of REDD+; others are vehemently opposed. While the Chiapas state government insists that no official REDD+ projects yet exist there, an impressive number of Chiapas state and Mexican federal government agencies, advisory bodies, technical committees, consultants, and local and international conservation organizations are actively involved in REDD+ planning and promotion, funded by grants and loans from the World Bank, the federal government, conservation NGOs based in the North, and a few private sector organizations.

It remains to be seen whether REDD+ in Chiapas can have a significant, positive impact on forest management in Chiapas without contributing directly to dispossessions of poor farmers and forest users. Less in doubt is that other factors are undermining the ability of the rural Chiapas majority to farm and live sustainably: anti-peasant government policies that promote export agriculture, ranching, and large-scale plantations of non-food crops including biofuels such as oil palm and jatropha; the expansion of mining and fossil fuel extraction; large-scale dams and wind-power projects; commercial logging; high-end tourism projects; infrastructure schemes aimed at further integrating southern Mexico into the global economy on terms determined by Mexican and transnational corporations (Bartra 2001). These trends are likely to cause far more forest and livelihood losses than the best-designed REDD+ programs can prevent.

Green-economy reasoning conceptualizes environment as distinguishable from society, making it possible to treat the climate crisis as analytically and programmatically separate from wider political-economic and political-ecological structures and trends. The most significant, underlying drivers of deforestation and of ecological degradation can therefore be set aside, postponed or ignored entirely as a focus of policy action. Daunting but crucial political choices can be avoided. The result of such reasoning is that green-economy logic and the market-based climate policy are likely to result in little more than greenwashing on a very large scale.

Notes

1. The author thanks the Rachel Carson Center, Munich, where this chapter was written in 2014, and the Office of Research and Sponsored Programs at San Francisco State University, which provided a small grant for research in Mexico in 2013. Portions of this chapter appear in *International Environmental Agreements: Politics, Law and Economics* (online first August 2015).
2. The term "green economy" is sometimes used in a broader sense to mean ecologically sustainable economy. As discussed in this chapter, the term refers to the explicitly market-oriented conceptualization of green economy that is emphasized today by major environmental policy-making agencies and policy-influencing institutions and academics.
3. Carbon credits are usually accounted for in $MtCO_2e$, or metric tons of carbon-dioxide equivalent, a device used to measure and compare the global-warming impact of CO_2, methane, nitrous oxides, ozone, and other GHGs.
4. By requiring enterprises worldwide to limit GHG emissions or purchase permits to pollute in the form of offsets, such a regime would increase the demand for offsets and thus raise their price. It is all the more remarkable that a discourse about the superiority of market-based instruments in conservation continues to predominate.
5. Recent research suggests that that the amount of private, profit-seeking investment in ecosystem-service and biodiversity offsets remains extremely low and heavily dependent on subsidized schemes (Dempsey and Suarez, forthcoming).
6. Biodiversity offsets and credits for wildlife conservation are also bought and sold internationally.
7. The majority of PES projects involve payments for carbon sequestration or for hydrological services financed at subregional and national scales.
8. According to the conventional economic precept of "gains to trade," market exchange benefits all parties involved. In theory, this should apply to markets in nature, including trade in material goods and environmental assets such as carbon sequestration services between the global North and South.
9. $I = PAT$ is meant to show that total human impact on the environment (I) can be calculated by multiplying population (P) by quantities indicating level of affluence (A) and technology (T).
10. Green economy was the most debated topic during preparatory conferences for the Rio+20 summit (UNCSD 2012a). Delegates from the G77+China bloc voiced concerns that green economy would undermine their sovereignty with burdensome new policy requirements and obstacles to their exports. In their Rio+20 plenary speeches, the presidents of Uruguay and Ecuador condemned green economy as merely the latest incarnation of colonialism. The emphasis on green economy in the negotiating text was whittled down to a brief mention in the summit's final outcome statement.
11. I have argued in more detail elsewhere that inequality is built into the framework and rationale for global carbon markets from the outset (McAfee 2012).
12. I have discussed elsewhere the diverse literature that documents these projects and debates the pros and cons of PES (McAfee 2012; McAfee and Shapiro 2010).
13. The oversupply of allowances in the Emissions Trading System (ETS) began when polluting firms overstated their past emissions in order to obtain more carbon credits. This has led to credit prices too low to spur the hoped-for emissions reductions.
14. One reason, in addition to the wide appeal of market-based scenarios, is simply that project proposals with climate-related themes are more likely to receive funding from development-aid agencies.
15. The observations in this paragraph are based on my field visits and interviews in October and November 2013.

References

Akram-Lodhi, A. Haroon. 2012. "Contextualising Land Grabbing: Contemporary Land Deals, the Global Subsistence Crisis and the World Food System." *Canadian Journal of Development Studies/Revue Canadienne D'études Du Développement* 33 (2): 119–142, doi: 10.1080/02255189.2012.690726.

Angelsen, A., M. Brockhaus, W. D. Sunderlin, and L. V. Verchot, eds. 2012. *Analysing REDD+: Challenges and Choices*. Bogor, Indonesia: Center for International Forestry Research (CIFOR). http:/www.cifor.org/fr/online-library/browse/view-publication/publication/3805.html.

Barbier, Edward. 2011. "The Policy Challenges for Green Economy and Sustainable Economic Development." *Natural Resources Forum* 35 (3): 233–245. doi:10.1111/j.1477-8947.2011.01397.x.

Bartra, Armando. 2001. *Mesoamérica, Los Ríos Profundos*. Mexico City: Instituto Maya.

Beymer-Farris, Betsy A., and Thomas J. Bassett. 2012. "The REDD Menace: Resurgent Protectionism in Tanzania's Mangrove Forests." *Global Environmental Change* 22 (2): 332–341. doi:10.1016/j.gloenvcha.2011.11.006.

Böhm, Steffen, and Siddhartha Dabhi, eds. 2009. *Upsetting the Offset: The Political Economy of Carbon Markets*. London: MayFly Books.

Boulding, Kenneth E. 1966. "The Economics of the Coming Spaceship Earth." *Environmental Quality in a Growing Economy* 2: 3–14.

Busch, Jonah. n.d. "World Bank's Carbon Fund Pioneers Multilateral Payments for Forest Carbon." *Forest Climate Change*. http://www.forestsclimatechange.org/forests-climate-change-finance/carbon-fund-payments-forest-carbon/.

CARB (California Air Resources Board). 2012. "Regulatory Guidance Document." http://www.arb.ca.gov/cc/capandtrade/guidance/guidance.htm.

Chichilnisky, Graciela, and G. M. Heal, eds. 2000. *Environmental Markets: Equity and Efficiency*. Economics for a Sustainable Earth Series. New York: Columbia University Press.

Chomitz, Kenneth M. 2007. *At Loggerheads?: Agricultural Expansion, Poverty Reduction, and Environment in the Tropical Forests*. World Bank Policy Research Report. Washington, DC: World Bank.

Coase, R. H. 1960. "The Problem of Social Cost." *Journal of Law & Economics* 3: 1–44.

Daly, Herman E., and Joshua C. Farley. 2003. *Ecological Economics*. Washington, DC: Island Press. http://search.ebscohost.com/login.aspx?direct=true&scope=site&db=nlebk&db=nlabk&AN=118237.

Dempsey, Jessica, and Daniel Suarez. forthcoming. "Arrested Development? The Promises and Paradoxes of 'Selling Nature to Save It.'" *Annals of the American Association of Geographer* 2106.

Fairhead, James, Melissa Leach, and Ian Scoones. 2012. "Green Grabbing: A New Appropriation of Nature?" *Journal of Peasant Studies* 39 (2): 237–261. doi:10.1080/03066150.2012.671770.

Ferraro, Paul J. 2008. "Asymmetric Information and Contract Design for Payments for Environmental Services." *Ecological Economics* 65 (4): 810–821. doi:10.1016/j.ecolecon.2007.07.029.

Forest Trends. 2013. "Covering New Ground: State of the Forest Carbon Markets." http://www.forest-trends.org/documents/files/SOFCM-full-report.pdf.

Forest Trends. 2014. "Sharing the Stage: State of the Voluntary Carbon Markets." http://www.forest-trends.org/vcm2014.php.

Funk, McKenzie. 2014. *Windfall: The Booming Business of Global Warming*. New York: Penguin Press.
Hamilton, Clive. 2013. *Earthmasters: The Dawn of the Age of Climate Engineering*. New Haven, CT: Yale University Press.
Hardin, Garett. 1968. "The Tragedy of the Commons." *Science New Series* 162 (3859): 1243–1248.
Intergovernmental Panel on Climate Change. 2014. "Fifth Assessment Report—Mitigation of Climate Change." *Climate Change 2014*. Accessed May 5. https://www.ipcc.ch/report/ar5/wg3/.
IPBES/UNEP. 2011. "Report of the First Session of the Plenary Meeting to Determine Modalities and Institutional Arrangements for an Intergovernmental Science-Policy Platform on Biodiversity and Ecosystem Services." K1173506. Nairobi, Kenya: Intergovernmental Science-Policy Platform on Biodiversity and Ecosystem Services/United Nations Environment Programme. http://ipbes.net/downloads/doc_download/501-report-of-the-first-session-of-the-plenary-meeting-final-en.html.
Janvry, Alain de, and Elisabeth Sadoulet. 2006. "Making Conditional Cash Transfer Programs More Efficient: Designing for Maximum Effect of the Conditionality." *World Bank Economic Review* 20 (1): 1–29, doi:10.1093/wber/lhj002.
Klooster, Dan. 2003. "Campesinos and Mexican Forest Policy during the 20th Century." *Latin American Research Review* 38 (2): 94–126.
Kossoy, Alexandre. 2014. "State and Trends of Carbon Pricing." *World Bank*. http://www.ecofys.com/files/files/world-bank-ecofys-2014-state-trends-carbon-pricing.pdf.
Kronenberg, Jakub, and Klaus Hubacek. 2013. "Could Payments for Ecosystem Services Create an 'Ecosystem Service Curse'?" *Ecology and Society* 18 (1). doi:10.5751/ES-05240-180110.
Lang, Chris. 2012. "NO REDD+! In RIO +20: A Declaration to Decolonize the Earth and the Sky." *REDD Monitor*. http://www.redd-monitor.org/2012/06/19/no-redd-in-rio-20-a-declaration-to-decolonize-the-earth-and-the-sky/.
Latour, Bruno. 2008. "It's Development, Stupid! Or How to Modernize Modernization?" *Espaces Temps.net*. May 29. http://www.espacestemps.net/en/articles/itrsquos-development-stupid-or-how-to-modernize-modernization-en/.
La Via Campesina. 2007. "Small Scale Sustainable Farmers Are Cooling Down the Earth." http://viacampesina.org/en/index.php/actions-and-events-mainmenu-26/-climate-change-and-agrofuels-mainmenu-75/437-small-scale-sustainable-farmers-are-cooling-down-the-earth.
Li, Tania Murray. 2010. "Indigeneity, Capitalism, and the Management of Dispossession." *Current Anthropology* 51 (3): 385–414. doi:10.1086/651942.
Lohmann, Larry. 2005. "Marketing and Making Carbon Dumps: Commodification, Calculation and Counterfactuals in Climate Change Mitigation." *Science as Culture* 14 (3): 203–235. doi:10.1080/09505430500216783.
Luks, Fred. 2010. "Deconstructing Economic Interpretations of Sustainable Development: Limits, Scarcity and Abundance." In *The Limits to Scarcity: Contesting the Politics of Allocation*, edited by Lyla Mehta, 93–108. London; Washington, DC: Earthscan.
Luttrell, Cecilia, Lasse Loft, Maria Fernanda Gebara, Demetrius Kweka, Maria Brockhaus, Arild Angelsen, and William D. Sunderlin. 2013. "Who Should Benefit from REDD+? Rationales and Realities." *Ecology and Society* 18 (4). doi:10.5751/ES-05834-180452.
Machin, Amanda. 2013. *Negotiating Climate Change: Radical Democracy and the Illusion of Consensus*. London: Zed Books.

Machin Sosa, Braulio, Adilén María Roque Jaime, Dana RocíoÁvila Lozano, and Peter Michael Rosset. 2008. *Agroecological Revolution: The Farmer-to-Farmer Movement of the ANAP in Cuba*. Havana: Asociación Nacional de Agricultores Pequeños and La Via Campesina.

McAfee, Kathleen. 2012. "The Contradictory Logic of Global Ecosystem Services Markets." *Development and Change* 43 (1): 105–131. doi:10.1111/j.1467-7660.2011.01745.x.

McAfee, Kathleen. 2014. "The Post- and Future Politics of Green Economy and REDD+." In *The Politics of Carbon Markets*, edited by B. Stephan and R. Lane, 237–260. New York: Routledge.

McAfee, Kathleen. 2016. "Green Economy and Carbon Markets for Conservation and Development: A Critical View". *International Environmental Agreements* 16: 333–353. doi:10.1007/s10784-015-9295-4.

McAfee, Kathleen, and Elizabeth N. Shapiro. 2010. "Payments for Ecosystem Services in Mexico: Nature, Neoliberalism, Social Movements, and the State." *Annals of the Association of American Geographers* 100 (3): 579–599. doi:10.1080/00045601003794833.

Moore, Jason W. 2010. "The End of the Road? Agricultural Revolutions in the Capitalist World-Ecology, 1450–2010." *Journal of Agrarian Change* 10 (3): 389–413. doi:10.1111/j.1471-0366.2010.00276.x.

Newell, Peter, and Matthew Paterson. 2010. *Climate Capitalism: Global Warming and the Transformation of the Global Economy*. Cambridge and New York: Cambridge University Press.

Newell, Peter, and Jon Phillips. 2011. "The Governance of Clean Development: CDM and Beyond." *GCD Briefing* (3). Tyndall Center for Climate Change Research. http://www.tyndall.ac.uk/gcd-CDM-and-Beyond.

Pagiola, Stefano. 2007. "Guidelines for 'Pro-Poor' Payments for Environmental Services." World Bank. http://siteresources.worldbank.org/INTEEI/Resources/ProPoorPES-2col.pdf.

Pattanayak, Subhrendu K., Sven Wunder, and Paul J. Ferraro. 2010. "Show Me the Money: Do Payments Supply Environmental Services in Developing Countries?" *Review of Environmental Economics and Policy* 4 (2): 254–274. doi:10.1093/reep/req006.

Point Carbon. 2014. "Global Carbon Market Contracts by 38% in 2013 as Prices and Volumes Drop." *Thomson Reuters*. http://www.metal.com/newscontent/56377_global-carbon-market-contracts-38-as-prices-and-volumes-drop.

Rights and Resources Initiative. 2014. Status of Forest Carbon Rights and Implications for Communities, the Carbon Trade, and REDD+ Investments. http://www.redd-monitor.org/wp-content/uploads/2014/03/ForestCarbon_Brief-for-web-16Mar14.pdf.

Rocheleau, Dianne. 2015. "Networked, Rooted and Territorial: Green Grabbing and Resistance in Chiapas." *Journal of Peasant Studies* 42 (3–4): 695–723. doi:10.1080/03066150.2014.993622.

Rosset, Peter. 2000. "The Multiple Functions and Benefits of Small Farm Agriculture in the Context of Global Trade Negotiations." *Development* 43 (2): 77–82. doi:10.1057/palgrave.development.1110149.

ROW. 2013. "California, Acre and Chiapas Partnering to Reduce Emissions from Tropical Deforestation." *REDD Offsets Working Group*. http://greentechleadership.org/documents/2013/07/row-final-report-executive-summary.pdf.

Shapiro-Garza, Elizabeth. 2013. "Contesting Market-Based Conservation: Payments for Ecosystem Services as a Surface of Engagement for Rural Social Movements in Mexico." *Human Geography* 6 (1): 134–150.

Stern, N. H. 2009. *The Global Deal: Climate Change and the Creation of a New Era of Progress and Prosperity*. New York: Public Affairs.

Stewart, Richard B., Benedict Kingsbury, and Bryce Rudyk, eds. 2009. *Climate Finance: Regulatory and Funding Strategies for Climate Change and Global Development*. New York: New York University Press, Abu Dhabi Institute.

Suarez, Daniel. 2013. "'You Cannot Manage What You Do Not Measure': Natural Capital Accounting at Rio+20." In *Blue and Green Economies 1: Paradigm Shift or Hegemonic Realignment in Environmental Discourse?* Los Angeles, CA. http://meridian.aag.org/callforpapers/program/AbstractDetail.cfm?AbstractID=50220.

Swallow, Brent, Beria Leimona, Thomas Yatich, Sandra J. Verlarde, and S. Puttaswamaiah. 2007. "The Conditions for Effective Mechanisms of Compensation and Rewards for Environmental Services: CES Scoping Study." Issue Paper No. 3. ICRAF Working Paper No. 38. Nairobi, Kenya: World Agroforestry Centre. http://www.worldagroforestrycentre.org/Sea/Publications/files/workingpaper/WP0081-07.PDF.

Tauli-Corpuz, Victoria, and Lars-Ander Baer. 2010. "The Copenhagen Results of the UNFCCC; 'Implications for Indigenous Peoples' Local Adaptation and Mitigation Measures." *Economic and Social Council*. http://www.google.com/url?sa=t&rct=j&q=&esrc=s&source=web&cd=1&ved=0CCsQFjAA&url=http%3A%2F%2Fwww.un.org%2Fesa%2Fsocdev%2Funpfii%2Fdocuments%2FE%2520C.19%25202010%252018.DOC&ei=yBZSUovHM-nuyQHLpICYBA&usg=AFQjCNE4DSjo6XKaaHtU6ZFZ_SGO7f75CA&bvm=bv.53537100,d.aWc.

TEEB. 2010. "The Economics of Ecosystems and Biodiversity Mainstreaming the Economics of Nature: A Synthesis of the Approach, Conclusions and Recommendations of TEEB." *Economics of Ecosystems and Biodiversity*. http://www.teebweb.org/mainstreaming-the-economics-of-nature-a-synthesis-of-the-approach-conclusions-and-recommendations-of-teeb-launch-of-final-teeb-report/.

UNCSD. 2012a. "The Future We Want—Zero Draft." *United Nations Conference on Sustainable Development*. http://www.uncsd2012.org/mgzerodraft.html.

UNCSD. 2012b. "Statements." *Rio+20 Statements*. https://rio20.un.org/rio20/records/page?field_meeting_conference_tid=81.

UNEP. 2011. *Towards a Green Economy: Pathways to Sustainable Development and Poverty Eradication*. Nairobi, Kenya: United Nations Environment Programme.

VCS. 2013. "Jurisdictional & Nested REDD+." *Verified Carbon Standard*. http://www.v-c-s.org/sites/v-c-s.org/files/FactSheet%20JNRI%202013%20-%20MidRes_2.pdf.

World Bank. 2010. *World Development Report 2010: Development and Climate Change*. Washington, DC: World Bank.

World Bank. 2012a. "Carbon Finance at the World Bank." *The World Bank: Carbon Finance Unit*. http://web.worldbank.org/WBSITE/EXTERNAL/TOPICS/ENVIRONMENT/EXTCARBONFINANCE/0,,menuPK:4125909~pagePK:64168427~piPK:64168435~theSitePK:4125853,00.html.

World Bank. 2012b. *Inclusive Green Growth: The Pathway to Sustainable Development*. Washington, DC: World Bank.

Wunder, Sven. 2007. "The Efficiency of Payments for Environmental Services in Tropical Conservation." *Conservation Biology* 21 (1): 48–58. doi:10.1111/j.1523-1739.2006.00559.x.

Wunder, Sven. 2013. "When Payments for Environmental Services Will Work for Conservation." *Conservation Letters* 6 (4): 230–237. doi:10.1111/conl.12034.

3

FOREST CARBON SINKS PRIOR TO REDD

A Brief History of Their Role in the Clean Development Mechanism

María Gutiérrez

Sinks under the CDM: Sinks or Toilets?

> Why should African governments let their land be used as a toilet for absorbing emissions from Americans' second cars?[1]

Already in 1977, in what is often considered the earliest paper on sinks and climate change, Dyson suggested that, provided they are planted on a sufficient scale, trees could offset the global annual increase in emissions in the face of an imminent ecological disaster from increasing CO_2 levels. To make this economically feasible, he suggested plantations be carried out "by labour intensive methods in countries where labour is cheap" (Dyson 1977, 290).

Twenty years later, arguments for including sinks in developing countries under the Kyoto Protocol had become more sophisticated. There was less explicit reference to cheap labor and instead, in line with prevalent neoliberal economic ideas (that each place should specialize in and produce what it can do best and trade the rest), the argument revolved mainly around the suitability of the land. The Special Report on Land Use, Land Use Change and Forestry (LULUCF) prepared by the Intergovernmental Panel on Climate Change (IPCC 2000) notes the great difference in the rate at which forests grow in the tropics and subtropics compared to temperate areas. The projected costs of establishing forestation schemes to act as sinks were generally assumed to be between $0.01 and $20.00 per ton of carbon absorbed in the tropics, compared to $20–100 per ton of carbon in non-tropical countries (see IPCC 2000, chapter 5.2.3).

This chapter aims to explain the context in which the idea of trees as offsets for greenhouse gas (GHG) emissions—so-called sinks—first arose under the UN climate change regime. It highlights key aspects of the concept of sinks and the trade that they were conceived to create, and briefly outlines the history of the negotiations on sinks under the Clean Development Mechanism (CDM) of the United Nations Framework Convention on Climate Change (UNFCCC). It also includes some notes on the UN and the nature of the negotiations. This is all based on the premise that to understand current discussions on REDD+, one has to attend to the history of the climate change negotiations on sinks. In this context, forests are only a small part of what multilateral climate change mitigation is about, and they are often used as a bargaining chip during negotiations. This helps explain why addressing deforestation in developing countries took off in the context of a market for GHG emission offsets and not in other UN processes dealing specifically with forests, and suggests that to understand REDD+ today it pays to keep this context in mind.

The chapter is based on personal observations and research carried out between 2000 and 2007 while covering the UNFCCC negotiations for the Earth Negotiations Bulletin (ENB), an independent reporting service summarizing proceedings at multilateral negotiations on environment and development[2] (see Gutiérrez 2007, 2011).

It is important to note that, as a history of the negotiations, this account focuses on nation-states as key actors. Indeed, it is the defining characteristic of the UN negotiations that only nation-states under the UN have a voice. What "sustainable development" actually means specifically, or any social and equity concerns, are deemed to be a matter of national sovereignty, and therefore beyond the purview of the UN body. Non-state actors or organizations with a clear idea of their interest have to lobby country delegates to have their position included, but cannot intervene in the UN negotiations on their own account.

The Problem with Sinks

Carbon sequestration appeared as a simple idea in principle. Trees and vegetation absorb carbon from the atmosphere through the process of photosynthesis. This carbon can be measured, albeit not always with great accuracy. With deforestation, the carbon is released back into the atmosphere, contributing to the greenhouse effect—as well as to loss of water retention by the soil and other processes often leading to environmental degradation. Because the problem of climate change was conceived "from the point of view of the atmosphere," it did not matter where in the world a reduction in emissions took place.[3] Thus trees and vegetative growth in one place could offset carbon emitted anywhere else. It also happens that areas primarily suitable for forests[4] in developing countries tend to be populated by the poorest peasant farmers, because they are isolated and often located on steep hillsides and are difficult to work, and because the soil cannot productively sustain intensive agriculture without a great amount of agricultural inputs; attempts to extract something out of this land usually lead to increased land degradation and poverty. So even though it was recognized that, in the context of climate change, carbon sequestration was largely a temporary solution of limited effect,[5] investing in carbon sinks appeared as a simple and inexpensive solution that addressed both the increase in greenhouse gas emissions and deforestation—two major environmental problems—at the same time.

Yet once the idea of accounting for carbon sinks in the context of a market was accepted, an interminable list of problems appeared. Simply creating a market for emission permits implies serious questions of equity and moral dilemmas. It requires assigning property rights to something that can be bought and sold. This, in practice, means assigning rights to pollute. How does one distribute rights to change the earth's climate for generations to come? Because climate change is the result of cumulative GHG emissions for which industrialized nations are historically responsible, it was not possible for starters to assume that all counties had equal responsibilities and therefore obligations. Per capita emission rights might sound like a fair and reasonable idea in principle, and was propounded most vocally by NGOs from India and other developing countries. But that would place limits on developing necessary infrastructure in areas with low population density, such as in small island states, and in countries with high emigration rates.

These difficulties are compounded in the specific case of assigning rights to carbon sinks. Since trees and vegetation absorb carbon dioxide as they grow and release it back into the atmosphere as they decay and die, how does one account for the non-permanent reductions of carbon? What if after 20 or 30 or 60 years (the time allowed for CDM projects), once the credits have been sold and used, the plantation catches fire or succumbs to a disease, releasing the stored carbon back into the atmosphere? Who is liable for this re-emission of carbon? Because agricultural and silvicultural economic activities do not take place in a vacuum, establishing a plantation in one area may lead to deforestation in another, resulting in an increase in carbon emissions (so-called leakage). How can one assure this does not happen, and if it does, how does one account for the possibility and deduct from the credits? Most basically, who owns the carbon that trees necessarily absorb in the process of photosynthesis? The answer varies even where land tenure and property rights are clear. Since this is part of a UN agreement that, it is mandated, should be public and transparent, the market has to be credible and legitimate. Real offsets in emissions have to be proven to have taken place, because for every credit bought in a developing country, an equivalent amount of greenhouse gas will be emitted into the atmosphere in an industrialized one.

Sinks as a Bargaining Chip

A key thing to understand about sinks as they developed under the UNFCCC is that they are mainly a negotiating tool in the wider context of overall emissions reduction commitments. Although the land use sector does account for emissions in many countries, for the largest emitters and other countries with important forestry or extensive agricultural sectors (the United States, Australia and Canada, as well as Russia, Norway and New Zealand, for example), including sinks in overall accounting of emissions allowed them to assume the commitment to cut emissions much more easily. For the United States, including broadly defined sinks meant that its reduction target under the Kyoto Protocol went down from 7 percent to 4 percent. Sinks were thus the ultimate flexible mechanism.

But sinks were also of interest to many Latin American countries and some African ones, which saw them as a potential source of income in a sector where foreign investment is scarce, and where the challenge is compounded by international pressure for forest conservation. For some poor African countries, whose emissions from fossil fuel burning account for less than 2 percent of the global total, they represented the only opportunity to participate in the Kyoto market (see Goetze 1999).

And precisely because they meant so much for some of the largest emitters, countries like the small island states (with no reduction commitments but with serious concerns for the effects of climate change) sometimes used sinks as a bargaining chip to press for more substantive action on other issues, such as emissions from aviation and maritime transport. A representative of the Alliance of Small Island States (AOSIS) referred to sinks as "the only leverage we have" (personal communication, November 2004).

In fact, sinks were used as a negotiating token when, at the Sixth UNFCCC Conference of the Parties (COP) in The Hague in 2000, in a last-minute attempt to save the negotiations, the United States offered the European Union (EU) to settle for no sinks under the CDM if they were allowed under Protocol Article 3.4 (additional LULUCF activities for developed countries)—even if in doing so the United States was betraying its Latin American partners (Agarwal et al. 2001, 256; Fry 2002, 167).

Yet in many ways this bargaining chip grew out of proportion to its importance in the overall climate change context, as people started the work of trying to delimit and make carbon sequestration from land use change quantifiable and verifiable. There is the widely held sense that very few people had any idea what they were getting into. Simply put, some of the technical challenges seemed insurmountable. Something as fundamental as clearly distinguishing direct human-induced changes from indirect and natural ones cannot be resolved scientifically, and people could only go back to the negotiating table to find a political solution.

Negotiating Sinks in the CDM[6]

Even before the adoption of the Kyoto Protocol in 1997, expectations on the inclusion of sinks in the CDM ran high. There was talk of billions of dollars going south for the sale of reduction emission credits. But these expectations were not shared by all. Many countries opposed sinks in the CDM based on the uncertainty of the reductions that, they claimed, could threaten the "environmental integrity" of the Protocol. This was the official position of the EU, AOSIS and others, as well as of a number of powerful international environmental NGOs (nongovernmental organizations) such as Greenpeace, the World Wildlife Fund, Friends of the Earth and others grouped under the Climate Action Network (CAN).[7]

Another persuasive and important argument against sink projects in the CDM was that they deliver little in terms of technology transfer to developing countries. Moreover, because the assumption was always that forest projects presented a cheap option to acquire reduction credits, it was feared that if they competed with energy projects under the CDM, they would divert investment that would otherwise be directed at more permanent improvements in energy efficiency and at developing renewable energy. China, for one, stood to gain much more from energy projects and was not really interested in sinks. AOSIS, knowing that they would be the most directly affected by the impacts of climate change and with few chances of implementing sink projects, also opposed sinks, wanting rather to ensure that emissions would be reduced at the source. The Indigenous Peoples Forum on Climate Change likewise opposed sinks in the CDM, seeing them as a threat to their worldview and land use, and called them "a worldwide strategy for expropriating our lands and territories and violating our fundamental rights that would culminate in a new form of colonialism."[8]

Brazil's position was mystifying to many people, who a priori assumed Brazil had most to gain from a market in sinks—in particular, from avoided deforestation. Yet while Brazil did support the inclusion of plantations and agroforestry under the CDM, it adamantly opposed including avoided deforestation. Sovereignty concerns are commonly used to explain this position, as Brazil, like Peru, resisted any strategy that would tie up land that could be used for the country's national development. Some have referred to the Brazilian government's fear of "'internationalization' of the Amazon using environmental protection as an excuse" (Fearnside 2001, 174). Moreover, there seemed to be a discrepancy of views within the government about this (differences between the Ministry of Foreign Relations and the Ministry of the Environment, as well as differences among and between state and federal governments [Fearnside 2001]). But all throughout the negotiations that led to the so-called Bonn Agreements and Marrakesh Accords in 2001, when the first basic rules on sinks under the CDM were adopted (see UNFCCC Decision 11/CP.7), the Brazilian delegation never wavered in presenting avoided deforestation as creating an enormous loophole for real reduction commitments in industrialized countries, knowing too well how precarious forest conservation is and how difficult it is to monitor (see ENB 1999, 2000a–d). In retrospect, it seems clear that Brazil foresaw a time when all countries would acquire emission reduction commitments and was not prepared to give up the cheapest option to developed countries under the CDM market when it could use it itself.

Yet for many other developing countries, in particular many Latin American ones, sinks appeared from the beginning as a development opportunity to address soil erosion, degradation of land and loss of forest cover. They welcomed what they saw as an opportunity to deal with natural resource conservation as free as possible from the intervention from international agencies and environmental lobby groups.[9] Sustainable land use policies and practices were basic sustainable development goals—insofar as they implied erosion prevention and improved water quality—and they thought the CDM could also provide inputs in research and capacity building activities to assist in this goal. They argued that assigning a price to carbon was one of the very few ways that sustainable forest management and conservation could compete with alternative land uses and avoid further land degradation. Costa Rica, Colombia and several other countries were already coming up with national policies recognizing so-called environmental services provided by forests, but as usual, lacked sufficient funding.

During the negotiations leading to the Marrakesh Accords at COP-7 in 2001, Latin American countries, including Bolivia, Colombia, Ecuador, Guatemala, Nicaragua, Uruguay and Costa Rica, were noteworthy in their support for a broad inclusion of sinks in the CDM.[10] Similarly, for many African countries with little prospect of investments in energy efficiency, sink projects represented the only possibility of participating in the Kyoto Protocol's new market mechanism. Meanwhile, some developed countries, in particular those in the so-called Umbrella Group (the United States, Canada, Australia, New Zealand, Iceland, Norway, the Russian Federation and Ukraine)[11] viewed sink projects under the CDM as a relatively inexpensive way to meet their reduction commitments, potentially while also meeting other international environmental goals. More to the point, since emissions from land use change and forestry account for an important part of emissions from developing countries, it was argued that improving forest and land management in developing countries would eventually contribute to stabilizing concentrations of GHGs—the ultimate objective of the Convention.

Other influential arguments for the inclusion of sinks under the CDM stemmed from the surge of public interest in forest conservation. This represented a large and extremely well-funded enterprise in developed countries. By contrast, most developing countries lacked the resources to undertake and maintain forest conservation, even if under pressure by international grant and loan-awarding organizations. Many had also bet on tourism as a source of foreign income, and had publicized natural resources as national attractions. At the time when sinks in the CDM were first being negotiated, the sale of emission reduction credits from sinks appeared to many developing countries as a potential source of sustained income in a sector where no strings attached investment was practically non-existent. The money so gained could be used to assist conservation efforts and even possibly contribute to rural development.

This view was very much promoted by what is known as the conservation lobby—a number of powerful environmental NGOs, mainly from the United

States, who had already invested in carbon sequestration projects even prior to the adoption of CDM rules on sinks. These organizations knew that they had a major role to play as intermediaries, and saw many potential benefits from the inclusion of sinks under the CDM. Although they were less vocal regarding tree plantations, groups such as The Nature Conservancy, the Union of Concerned Scientists, Conservation International, Environmental Defense Fund and others argued that, with clear rules, sink projects in the CDM could deliver social and environmental benefits, promote biodiversity and help developing countries in the conservation of forests (see Boyd et al. 2008).

But the technical details and unresolved questions on accounting for sinks appeared more daunting the more negotiators got into it. Some of the basic questions that had to be resolved included: how does one define a forest in a way that applies to all countries and contexts? Even the common definition of forests had changed in the last few years, having evolved from notions that included the idea of ecological climax, to references to ecosystems, and then images of an "erratic, shifting mosaic of trees." But then what about savannas and woodlands? And short shrubs, mangroves, or marshland? And forest transitions? And similarly, how does one define afforestation, reforestation and deforestation?[12]

Other equally tricky questions were: What makes the removal by sinks "human-induced"—which all CDM activities were required to be? How does one set apart natural regeneration or fertilization from other natural processes or past practices? What happens when the carbon reverts back to the atmosphere as a result of forest fires, pests and other natural disturbances? Will setting aside land as carbon stock result in other areas being affected from the displacement of activities in these land to other locations? How does one interpret "since 1990"? Is it fair to use the same baselines for all countries? How does one verify and ensure accuracy? Clearly these questions were hardly just technical. In fact, any call to distinguish between political and technical issues on matters related to land use and forestry was quickly considered politically motivated.[13]

In order to sort out these problems of accounting and assure the credibility of the market, delegates at the UN came up with an incredibly complex set of rules for sink projects under the CDM, first adopted at COP-7 in Marrakesh in 2001 and then at COP-9 in Milan in 2003. A deal was struck to exclude avoided deforestation from the CDM and allow only afforestation and reforestation (A/R) project activities.[14] A limit was placed on the amount of emission reductions an industrialized country with reduction commitment under the Protocol (so-called Annex I country) could claim from these projects under the CDM, amounting to a maximum of 1 percent of the country's base year emissions for each of the 5 years of the commitment period (UNFCCC Decisions 5/CP.7 and 19/CP.9). To enter the CDM market, project developers had to submit an application to the UN, for which they had to pay up front, and which included a complex carbon accounting methodology.

Since the definitions and rules were intended to apply to the whole world, they hardly fit the diverse reality on the ground. An analysis by ENCOFOR (Zomer et al. 2005) showed that much of the land area technically eligible for CDM projects in developing countries had a high density population; was primarily under 1,000 m in altitude; was for the most part already agriculturally productive; and 75 percent of it was in Asia. So while the sale of emission credits could have been used to reforest poor areas of primarily forest altitude, where land degradation is a serious problem and economic opportunities scarce, the areas that applied were those that were already agriculturally productive and comparatively prosperous.

Moreover, given the complexity of the process and the rules, the transaction costs became so high that only the largest projects that already had the surplus money to invest, or those that managed to have the transaction costs covered by an NGO, were able to participate. And because the rules require highly specialized technical knowledge and methods, only a few entities have been accredited to review CDM projects—almost all of which are based and staffed, predictably, in the wealthier industrialized countries. As a simple commodity chain could show, the biggest profit is bound to be made by the few accredited overseeing agencies that verify and validate the product. And yet it was precisely these complicated rules and restrictions that enabled the creation of the scarcity conditions necessary for the market to have some value—for those who could afford it.

On the UN and the Nature of Negotiations

> A true bureaucracy (in the pejorative sense of that word) will never be interested in the validity of the results, but in the validity of the process producing the results.
> (Galtung 1986, 6)

The complexity of climate change is compounded. Like other environmental problems, it knows no geographic boundaries and needs cooperative and coordinated action by states to limit its effects, yet this goes against the foundations of international law, which is based on "concepts of state responsibility, sovereign equality, and the paramountcy of state consent" (Yamin and Depledge 2003, 2). But in addition, climate change cuts across scales not only of space but also of time, and is characterized by scientific uncertainty (involving complex interactions between the Earth's atmosphere, the biosphere and the oceans over time, and the impact of human activities upon them). This implies potentially huge damages and costs (such as cultural loss and human mortality) that are impossible to calculate, and entails vast discrepancies between those responsible for the damage and those mostly affected by it.

The international community has managed to negotiate two treaties in less than a decade to deal with the issue: the UNFCCC in 1992 and the Kyoto Protocol in 1997 (the latter binding). These agreements, and particularly the Kyoto Protocol, have come to share the complexity of the issue itself and become inaccessible to anyone except those involved in the negotiations. The rules are increasingly technical, and have produced specialized experts who know in detail only particular topics. Very few people have an understanding of the whole picture. Even assuming one could read all the documents, they are impossible to understand without knowledge of the institutional practices and procedures, both formal and informal, behind the negotiations. As Bodansky notes regarding the difficulty of interpreting the UNFCCC:

> Words are debated and selected as much for their political and for their legal significance. Indeed, proposed formulations often took on a talismanic quality, only distantly connected to the actual meaning of the words. Linguistic debates became a proxy for political confrontation, with success or failure measured not just by the substantive outcomes, but also by the inclusion or exclusion of particular terms. For example, developing and developed countries argued for hours over whether economic development should be characterized as "essential" or a "prerequisite" for developing countries' response measures. Delegations often sought to introduce identical language in different parts of the Convention or to move language from one part of the Convention to another, not to effect particular legal consequences, but to highlight certain provisions for political reasons.
> *(Bodansky 1993, 492–493)*

The Elusive Fairness

All men are equal before the law, but they are no longer equal after it.
(Dahrendorf 1968, quoted in Collier 1975, 126)

In principle, the UNFCCC and Kyoto Protocol negotiations, like most other UN processes, are set up to ensure equality of representation and opportunities. So, at least in the context of formal negotiations, the United States has exactly the same rights and responsibilities as Niue or Bhutan. It is one of the few fora where Parties' interests are equally heard and, as the nation-state of Tuvalu has proven several times, individuals can exert an influence in the process not proportional to the economic or political power of the nation they represent. Yet as anthropologists have noted, redistribution and reciprocity can also reinforce inequality, when the exchanges made result from different valuations of goods exchanged among partners with different needs (see Mauss [1950] 1990; see also Orlove 1977 in Wolf 2001, 164). This inequality is reinforced and reproduced in myriad ways—starting with the most insidious and obvious one, that of the use of English as the common language for drafting the texts, and for negotiations when official translation to the six UN languages is not available.[15]

UN institutional practices and procedures are highly ritualistic and self-referential. And under the rituals it is easier to hide real inequalities.[16] As noted by Gupta (2001), although the process is political, the manner is by default legal. This international legal system generates frustration among delegates from developing countries, because

> while promising the rule of law in terms of procedural issues, [it] does not provide substantive guidance. There is a fear that legal principles of justice and fairness are not being developed, and instead the international legal system provides an arena for *realpolitik*, which then gets institutionalized by the legal power of precedents.[17]

Thus, the international legal process appears to merely ensure a "'polite order' within which the 'rules of the jungle' operate" (Gupta 2001, 142; see also Gupta 2000a, 2000b).

Furthermore, the highly formal and ritualistic manner of international negotiations in the UN fora is eccentric and complicated to learn. Delegates from poor countries are "socialized" into the process, but often their involvement remains a formality (Gupta 2001). During the negotiations, the less compromising option for the delegates, and often the only one available when they lack explicit guidelines from the home government, is keeping quiet. Even when a certain general mandate has been approved by their government, because the negotiations proceed rapidly and often unpredictably (given the large number of parties involved and complexity of issues addressed), it is very hard to have a position on everything. Delegates thus often accept things by default to avoid being further exposed. Moreover, developing country delegates dedicate most efforts toward analyzing developed countries' proposals, suspicious of what they entail, leaving them little time to elaborate their own proposals. As one delegate told Gupta, "We mistrust the North and we spend all our time analyzing their agenda rather than preparing ours" (Gupta 2001, 142). This attitude often generates frustration on the other side of the table: developed country delegates often lose patience and faith in frequently ill-prepared negotiators who respond most often on the defensive. Bias is thus reproduced, and mistrust bred.

A lot has to do with the importance accorded to climate change and environmental issues in general in most developing countries: *it is not high on the agenda given other more pressing social problems and few resources*. The sense is that, unlike the perception in industrialized countries of climate change as a common, scientific and technological problem, climate change in most developing countries is for the most part perceived as symptomatic of a more systemic problem to do with unequal distribution and development, caused largely by industrialized countries (Gupta 1997).

Inequality Reproduced

Perhaps the most important expression of a market is the creation of conditions of scarcity that ensure that the goods have value. Although vital, this has proved especially complicated in the case of carbon sequestration carried on by trees and vegetation, which as noted earlier happens anyway, everywhere. A number of measures were introduced in the CDM rules and definitions to ensure that sink credits would not "flood the credits market." These included placing a cap on CDM sink credits that can be used for compliance with countries' emission reduction commitments, and the allowance of only A/R project activities. But scarcity was also achieved by making the rules complex, the modalities difficult to apply to projects, and the transaction costs high.

Thus the inbuilt inequality in the creation of a market that Marx, Polanyi and countless others have described could be seen slowly but clearly emerging, as carbon sequestration went from the novel (if also controversial) concept of a payment by industries for the environmental service that peasant farmers in developing countries provide in maintaining forests, to the present state where carbon accounting and a safe investment take precedence over every other social or environmental consideration. In the process, a new enclosure of the environment was conceived, so that a new commodity could be sold.

REDD+ started out as a way to address many of the problems associated with A/R projects under the CDM. It was proposed by developing countries, meant to be undertaken at the national level with state support and resources, and to focus primarily on a reduction in emissions from deforestation and forest degradation. Ten years and many millions of dollars later, some of the same problems that made the CDM projects so intractable have developed under REDD+ (including, for example, the increased emphasis on project-level interventions) (see Angelsen et al. 2012). It is particularly problematic to see REDD+ in the context of climate change (on which this chapter insists) as dependent on payment for verified emission reductions under the UNFCCC, which is unlikely to materialize given the dismal level of climate change mitigation ambition at the nation-state level and the lack of legally binding commitments under that process.

Still, there is no denying that in many countries there is greater awareness and information on deforestation and forest degradation, as well as national-level policy and institutional development, and even in many instances local engagement, that would most likely not exist were it not for this process. While there is much to improve on and watch out for, compared to business as usual, where forests and land are cleared for mining or extensive large-scale farming without regard for the people or the environment, REDD+ is probably a problem worth having.

Notes

1. The metaphor of the toilet, as recounted by Grubb in noting the "depth of feelings" and complexities surrounding the issue, was used by an African attendant to

a meeting in the early 1990s. It was in response to a talk by an economist from a US environmental NGO, who elaborated on the advantages of Activities Implemented Jointly (the pilot phase for flexible mechanisms before the creation of the CDM and Joint Implementation) and the reduced costs of absorbing CO_2 in Africa compared to limiting emissions in the United States. "Shaking with anger, an African present arose and asked 'why should African governments let their land be used as a toilet for absorbing emissions from Americans' second cars?'" (Grubb et al. 1999, 99).
2. ENB is published by the International Institute for Sustainable Development (IISD), a non-profit organization based in Canada. See http://www.iisd.ca.
3. "From the point of view of the atmosphere" is an expression often heard in the negotiations to this day. Key to this understanding of the problem is the "comprehensive approach," which entails collectively accounting for removals as well as emissions of different greenhouse gases and measuring them according to a single metric—that of their global warming potential. Until the United States brought up this approach in 1989, sources of emissions and sinks had been considered separately. But as negotiations were launched for the UNFCCC in 1990, the US government presented sinks as "negative emissions" (Bodansky 1993, 517).
4. This refers to land most aptly and naturally suited to forest and where any other land use would require substantial amounts of inputs and transformation.
5. Even large areas of forests have relatively small effects in relation to the amount of daily carbon emissions. In 1994 the UN Conference on Trade and Development (UNCTAD) calculated that an area of forest larger than France would need to be planted every year to compensate for the existing rate of fossil fuel emissions—and emissions have only increased since then.
6. To give an idea of the negotiations at that time, this section provides a simplified picture and generalizes countries' positions, taking at face value their claims during the discussions in the negotiation process. It does not question individual governments' intentions or try to uncover whether what they claimed was meant.
7. For a general overview of the issues and views, see ENB (1999). See also Boyd et al. (2008).
8. See the Declaration of the First International Forum of Indigenous Peoples on Climate Change, http://amazonwatch.org/news/2000/0906-declaration-of-the-first-international-forum-of. See also http://www.c3.hu/~bocs/eco-a-1.htm. For a more elaborate account of their position, see Boyd (2003).
9. In contrast with other options, such as debt-for-nature swaps.
10. As a group, Latin American countries had hosted 15 out of 20 LULUCF projects under the Activities Implemented Jointly (AIJ) pilot phase (see http://unfccc.int/kyoto_mechanisms/aij/activities_implemented_jointly/items/2094.php).
11. The Umbrella Group is a loose coalition of countries. Membership varies, and although they rarely make statements together, they usually share information and meet during UNFCCC negotiating sessions.
12. For a general sense of these discussions and issues at the time, see ENB (2000b), http://www.iisd.ca/download/pdf/enb12141e.pdf. See also Gutiérrez (2000) for further elaboration on the various definitional questions.
13. See ENB (1999), http://www.iisd.ca/download/pdf/enb12123e.pdf.
14. For a more detailed discussion on how avoided deforestation was excluded from accruing emission credits under the CDM, see Boyd (2003).
15. In regards to CDM sink projects, the working language for the CDM executive board is English-only, and the project design document (PDD) has to be filled out in English (even if some guidance and methodologies are available in the six official UN languages). See http://cdm.unfccc.int/Reference/index.html.
16. There is a vast literature in anthropology on this role of rituals. A classic example is Turner (1967).

17. Note the importance of precedent. This aspect might be well known to legal historians and practitioners, but its relevance is not self-evident to the layperson. The concern with precedent affects the negotiations through and through. In it lies the explanation for many obscure positions—or at least it is so argued. Pieces of text are taken from other pieces of text, and decisions are made based on other decisions.

References

Agarwal, Anil, Sunita Narain, and Anju Sharma. 2001. *Poles Apart: Global Environmental Negotiations 2*. New Delhi: Centre for Science and Environment.

Angelsen, A., M. Brockhaus, W.D. Sunderlin, and L.V. Verchot, eds. 2012. *Analysing REDD+: Challenges and Choices*. Bogor, Indonesia: CIFOR.

Bodansky, D. 1993. The United Nations Framework Convention on Climate Change: A Commentary. *Yale Journal of International Law* 18(2): 451–558.

Boyd, Emily. 2003. *Forests Post-Kyoto: Global Priorities and Local Realities*. Unpublished PhD thesis, University of East Anglia.

Boyd, Emily, Esteve Corbera, and Manuel Estrada. 2008. UNFCCC Negotiations (Pre-Kyoto to COP-9): What the Process Says about the Politics of CDM-Sinks. *International Environmental Agreements: Politics, Law and Economics* 8(2): 95–112.

Collier, Jane E. 1975. Legal Processes. *Annual Review of Anthropology* 4: 121–144.

Dahrendorf, R. 1968. *Essays in the Theory of Society*. Stanford, CA: Stanford University Press.

Dyson, F.J. 1977. Can We Control the Carbon Dioxide in the Atmosphere? *Energy* 2: 287–291.

Earth Negotiations Bulletin (ENB). 1999. Fifth Conference of the Parties to the UNFCCC: 25 October–5 November. Vol. 12, no. 123. IISD.

ENB. 2000a. Twelfth Session of the Subsidiary Bodies of the UNFCCC: 5–16 June. Vol. 12, no. 137. IISD.

ENB. 2000b. Summary of the Workshop on Land Use, Land-Use Change and Forestry: 10–13 July. Vol. 12, no. 141. IISD.

ENB. 2000c. Sixth Conference of the Parties to the Framework Convention on Climate Change. Vol. 12, no. 163. IISD.

ENB. 2000d. Thirteenth Session of the Subsidiary Bodies of the UN Framework Convention on Climate Change: 4–15 September. Vol. 12, no. 151. IISD.

Fearnside, P.M. 2001. Saving Tropical Forests as a Global Warming Countermeasure: An Issue That Divides the Environmental Movement. *Ecological Economics* 39(2): 167–184.

Fry, Ian. 2002. Twists and Turns in the Jungle: Exploring the Evolution of Land Use, Land-Use Change and Forestry within the Kyoto Protocol. *Review of European Community & International Environmental Law (RECIEL)* 11: 2.

Galtung, Johan. 1986. On the Anthropology of the UN system. In *The Nature of United Nations Bureaucracies*, David Pitt and Thomas G. Weiss, eds., 1–22. Boulder, CO: Westview Press.

Goetze, Darren. 1999. Report on the 5th Meeting of the Conference of the Parties to the Climate Convention. *Union of Concerned Scientists*. November.

Grubb, Michael, Christiaan Vrolijk and Duncan Brack. 1999. *The Kyoto Protocol: A Guide and Assessment*. London: Royal Institute of International Affairs.

Gupta, Joyeeta. 1997. *The Climate Change Convention and Developing Countries: From Conflict to Consensus?* Dordrecht: Kluwer Academic.

Gupta, Joyeeta. 2000a. 'On Behalf of My Delegation, . . .': A Survival Guide for Developing Country Climate Negotiators. In *Climate Change Knowledge Network, Center for Sustainable Development in the Americas and International Institute for Sustainable Development.*

Gupta, Joyeeta. 2000b. North-South Aspects of the Climate Change Issue: Towards a Negotiating Theory and Strategy for Developing Countries. *International Journal of Sustainable Development* 3(2): 115–135.

Gutiérrez, María. 2007. *All That Is Air Turns Solid: The Creation of a Market for Sinks under the Kyoto Protocol on Climate Change.* Unpublished PhD thesis, City University of New York.

Gutiérrez, María. 2011. Making Markets Out of Thin Air: A Case of Capital Involution. In *The New Carbon Economy: Constitution, Governance and Contestation*, Peter Newell, Max Boykoff and Emily Boyd, eds., 41–64. West Sussex: Wiley-Blackwell.

Intergovernmental Panel on Climate Change (IPCC). 2000. *Land Use, Land Use Change and Forestry: Special Report*, Robert T. Watson, Ian R. Noble, Bert Bolin, N. H. Ravindranath, David J. Verardo and David J. Dokken, eds. Cambridge: Cambridge University Press.

Mauss, Marcel. [1950] 1990. *The Gift: Forms and Functions of Exchange in Archaic Societies.* London: Routledge.

Newell, P., M. Boykoff and E. Boyd, eds. Special Issue. In *Antipode* 43(3): 639–661. Wiley-Blackwell.

Orlove, B. 1977. *Alpacas, Sheep and Men: The Wool Export Economy and Regional Society in Southern Peru.* New York: Academic Press.

Turner, Victor. 1967. *The Forest of Symbols: Aspects of Ndembu Ritual.* New York: Cornell University Press.

Wolf, Eric R. 2001. *Pathways of Power: Building an Anthropology of the Modern World.* Berkeley: University of California Press.

Yamin, Farhana and Joanna Depledge. 2003. *The International Climate Change Regime: A Guide to Rules, Institutions and Procedures.* Cambridge: Cambridge University Press.

Zomer, Robert, Antonio Trabucco, Oliver Van Straaten, Lou Verchot and Bart Muys. 2005. Global Analysis of the Implications of the Definition of Forests on Land Area Eligible for Afforestation and Reforestation Activities in the CDM. *ENCOFOR.* Consultative Group on International Agricultural Research (CGIAR). http://csi.cgiar.org/encofor/forest/.

4

JUSTICE AND EQUITY IN CARBON OFFSET GOVERNANCE

Debates and Dilemmas

Mary Finley-Brook

Justice and Equity in Carbon Offset Governance

Trade-offs complicate development interventions so that benefits for one group or area often imply costs for another; large-scale projects deemed highly efficient in economic terms may generate harmful environmental or social externalities. This chapter explores issues of justice in carbon trading in terms of decision-making power and the subsequent distribution of positive and negative impacts. It explores whether offset governance can help resolve widespread problems, such as racial or income inequality and environmental injustice.

A spectrum of offset governance structures exists in regulated compliance markets organized through the United Nations Framework Convention on Climate Change (UNFCCC) and in voluntary markets, ranging from highly rigorous approaches to informal exchanges. To contextualize offsets, it is helpful to identify key actors and assess prior initiatives, including national-level payments for environmental services (PES). Lessons can also be drawn from the UNFCCC's Clean Development Mechanism (CDM) and first-generation initiatives for reducing emissions from deforestation and forest degradation (REDD). While offset impacts are site-specific, the study of cross-national and multisite patterns can help identify nascent trends (Caplow et al. 2011; Murdiyarso et al. 2012; Sunderlin et al. 2014). To achieve REDD+, as opposed to basic REDD, is something that few projects have done—it requires sustainable forest management and enhancement of carbon stocks as well as 3E+ criteria (effectiveness, efficiency, equity, social and environmental co-benefits) (Sunderlin et al. 2014).

The analysis in this chapter draws from civic, state, and private sector accounts, scholarly publications, and fieldwork since the mid-1990s assessing a wide range of community development models and their intersections with state governmentality, donor interventions, market trends, and nongovernmental organization (NGO) initiatives. Important insights arise from annual fieldwork since 2009 analyzing CDM projects in Costa Rica, the Dominican Republic, Nicaragua, and Panama, as well as REDD readiness initiatives in Panama and Peru (2013 and 2014).[1]

Geographic Tools for Analyzing Ecological Change, Complexity, and Power

Geographers view climate change mitigation and adaptation as important examples of dynamism and complexity in human-environment interactions. Topics that integrate political, economic, cultural, and ecological knowledge generally benefit from geographic analysis; and PES and offset cases throughout this book demonstrate the utility of several common practices in geography, including:

1. Simultaneous focus on multiple governance scales and various social groups, with the recognition that perceptions and priorities can shift based on scale and sector;
2. Attention to economic flows and other connections between locations as well as resulting changes to landscape patterns and livelihood strategies;
3. Documentation of uneven development through patterns of ecological, spatial, and social distributions of costs and benefits;
4. Analysis of power and its linkages to territoriality and claims to space.

Offset Governance

Environmental governance has broadened from the centralized government mandates of former decades as decision-making shifted upward to multilateral and transnational organizations, downward to regional and local institutions, and outward to indigenous federations, NGOs, and the private sector (Swyngedouw 2000; Thompson et al. 2011). Changes often resulted from intense political struggle, and governance processes remain marred by power inequalities and vested interests (Swyngedouw 2000). Offset governance research must examine decision-making power (procedural equity) as well as implications for material outcomes, such as access to resources or profits (distributional equity) (Larson and Petkova 2011). It is also essential to determine if there are historical or preexisting conditions that limit or facilitate access to procedural and distributional equity (contextual equity) (McDermott et al. 2013). For example, land tenure insecurity will negatively impact both distributional and procedural equity.

The ways powerful groups define environmental issues legitimizes certain potential solutions while marginalizing others (Thompson et al. 2011). The Kyoto Protocol sought to mitigate climate change through carbon offsets in developing countries without achieving binding commitments to reduce domestic emissions in industrialized nations. Offsets build from the assumption that valuation in monetary terms will correct ecological imbalance with more efficient resource allocation (Corbera 2012). Using market instruments to rectify environmental problems is limited because it breaks nature into measurable components, while financial mechanisms protect only the parts capable of generating income (Corbera 2012; COICA 2014).

Carbon trading is only one form of climate change mitigation, but it has broad support among multilateral agencies, development banks, private firms, and big NGOs. As networks of powerful actors promote carbon trade, developing countries with high potential for forest offsets may experience greater supply of funding for REDD readiness than actual demand (Gustavo Suárez de Frietas, interview, July 7, 2014). Meanwhile, dozens of voluntary sector forest carbon projects began in many countries before national policies or procedures for avoided deforestation could be defined.

Climate governance regimes are becoming increasingly complex (Chhatre et al. 2012; Suárez de Frietas, interview, July 7, 2014) through the growth of entities such as the UNFCCC, voluntary market associations, consulting agencies, and carbon brokers. At the same time, bilateral and multilateral institutions like the United Nations and World Bank have assumed leadership roles in carbon finance and climate policy. Offset mechanisms can be ill-suited for governance in national or local arenas because they are based on external priorities and build from development networks with predefined ideals (Clements 2010; Corbera 2012). In order to limit corruption, donors' reporting standards and accountability requirements are demanding and thus may impede or complicate customary governance processes within and among local and regional bodies. Changes required by donors might not respect the autonomy of local organizations, institutions, and government entities. For example, a lack of bylaws impeded Panama's National Coordinator of Indigenous Peoples (COONAPIP) from directly receiving UN-REDD funds, but subsequent pressure to convert the indigenous federation into an NGO violated a legal right to define self-governance (Mezúa, interview, July 9, 2014). This example demonstrates the complexity of defining authority within multiscale, multisector offset governance as stakeholders have different priorities, yet the ideals of those who control funds may be disproportionately privileged.

Tenure insecurity, widespread in developing countries and particularly in indigenous territories, is a contextual or historical inequality that often constrains local decision-making toward and benefits from avoided deforestation projects (Chhatre et al. 2012). Tenure rights should be clarified before starting resource management projects. Nevertheless, even after titling, rights sometimes exist only on paper: corruption and state failure to defend local rights against intrusions or competing interests contribute to ongoing insecurity (Larson 2011). Further, titleholders may not be considered owners of natural resource rights above and below the ground, including trees and carbon (Sunderlin et al. 2014), which could limit community benefit from REDD and encourage conflict during project creation and implementation.

Learning from Prior Conservation Initiatives

For more than two decades, integrated conservation and development programs (ICDPs) have attempted to move beyond the fortress-style environmental protection stemming from a false dichotomy of "people versus parks" in order to integrate social development into conservation. By 2000 an estimated 300 ICDPs utilizing millions of dollars of investment existed worldwide (Hughes and Flintan 2001). Nevertheless, most ICDPs struggled to balance both conservation and development and some did not achieve either (Blom et al. 2010), in part because project-scale interventions did not adequately address broader drivers of ecosystem degradation or lacked effective involvement from local communities (Sunderlin et al. 2014). ICDPs are similar to avoided deforestation initiatives because they share a goal of promoting sustainable livelihoods and poverty reduction within subnational environmental protection projects; thus the failures of ICDPs can provide important insights on how to improve the design of REDD (Blom et al. 2010). While REDD readiness programs generally promote institutional and policy reforms to encourage broader conservation, subnational avoided deforestation projects often still cannot address pressures exerted through tax policies or foreign investments that facilitate resource extraction and agro-industrial expansion (Sunderlin et al. 2014; Crippa and Gordon 2013). Moreover, government agents frequently remain unwilling or unable to impede illegal extraction in logging, mining, and other economic sectors that will negatively impact the viability and broader social and ecological resilience of project-based interventions.

As the implementation of many REDD initiatives continues to be at the project level, existing community-based natural resource management efforts, including ICDPs, have begun to target the inclusion of emissions reductions to take advantage of economic incentives (Blom et al. 2010; Sunderlin et al. 2014). Building from established ICDPs could benefit REDD projects, since relationships and governance structures already exist. Nevertheless, standing agreements and practices with communities could also create barriers to the adoption of new performance-based measures, whereby participants are compensated only after verification of emissions reductions, which is a foundational element of REDD.

Other pre-existing community forestry projects that might provide a base for REDD sell certified-sustainable timber. In these hybridized projects, extraction would complement avoided deforestation and carbon credits would be supplemental to logging income (Danis Saavadra, interview, June 30, 2014); however, the total amount of carbon captured per acre would decrease. Income diversification is important for poverty alleviation, especially because forest-based communities usually share proceeds from carbon sales with brokers, intermediaries, and consultants. If logging is curtailed, avoided deforestation projects need to recover opportunity costs so as not to intensify poverty. Potentially complementary activities include agro-forestry, sales of non-timber and artisanal products, ecotourism, and income from additional ecosystem services.

Avoided deforestation project designers have much to learn from prior PES initiatives due to concerns about procedural and distributive equity in PES programs around the world (Pascual et al. 2010; Corbera 2012; McDermott et al. 2013). PES efforts in national programs in places like Costa Rica, Mexico, Tanzania, Uganda, and Vietnam suggest the need to improve engagement of marginalized populations (Corbera 2012; McElwee 2012; McDermott et al. 2013). Without targets for increasing income equity and reducing poverty gaps, most PES arrangements are unlikely to have strong anti-poverty implications. For example, early Costa Rican PES programs demonstrated a tendency toward capture of benefits by businesses, large landholders, and farmers with higher levels of education (Kaimowitz 2008; McDermott et al. 2013). Farmers without land are unable to participate in some locations, while in Mexico they could participate but received fewer benefits, potentially widening inequalities between land holders and landless (Rico García-Amado et al. 2011). Furthermore, in some PES programs, intermediaries obtained as much as 50 percent of total income (Hajek et al. 2011). Even though financial mechanisms in PES and REDD have been identified in the literature as "game changers" because they provide direct incentives for conservation-oriented behavior (Murdiyarso et al. 2012), according to research done by the Center for International Forestry Research (CIFOR), provision of conditional incentives provided to participants after conservation success has become less of a priority in recent years (Sunderlin et al. 2014). International CIFOR research suggests many initial REDD projects are currently backing away from promising direct payments for avoided deforestation because of the lack of reliable and predictable future funding.

Learning from Clean Development Mechanism Offsets

For forestry projects to enter into the CDM they could not merely involve the preservation of standing forests and instead had to increase afforestation or reforestation (A/R). Due to high costs and administrative problems, A/R forestry offsets were seldom included in the CDM (Lederer 2011), but it is still important to evaluate clean development as a precursor to REDD in terms of social justice impacts. The first decade of CDM project experience highlights a number of risks for host communities, as well as negative ramifications for indigenous peoples and economically marginalized populations (Indigenous Environmental Network 2008; Finley-Brook and Thomas 2011; Savaresi 2012). For example, since 2007, a series of wind projects have been registered in Mexico under the CDM in spite of three interrelated concerns: (1) insufficient compensation to local populations; (2) disingenuous and illegal procedures for obtaining land contracts; and (3) the erosion of indigenous common property governance institutions (Baker 2012). Members of host communities often complain they were not formally consulted about projects and when they received information it was inaccurate, misleading, or partial (Pasqualetti 2011; Baker 2012). Project developers did not allow community decision-making about the size and number of turbines, construction of access roads, uses of water, irrigation access, and other issues impacting local economies. Local ranchers and farmers have found wind farm rents too low to compensate for production losses (Hawley 2009; Baker 2012).

Driven by methodological concerns related to leakage and verification, CDM validation processes are often perceived as cumbersome and costly. Nonetheless, public consultation procedures are frankly minimal and usually only involve a brief informational meeting where stakeholder comments are recorded; this is usually held in a semi-urban or town location and requires travel for impacted community members living at or near project sites. Second, the public consultation period has a 30-day window to record comments via the Internet (Finley-Brook and Thomas 2011). CDM proponents interpret low participation in consultation processes as "affirmative silence" (Lederer 2011); an alternative interpretation is that the consultation methods used are ineffective or inaccessible.

To assure CDM integrity, social safeguards should go beyond a letter of support from a host government, the current CDM application requirement (du Monceau and Brohé 2011). National authorities cannot be considered neutral parties because they are biased in favor of projects that build infrastructure such as roads, ports, and electrical lines or encourage foreign investment (Finley-Brook and Thomas 2011; Savaresi 2012). Nonetheless, UNFCCC verification procedures rely heavily on national-level determinations of CDM project suitability because they (1) reduce UN oversight responsibilities and (2) uphold principles of sovereignty and non-intervention.

CDM project developers have relatively unchecked power in terms of how they treat host communities. A CDM project has not been rejected for purely social reasons (du Monceau and Brohé 2011; author's analysis),[2] even when the CDM Executive Board was aware of alleged violations of human rights. For example, in 2011, the Honduran oil palm company Exportadora del Atlántico received CDM registration following the murder of dozens of local farmers and the displacement of hundreds (Directorate-General for External Policies 2012). Although describing events in Honduras as "deplorable," the chairman of the Executive Board asserted, "the Board is not equipped" to address or investigate human rights abuses (Neslen 2011).

Indigenous populations living near the Barro Blanco CDM hydroelectric project in Panama were not properly consulted before the initiation of construction (Anaya 2013). Barro Blanco was registered under the CDM even after impacted Ngäbe submitted letters arguing project developers violated CDM stakeholder consultation rules. The secretary to the CDM Executive Board replied to the letter submitters that no comments were received "during the global stakeholder commenting period (27 June 2009–26 July 2009)" (Howard 2011, 1). Since this one-month window was missed, the host community's letters, although received "in a timely matter" and "considered in full," did not influence the decision of the Executive Board (Howard 2011, 1). Howard added, "It may be mentioned here that the Board at its next meeting will consider means of addressing significant deficiencies in validation and verification reports which lead to registration/issuance." This response suggested Barro Blanco's shortcomings might influence broader CDM procedural reform, but in subsequent years the Board did not change policies to impede recurrence.

In 2012, Ngäbe communities near Barro Blanco experienced intimidation campaigns and violent repression leading to three deaths (Bill et al. 2012). With dam construction nearly complete, villagers face relocation and flooding of ancestral lands. UN bodies and host governments must take responsibility for human rights protections throughout the life cycle of CDM projects and develop impartial procedures to investigate and redress complaints. With reforms to CDM procedures under discussion since 2012, prevention of human rights violations was overshadowed by concerns about procedural efficiency and streamlining (see, for example, CDM Executive Board 2013).

REDD Debates and Dilemmas

Offsetting policies and markets are dynamic and disputed. Experts generally agree there are contradictory possibilities in terms of positive and negative governance and equity impacts from REDD (Table 4.1), meaning each project has potential to create opportunities, threats, or some combination of both.

REDD+ has been promoted as a method to achieve what previous conservation and climate change mitigation initiatives could not. Forest carbon preservation was initially highlighted as a means to mitigate more cheaply and easily than industrial offsets (Sunderlin et al. 2014), perhaps suggesting naiveté about the complexity of the institutional transformations and policy reforms necessary. More recently, it has been argued that REDD+ can generate transformative change through the capacity building required in preparatory REDD readiness programs (Murdiyarso et al. 2012). However, in spite of millions of dollars transferred to state agencies as part of readiness investments, pitfalls are likely if institutional reforms are superficial and true participatory collaboration is lacking (Hagen 2014).

Successful offsets, defined here as providing procedural and distributional equity as well as contributing to ecological resilience in addition to greenhouse gas reductions, often emerge from unique windows of opportunity, such as in a partnership between the Smithsonian Tropical Research Institute (STRI) in Panama and the Ipeti-Emberá community in eastern Panama. STRI paid indigenous families to plant native tree species on deforested areas and maintain stands of

TABLE 4.1 Potential for contradictions in REDD results

Opportunities	Sources	Threats	Sources
Reform and strengthen forest policy and governance in developing countries	Clements (2010); Larson and Petkova (2011)	Create a new form of imperialism	Clements (2010); Corbera (2012)
Establish new sources of employment and income; food security with reforestation of agroforestry species	Caplow et al. (2011); Larson and Petkova (2011)	Create restrictions on local access to resources; threats to food production	Caplow et al. (2011); COICA (2014)
Recognition and valuation of numerous ecosystem services	COICA (2014)	Carbon focus could undermine ecology and other forest benefits	Clements (2010); Corbera (2012); COICA (2014)
Decentralization of development decision-making	Larson and Petkova (2011)	Recentralization; tensions between international, national, and local scales	Larson and Petkova (2011); Corbera (2012)
Improvements in legal and financial support to define and defend property rights	Larson (2011); Crippa and Gordon (2013)	Loss of local access to land; land grabbing	Larson (2011); Chhatre et al. (2012); Crippa and Gordon (2013)

old growth forest in order to offset the carbon produced from their operations, determined to be equivalent to 4,000 tons annually. Twenty families committed a quarter of their land to the project and in exchange received direct payments for the carbon sequestered as well as a pool of shared funds to benefit of the whole community. After a successful 3-year pilot, preparation of a second longer-term contract involving more families is underway (Catherine Potvin, personal communication, November 2, 2015). A McGill University scientist worked with STRI personnel to train local populations in participatory mapping and carbon inventories, but if technical support were purchased at market value, would projects like this be economically viable, particularly if intermediaries commonly found in the voluntary market were also paid transaction fees? Furthermore, it is often difficult for avoided deforestation projects working with local communities to pay participants an adequate compensation to cover opportunity costs, meaning the equivalent a producer or landowner could earn from timber, palm oil, or other crops on the same land (Rico García-Amado et al. 2011; Sunderlin et al. 2014).

Analyses of early offset projects suggest that improvements to procedural equity can increase distributional equity and, alternatively, that linking marginalized populations to markets without identifying and addressing power inequalities can contribute to exploitation and harm (Tienhaara 2012). The designation of some exploitative projects as carbon piracy—a form of green grabbing—is accurate (Espinoza Llanos and Feather 2011). Even UN agencies did not comply with standards in the UN Declaration on the Rights of Indigenous Peoples during the initiation of the UN-REDD program in Panama (Cuéllar et al. 2013; Feiring 2013): state and donor agencies were unwilling to spend the time and resources necessary to consult regional and local indigenous governance bodies (Cuéllar et al. 2013). In Peru, delays in sharing information and power contributed to frustration and strong criticisms from indigenous federations (Espinoza Llanos and Feather 2011; Feiring 2013; White 2014). Once conflicts arose in Panama and Peru, officials promised to improve participation, but state institutions may not have the capacity to effectively consult and share governance (Feiring 2013). Meanwhile, REDD programs can foment conflict within indigenous organizations. For example, the indigenous federation COONAPIP's involvement in UN-REDD in Panama led to two *comarcas* (semi-autonomous territories) withdrawing from the federation to demonstrate disapproval (Cuéllar et al. 2013; Feiring 2013).

Indigenous Responses to REDD and Justice Concerns

Donors often respond to emerging critiques about procedural and distributional inequity in avoided deforestation initiatives by creating additional training manuals, safeguard policies, reporting tools, accountability assessments, and workshops. Although information-sharing strategies are essential, written documents and brief informational gatherings cannot replace consultation and dispute resolution based on standards for free, prior, and informed consent (FPIC). One-time training events, although popular, have limited impact (Hagen 2014).

Since 2008, indigenous federations such as the Inter-Ethnic Association for the Development of the Peruvian Amazon (AIDESEP) have criticized the exclusion of indigenous peoples from decision-making in REDD readiness initiatives (Espinoza Llanos and Feather 2011; White 2014). In 2011, AIDESEP circulated an alternative proposal they call "Indigenous REDD+" (Espinoza Llanos and Feather 2011; Abate and Kronk 2013), which is essentially incompatible with the Peruvian state approach as designed with support from the World Bank and multilateral agencies (White 2014). Key conditions for Indigenous REDD+ are (1) industrialized countries need to target domestic emission reductions and (2) specific fiscal mechanisms to trade carbon in forest offsets must be eliminated, while still finding methods for wealthy countries to pay the ecological debt they owe. According to advocates, Indigenous REDD+ arrangements are more likely to support holistic forest protection if they remain outside of international carbon markets seeking to provide cheap offsets (Abate and Kronk 2013).

In 2013, AIDESEP agreed to co-coordinate Peru's Indigenous Platform for REDD readiness (Feiring 2013). Feiring wonders if the Peruvian state created this platform to subdue criticisms of exclusionary planning rather than as a sincere commitment to collaboration. Can AIDESEP use this role to advance an indigenous agenda?

AIDESEP is collaborating with the Council of Indigenous Organizations of the Amazon Basin (COICA) to forward an integral conservation plan called Indigenous Amazonian REDD+ (RIA). RIA recognizes a minimum of 22 ecosystem services beyond carbon sequestration and aims to improve coordination between strategies for adaptation, mitigation, and resiliency. COICA's (2014) proposed front line of defense to stop rainforest destruction is indigenous self-governance and land rights. Yet, as noted, indigenous resource rights are frequently challenged.

Indigenous peoples around the world have criticized exclusionary development of climate policy and worked to development alternative carbon mitigation strategies that acknowledge harm caused by historical political and economic marginalization and that aim to provide emancipatory and equitable means to reduce environmental degradation and avoid deforestation without risk of additional land dispossession (Doolittle 2010; Johnstone 2010; Baez 2011). Although indigenous peoples have often struggled to gain agency to influence state climate policy at national scales, they have increasingly formed alliances across borders to advocate for change at international scales thus increasing their leverage in local, regional, and national policy arenas (Johnstone 2010; Schroeder 2010; Baez 2011).

Conclusion

Some environmental justice advocates remain critical of emissions markets (Indigenous Environmental Network 2008; Burnham et al. 2013). Increasingly, researchers point out that carbon mitigation schemes in some locations have been shown to threaten and cause harm to marginalized groups (Baez 2011; Finley-Brook and Thomas 2011; Burnham et al. 2013; Crippa and Gordon 2013). For many critics, the root cause of runaway climate change is capitalism, an inequitable, growth-oriented economic system unlikely to fix the crisis it created. Carbon trading allows the wealthy to pay to pollute, does not reduce consumption, and supports privatization of global and local commons (Durban Group for Climate Justice 2004).

Even indigenous leaders who engage in REDD readiness processes may still remain skeptical about the willingness or ability of the state to reform, although they hope for opportunities to reinforce self-governance (Faquin Fernandez 2014; Mezúa 2014). The Indian Law Resource Center[3] has called for reorientation of carbon trading to defend the rights of indigenous peoples by cementing procedures for FPIC and self-determination (Crippa and Gordon 2013). A message for policy-makers is that placing indigenous peoples in broad categories such as "vulnerable groups" will not protect their distinct land, resource, and self-governance rights. There is a need for rigorous, enforceable, and specific safeguards for indigenous populations in alignment with FPIC standards as defined in the UN Declaration on the Rights of Indigenous Peoples (Doolittle 2010; Johnstone 2010; Crippa and Gordon 2013). Equally necessary are procedural reforms creating spaces for indigenous peoples to become directly involved in the formulation of climate policies at all governance scales, including within UNFCCC proceedings (Doolittle 2010; Schroeder 2010; Abate and Kronk 2013).

McDermott et al. (2012) define forest carbon trade as plastic and open to interpretation, since many details remain to be defined, and scenarios for time periods, objectives, market prices, and even payment strategies diverge widely. The unsettled structure contributes to ambiguity and even confusion. Furthermore, performance-based financial mechanisms complicate cost-benefit calculations, as payment transfers may not occur for years or decades. In the meantime, geographic frameworks help organize and understand complex, uneven distributions of social, political, economic, ecological, and cultural costs and benefits between various scales, sectors, and spaces.

As researchers slowly record offset results, social justice advocates and environmentalists should not wait passively. Based on CDM results, there is strong evidence of potential risks from emissions trading for indigenous peoples and low-income populations. Moreover, policy objectives must go beyond prevention of human rights violations to assure poverty eradication and build socio-ecological resilience. Broad political transformations are necessary; otherwise business-as-usual practices are likely to impede emissions reduction on the one hand, and expand environmental injustice on the other.

Notes

1. A Carole Weinstein International Grant supported fieldwork in the Dominican Republic, a Higher Education for Development and United States Agency for International Development-sponsored project, Building Conservation Capacity for a Changing Amazon financed research in Peru, and the University of Richmond funded travel to Costa Rica, Nicaragua, and Panama.
2. This conclusion was drawn after reading ruling notes on rejected projects available from https://cdm.unfccc.int/Projects/rejected.html. Central reasons for rejection are (1) concerns about *additionality*—proving the project required CDM support and (2) lack of evidence of progress toward CDM registration throughout project implementation. By late 2015, 7,680 projects had been registered and only 335 were rejected or withdrawn, showing a propensity toward approval.
3. See http://www.indianlaw.org/climate for more information.

References

Abate, Randall S. and Elizabeth Ann Kronk. 2013. Commonality among unique indigenous communities: An introduction to climate change and its impacts on indigenous peoples. *Tulane Environmental Law Journal* 26: 1–17.

Anaya, S. James. 2013. *Declaración del Relator Especial sobre los derechos de los pueblos indígenas al concluir su visita oficial a Panamá.* http://unsr.jamesanaya.org/statements/declaraciondel-relator-especial-sobre-los-derechos-de-los-pueblos-indigenas-al-concluir-su-visita-oficial-a-panama (accessed July 13, 2014).

Baez, Stephanie. 2011. The "right" REDD framework: National laws that best protect indigenous rights in a global REDD regime. *Fordham Law Review* 80: 821–875.

Baker, Shalanda H. 2012. Why the IFC's free, prior, and informed consent policy doesn't matter (yet) to indigenous communities affected by development projects. *University of San Francisco Law Research Paper No. 2012–16*.

Bill, Doris, Marciela Arce, Félix Wing Solís, Washington Lum Sandoya and David de Leon. 2012. *Informe de gira de observación de derechos humanos luego de las protestas contra la minería e hidroeléctricas en la Comarca Ngäbe-Buglé*. http://cdn.otramerica.com/OTRAMERICA_web/48/posts/docs/0448033001330859048.pdf (accessed July 13, 2014).

Blom, Benjamin, Terry Sunderland, and Daniel Murdiyarso. 2010. Getting REDD to work locally: Lessons learned from integrated conservation and development projects. *Environmental Science & Policy* 13, 2: 164–172.

Burnham, Morey, Claudia Radel, Zhao Ma, and Ann Laudati. 2013. Extending a geographic lens towards climate justice, part 2: Climate action. *Geography Compass* 7: 228–238.

Caplow, Susan, Pamela Jagger, Kathleen Lawlor, and Erin Sills. 2011. Evaluating land use and livelihood impacts of early forest carbon projects: Lessons for learning about REDD+. *Environmental Science and Policy* 14: 152–167.

CDM (Clean Development Mechanism) Executive Board. 2013. *Compilation of Inputs Considered by the Board in Its Review of the CDM Modalities*. https://cdm.unfccc.int/filestorage/l/q/F02QOJZ359UME6WARCBYX7GI1NPTSL.pdf/eb72_repan01.pdf?t=UFp8bjNkaWs0fDAy5HBAm2ZIn1QTqGLp7lgd (accessed July 13, 2014).

Chhatre, Ashwini, Shikha Lakhanpal, Anne Larson, Fred Nelson, Hemant Ojha, and Jagdeesh Rao. 2012. Social safeguards and co-benefits in REDD+: A review of the adjacent possible. *Current Opinion in Environmental Sustainability* 4: 654–660.

Clements, Tom. 2010. Reduced expectations: The political and institutional challenges of REDD+. *Oryx* 44, 3: 309–310.

COICA (Coordinadora de las Organizaciones Indígenas de la Cuenca Amazónica). 2014. *Integralidad de Redd+ Indígena Amazónico*. http://www.coica.org.ec/index.php/es/noticias/241-redd-indigena-amazonico (accessed July 13, 2014).

Corbera, Esteve. 2012. Problematizing REDD+ as an experiment in payments for ecosystem services. *Current Opinion in Environmental Sustainability* 4: 612–619.

Crippa, Leonardo A. and Gretchen Gordon. 2013. *International Law Principles for REDD+: The Rights of Indigenous Peoples and the Obligations of REDD+ Actors*. Washington, DC: Indian Law Resource Center.

Cuéllar, Nelson, Susan Kandel, Andrew Davis, and Fausto Luna. 2013. *Indigenous Peoples and Governance in REDD+ Readiness in Panama*. San Salvador: PRISMA.

Directorate-General for External Policies. 2012. *Human Rights and Climate Change: European Union Policy Options*. Brussels: European Parliament.

Doolittle, Amity A. 2010. The politics of indigeneity: Indigenous strategies for inclusion in climate change negotiations. *Conservation and Society* 8: 286–291.

du Monceau, Tanguy and Arnaud Brohé. 2011. *Study on the Integrity of the Clean Development Mechanism*, AEA/ED56638/Issue 1. Oxford: AEA Technology.

Durban Group for Climate Justice. 2004. *Durban Declaration*. http://www.durbanclimatejustice.org/durban-declaration/english.html (accessed July 13, 2014).

Espinoza Llanos, Roberto and Conrad Feather. 2011. *The Reality of REDD+ in Peru: Between Theory and Practice: Indigenous Amazonian Peoples Analyses and Alternatives*.

Lima: Interethnic Association for the Development of the Peruvian Amazon (AIDESEP) and Forest Peoples Program (FPP).

Faquin Fernandez, Josue. July 3, 2014. President, Regional AIDESEP Organization-Ucayali. Pucallpa, Peru. Interview with author.

Feiring, Birgitte. 2013. Participation and Consultation Standards, Guidelines and Country Experiences. *Expert Workshop on Practical Approaches to Ensuring the Full and Effective Participation of Indigenous Peoples in REDD+*. Weilburg, Germany, September 10–12.

Finley-Brook, Mary and Curtis Thomas. 2011. Renewable energy and human rights violations: Illustrative cases from indigenous territories in Panama. *Annals of the Association of American Geographers* 101, 4: 863–872.

Hagen, Roy. 2014. *Lessons Learned from Community Forestry and Their Relevance for REDD+*. Arlington, VA: Forest Carbon, Markets and Communities Program.

Hajek, Frank, Marc J. Ventresca, Joel Scriven, and Augusto Castro. 2011. Regime-building for REDD: Evidence from a cluster of local initiatives in South-Eastern Peru. *Environmental Science & Policy* 14, 2: 201–215.

Hawley, Chris. 2009. Clean energy windmills: A 'dirty business' for farmers in Mexico. *USA Today*. http://www.usatoday.com/money/industries/energy/environment/2009-06-16-mexico-wind-power_N.htm (accessed September 6, 2014).

Howard, Andrew. December 20, 2011. Secretary CDM Executive Board. *Re: CDM Project Application #3237: Barro Blanco Hydroelectric Project*. http://carbonmarketwatch.org/wp-content/uploads/2011/02/EB60-3_EB60-16_CDMWatch_3237_Response.pdf (accessed July 13, 2014).

Hughes, Ross and Fiona Flintan. 2001. *Integrating Conservation and Development Experience: A Review and Bibliography of the ICDP Literature*. London: International Institute for Environment and Development.

Indigenous Environmental Network. 2008. *Indigenous Peoples' Guide: False Solutions to Climate Change*. http://www.earthpeoples.org/CLIMATE_CHANGE/Indigenous_Peoples_Guide-E.pdf (accessed July 13, 2014).

Johnstone, Naomi. 2010. Indonesia in the 'REDD': Climate change, indigenous peoples and global legal pluralism. *Asian-Pacific Law and Policy Journal* 12: 93–123.

Kaimowitz, David. 2008. The prospects for Reduced Emissions Deforestation and Degradation (REDD) in Mesoamerica. *International Forestry Review* 10, 3: 485–495.

Larson, Anne M. 2011. Forest tenure reform in the age of climate change: Lessons for REDD+. *Global Environmental Change* 21: 540–549.

Larson, Anne M. and Elena Petkova. 2011. An introduction to forest governance, people, and REDD+ in Latin America: Obstacles and opportunities. *Forests* 2: 86–111.

Lederer, Markus. 2011. From CDM to REDD+: What do we know for setting up effective and legitimate carbon governance? *Ecological Economics* 70: 1900–1907.

McDermott, Constance L., Lauren Coad, Ariella Helfgott, and Heike Schroeder. 2012. Operationalizing social safeguards in REDD+: Actors, interests and ideas. *Environmental Science and Policy* 21: 63–72.

McDermott, Constance L., Sango Mahanty, and Kate Schreckenberg. 2013. Examining equity: A multidimensional framework for assessing equity in payments for ecosystem services. *Environmental Science and Policy* 33: 416–427.

McElwee, Pamela D. 2012. Payments for environmental services as neoliberal market-based forest conservation in Vietnam: Panacea or problem? *Geoforum* 43: 412–426.

Mezúa, Cándido. July 9, 2014. President, National Organization of Indigenous Peoples of Panama. Lima, Peru. Interview with author.

Murdiyarso, Daniel, Maria Brockhaus, William D. Sunderlin, and Lou Verchot. 2012. Some lessons learned from the first generation of REDD+ activities. *Current Opinion in Environmental Sustainability* 4: 678–685.

Neslen, Arthur. October 3, 2011. EU carbon credits scheme tarnished by alleged murders in Honduras. *Guardian*. http://www.guardian.co.uk/environment/2011/oct/03/eu-carbon-credits-murders-honduras (accessed July 13, 2014).

Pascual, Unai, Roldan Muradian, Luis C. Rodrígues, and Anantha Duraiappah. 2010. Exploring the links between equity and efficiency in payments for environmental services: A conceptual approach. *Ecological Economics* 69: 1237–1244.

Pasqualetti, Martin J. 2011. Social barriers to renewable energy landscapes. *Geographical Review* 101: 201–223.

Rico García-Amado, Luis, Manuel Ruiz Pérez, Felipe Reyes Escutia, Sara Barrasa García, and Elsa Contreras Mejía. 2011. Efficiency of payments for environmental services: Equity and additionality in a case study from a biosphere reserve in Chiapas, Mexico. *Ecological Economics* 70: 2361–2368.

Savaresi, Annalisa. 2012. The human right dimensions of REDD. *Review of European Community and International Environmental Law* 21, 2: 102–113.

Schroeder, Heike. 2010. Agency in international climate negotiations: The case of indigenous peoples and avoided deforestation. *International Environmental Agreements* 10: 317–332.

Sunderlin, William D., Andini Desita, Ekaputri, Erin O. Sills, Amy E. Duchelle, Demetrius Kweka, Rachel Diprose, Nike Doggart, Steve Ball, Rebeca Lima, Adrian Enright, Jorge Torres, Herlina Hartanto, and Angélica Toniolo. 2014. *The Challenge of Establishing REDD+ on the Ground: Insights from 23 Subnational Initiatives in 6 Countries*. Bogor, Indonesia: Center for International Forestry Research.

Swyngedouw, Erik. 2000. Authoritarian governance, power, and the politics of re-scaling. *Environment and Planning D: Society and Space* 18: 63–76.

Thompson, Mary, Manali Baruah, and Edward R. Carr. 2011. Seeing REDD+ as a project of environmental governance. *Environmental Science and Policy* 14: 100–110.

Tienhaara, Kyla. 2012. The potential perils of forest carbon contracts for developing countries: Cases from Africa. *Journal of Peasant Studies* 39, 2: 551–572.

White, Douglas. 2014. A perfect storm? Indigenous rights within a national REDD readiness process in Peru. *Mitigation and Adaptation Strategies for Global Change* 19, 6: 657–676.

SECTION II
Accounting and Accountability

5
THE LIMITATIONS OF INTERNATIONAL AUDITING

The Case of the Norway-Guyana REDD+ Agreement

Janette Bulkan

Introduction

This chapter focuses on the practices of international auditing firms sponsored directly or indirectly by the government of Norway to verify Guyana's compliance with indicators intended, inter alia, to improve forest governance as part of a global Reducing Emissions from Deforestation and forest Degradation (REDD+) strategy. I argue that by dealing principally with the executive arm of the Guyana government, and not including the legislative arm (Parliament), the Norwegian government defeated its own declared intention to advance governance in REDD. I show that both governments have a vested interest in positive audit reports and appear to have used their power accordingly to reward or dismiss auditors. I argue that the progressive watering down of the compliance indicators makes such weakly monitored REDD+ schemes ineffective as one mechanism for slowing global warming. I recommend that credible auditing should be carried out by mixed teams of global and local auditors with strong terms of reference, recruited independently of the client governments, and required to report back in culturally appropriate language to key affected constituency groups.

A political ecology perspective, which links political economy to research on different social and ecological scales, frames this chapter. Political ecology pays attention to context, the workings of power and "the general ecological impacts of the state, interstate relations and global capitalism" (Bryant 1997). Political ecology provides a lens through which I examine a specific case of the performance of international auditors contracted to verify whether agreed actions or processes had taken place. Logically, the need for reputational credibility should ensure that auditing firms maintain their independence against powerful clients. The anthropologist Strathern argues that with the growth of audit regimes, "the apparently neutral 'market' provides a ubiquitous platform for individual interest and national politics alike" (2000, 2). For auditing to retain credibility, robust transparency is required.

My analysis is based on long-term fieldwork in Guyana complemented by scholarly research. I have actively engaged with international auditing systems and practices for over 15 years. I am a member of the six-person Policy and Standards Committee that is mandated to advise the international board of directors of the Forest Stewardship Council (FSC), globally the leading voluntary, independent, third-party forest certification scheme.

In November 2009, the governments of Norway and Guyana signed a "Memorandum of Understanding (MoU) regarding cooperation on issues related to the fight against climate change, the protection of biodiversity and the enhancement of sustainable development," with an associated Joint Concept Note (JCN) on REDD+ cooperation (Government of Guyana and Government of Norway 2009). Norway agreed to provide support of up to US$250 million between 2010 and 2015 to a Low Carbon Development Strategy (LCDS) developed within the Office of the President (OP) of Guyana (OP 2009).

Two years earlier, in April 2007, Norway had pledged that it would become carbon neutral by 2050.[1] Less than a year later, the government brought forward the date for achieving carbon neutrality to 2030.[2] That national policy commitment was achievable only if large amounts of emission reduction units (ERUs) could be acquired to offset Norway's petroleum production and its industrial and shipping emissions. That national commitment was expanded into an ambition in which Norway would develop a global model for REDD. From Norway's point of view, if other countries generated large amounts of ERUs, oversupply of ERUs would drive down unit costs, which would benefit Norwegian purchases (Thomas 2009; Lohmann 2011, 94).

Other beneficiary countries selected by Norway were Brazil and Indonesia (US$1 billion each), the Congo Basin Fund (US$250 million) and Tanzania. An evaluation report commissioned by the Norwegian Agency for Development Cooperation (NORAD) opined that "leveraging of political momentum within climate change negotiations through the provision of large scale funds to selected countries appears to be a risky and inefficient approach, from an advocacy perspective" (LTS International et al. 2011, xviii). In other words, if those few beneficiary countries did not reduce their forest carbon emissions, then Norway's strategy would have failed. Auditors and consultants could be, and were, enlisted to reduce the risk.

The Norwegian commitment was equivalent to an additional 7–9 percent of Guyana's declared annual budget.[3] For Norwegians, Guyana was a little known flyover country—not a destination (Bade 2012, 13). However, the agreement with Guyana allowed Norway to claim progress on REDD+ even though global negotiations for a successor agreement to the Kyoto Protocol were glacially slow (Solheim 2009). Norwegian policy has been to dedicate some of its sovereign wealth to generate support for its position from other countries. For example, the Norwegian government paid US$5 million to a think tank, the Center for Global Development (CGD), based in Washington, DC, "to persuade Washington officials to double United States spending on global forest protection efforts to $500 million a year" (Lipton et al. 2014). REDD-Monitor noted that while the Norwegian side recommended appropriate disclosure of its financial support, CGD only mentioned that support once in five reports published between March and August 2014 (REDD-Monitor 2014). The "troubling questions about intellectual freedom" of think tanks explored by Lipton et al. (2014) apply equally to the international auditors contracted by Norway, as described later.

Positive action on climate change mitigation seemed favorable in Guyana as its president also held the portfolio of Minister of Forests, and until November 2011, the governing party controlled a majority of seats in the National Assembly (Parliament).

In Guyana, the routine denials of research permits and official restrictions on access to information meant that the international consultants would become principal sources for information. For example, a consultant in the Project Management Office and the Head of the Office of Climate Change (OCC) of Guyana were named as the principal sources of information for a complementary report on the Guyana REDD+ Investment Fund (GRIF) (Gruning and Shuford 2012). A NORAD-commissioned consultancy on REDD in Guyana interviewed Winrock and Indufor, global consultancy firms, also paid by Norway for work in Guyana (LTS International et al. 2013). Those statements of government officials and consultants were at variance with those of other constituency groups, as reported in the local independent newspapers and on websites like REDD-Monitor.

Guyana Context

The REDD+ agreement was also self-serving for the government of Guyana. The government holds administrative control over the majority of the forests that cover over 87 percent of the country. Yet Guyana made no policy changes to reduce carbon emissions during the period of its agreement with Norway. Declared gold production increased from 305,000 in 2009 to 458,000 troy ounces in 2013; and forest production from 449,000 m^3 to 489,000 m^3 in the same period.[4] Under-declaration was acknowledged: government staff estimated that between one-third and two-thirds of gold produced was not declared (Harvard Law School 2007, 22n149, 35n240). From 2007 the Guyana Forestry Commission (GFC) stopped releasing disaggregated information on the names and quantities of the timber species exported. In 2011 the GFC reduced the requirement that loggers leave a distance between harvestable trees of the same species, from 10 m to 8 m (GFA 2011, 37). In other words, the GFC restricted information that would show increasing forest degradation resulting from the selective logging of a few species only, contrary to GFC's Code of Practice for Timber Harvesting—another example of non-compliance with the Norwegian agreement. The GFC also refused to disclose the names of concession holders and log exporters, which would confirm that Asian loggers controlled over 75 percent by area of all large-scale, long-term forest concessions and that the bulk of log production was exported to Asia (Bulkan 2014a).

For then president Jagdeo, the memorandum of understanding with Norway (2009–2015) was the culmination of a personal quest begun in 2006 to use Guyana's natural tropical forests to generate cash under an international regime in the service of climate change mitigation (Howden and Brown 2007). As Guyana's Constitution concentrates political power in an Executive Presidency, there was no requirement for prior presidential consultation with the National Assembly (Parliament) or other constituency groups before signing a government-to-government bilateral REDD+ agreement.

Guyana is marked by inequality and poverty and was ranked second lowest (118th of 186 countries surveyed) in the Caribbean (only Haiti was lower) and lowest in South America in the United Nations Human Development Index 2013.[5] The indigenous peoples (Amerindians, as identified in Guyana's decennial census) who constitute 10 percent of the national population are also the poorest, experiencing "about three times greater than the national average for both extreme and moderate poverty" in 2011 (Thomas 2014). Extreme poverty was defined as those persons with access to less than US$1.25 per day at 2006 prices, equivalent to the cost of 2,400 calories in that year. Moderate poverty was fixed at US$1.75 per day. Amerindians are the majority forest populations so should be among the principal beneficiaries of a REDD+ scheme. However, there was no evidence that the government had observed the protocols of free, prior, and informed consent (FPIC) before signing the REDD agreement that in principle included Amerindian customary lands currently administered as state lands (La Rose 2013; Dooley and Griffiths 2014).

Corruption is a major feature in Guyana's political life and civil society. Guyana ranked 133rd out of 176 countries surveyed in Transparency International's (TI) Corruption Perception Index (CPI) of 2011[6] and scored 28 out of a possible 100 points. In 2013, Guyana's score was 27 out of 100 points. Between 2001 and 2008 an estimated 42–52 percent of the national economy was derived from illegal activities, principally cocaine trafficking (Thomas et al. 2011). Thomas, a leading Caribbean economist, calculated that that the illegality component of the national economy had increased to 60 percent in 2014 (Wilburg 2014).

On paper, Norway's monetary commitment presented a substantial opportunity to address hinterland poverty and restart stalled processes, like the National Development Strategy, and national land use planning in partnership with indigenous peoples and other constituencies. In practice, the Office of the President alone decided what projects would be submitted for REDD+ funding. The 2011 NORAD-commissioned report on Guyana noted:

> The apparent mechanism by which any REDD funds accruing to communities would be spent, essentially through approval by the Ministry of Amerindian Affairs, is inappropriate. Efforts to build capacity and leave communities in charge of their own affairs, while meeting appropriate transparency and accountability standards should be prioritized.
> *(LTS et al. 2011, xvi)*

That recommendation went unheeded and Norwegian funding for Amerindian communities continued to be disbursed through the government of Guyana's patronage networks. In other words, so-called REDD funding supported the status quo, in which Amerindians were cast as clients of state patronage through ex gratia payments.

Rule of Experts

The global South governments that are in the forefront of REDD rely in part on a network of experts from the North, many serving as consultants to the World Bank, global North countries, non-profit research organizations and large conservation NGOs (nongovernmental organizations) (Cadman and Maraseni 2011). Successful consultants have developed their own internal mechanisms of self-discipline, in line with the expectations of their paymaster. An early example of this phenomenon is described in a meta-study of consultancy reports on deforestation in Nepal:

> There are two ways of approaching the environmental problems of the Himalaya. You can ask "What are the facts?" or you can ask "What would you like the facts to be?" . . . The management scientists . . . tend to work in (or as consultants to) organisations, and they tend to work with those who are towards the top of those organisations. In such culturally homogenous settings the chances are that everyone will see the problem in much the same way (if only because anyone who does not soon finds himself in the outside of those organisations).
> *(Thompson et al. 1986, 1–2)*

The political scientists, Paterson et al., used social network analysis to lay out "the 'emissions trading mycelium,'" which proliferated in the late 1980s and early 1990s, as individuals and organizations discussed and promoted carbon emissions trading in climate change debates across a broad range of places, and which bore its first fruits in the Kyoto Protocol. On the basis of those network linkages, emissions trading emerged in different spaces when conditions were ripe, without direct causal pressure or other sorts of connecting links on the surface, that is, through intergovernmental agreement or international convention or law (Paterson et al. 2011). Consultancy and audit reports provide the linkages among the fragmented global governance structures that seek to regulate national and international responses to anthropogenic global warming. The networked elites—both international bureaucrats and consultants—serve as translators between global South political directorates and international project funders or global North governments. In general their reports address only the vague or undisclosed terms of reference (TORs) of any particular project, and assiduously avoid any contradictory evidence. In the case of Guyana, there are no references in any of the auditors' reports to any article or report critical of Guyana's so-called REDD scheme that are readily accessible on websites such as REDD-Monitor or via an Internet search.

The remuneration of auditors and consultants with REDD contracts in Guyana, including and apart from Norway's program, represent a considerable part of agency budgets. In a 6-year period (2005–2010), the GFC paid out an estimated US$2.5 million, or 10 percent of its total income, in professional fees (Ram 2014). The GFC has not responded to public calls to identify the recipients and the purpose of the payments (Stabroek News 2014). The reported operating costs of the Guyana monitoring, reporting and verification (MRV) system were in the region of US$500,000 per year (LTS et al. 2013, 57). Direct consultancy payments made up 48 percent of the Inter-American Development Bank's budget for its Forest Carbon Partnership Facility (FCPF) Guyana project (GY-T1097); Norway is one of the principal funders of the FCPF. Additional consultancy payments are included in the 51 percent allocated to "goods and services" in that project (IADB 2013).

Individual experts can move through a revolving door, between consultancies and academia, each reinforcing the other. For example, the second author in reports on REDD+ for the government of Norway produced by the US-based Meridian Institute (Angelsen et al. 2011a; Angelsen et al. 2011b), was also the lead writer of a glowing report on REDD+ in Guyana issued by the Union of Concerned Scientists (Boucher et al. 2014). A Winrock staff member was a collaborator in the aforementioned reports for Norway, and in academic papers that include positive references to REDD+ in Guyana (Braswell et al. 2013; Pearson et al. 2014). That Winrock staffer was also a member of the Technical Advisory Panel for Guyana of the World Bank's FCPF and hired by the GFC for its REDD readiness work.

The US-based Center for Global Development was not the only not-for-profit that benefited from this new industry. Before the signing of the Norway-Guyana MoU, the Meridian Institute, a US-based not-for-profit, was hired by Norway as an organizer for the Guyana work (GINA 2009) and one of its REDD reports was approvingly quoted in Guyana's Low Carbon Development Strategy (LCDS)

(OP 2009, 15). In 2011, Guyana hired the Meridian Institute to provide an independent review of the Guyana REDD+ Investment Fund (GRIF) management (RA 2012, 42), a holding mechanism for Norway's funds for Guyana. Shortly after, from March 2012, the Meridian Institute began to serve as the interim secretariat of the GRIF.

Auditing Guyana's REDD+ Scheme

Between 2009 and 2014, Norway commissioned a succession of consultancies to assess, scope, monitor or verify Guyana's performance in its MRV or compliance with the indicators set out in the JCN, itself revised at least four times (Government of Guyana and Government of Norway 2009, 2011, 2012, 2014). In some cases, the same consultants or consultancy firms that helped Guyana set up an MRV system later returned in the capacity of auditors. For example, the GFC "hired two technical experts to aid the development of the Guyana MRV system (Poyry/Indufor Asia Pacific and Winrock International) and is extremely happy with the quality of support provided" (LTS et al. 2013, 31). Poyry's principal consultant in the 2011 assessment returned as Indufor's principal in the 2012 and 2013 assessments. One might reasonably conclude that for the clients—Norway and Guyana—that consultant was key.

Norway sponsored 10 assessments of areas of Guyana's forests and non-forest during the period of its agreement (a critical review of those reports would require more space). Norway had commissioned an earlier governance consultancy from CIFOR (Trevin and Nasi 2009) and three consultancies on independent forest monitoring (IFM) by the GFA Consulting Group based in Germany. It is not surprising that the GFC, which also served as the counterpart agency in most cases, would become practiced at managing consultants.

Here I look at two consultancies in 2012 and 2013, carried out by the Rainforest Alliance (RA) and Indufor respectively, to verify Guyana's compliance with the governance indicators set out in the JCN. Indufor, which carried out MRV assessments of forest cover in 2012 and 2013, was also awarded the contract to provide independent verification of progress set out in the JCN in 2013 (Year 3 of the MoU period). The auditing company, Rainforest Alliance (RA), had held that contract in Years 1 and 2. However, Guyana had registered its disapproval of RA's Year 2 Draft Report delivered on September 7, 2012. RA's summary of the events that followed is instructive:

> Extensive comments were received and were addressed by the team. . . . A final report that incorporated many revisions that responded to the comments from GoG and GoN was delivered . . . on November 11, 2012 . . . The GoG requested time to prepare additional evidence that had not been delivered to the audit team . . . Ninety-six documents were subsequently delivered by the GoG to the audit team on December 5, 2012 . . . Some additional

minor changes and additional pieces of information were added to the report as a result of the additional review. The fundamental Audit Conclusions for all 10 indicators remain unchanged from the November 11, 2012 version.
(RA 2012, 14–15)

My review of a sample of the auditors' reports commissioned by Norway showed that each team spent 10–12 calendar days on average in Guyana. Few of the consultants ventured beyond contacts with government officials. The brevity of the visits meant that there was little field sampling of the veracity of what government reports claimed. Only Rainforest Alliance in its second verification fanned out across the hinterland, and without government handlers, in response to criticism of its office-based approach in its first audit in 2011: "The three team members made five different field trips and visited with members of 16 different Amerindian communities in six different regions . . . spoke with approximately 264 community members" (RA 2012, 14).

In its 2012 assessment, RA examined 10 enabling indicators to measure Guyana's REDD+ performance and found that three of the indicators were met, four were partially met and three were not met—in short, a 30 percent pass rate. Predictably RA's report did not please either Norway or Guyana and RA's contract was not renewed for a third year. Eight months later, in August 2013, the Indufor team conducted its assessment against 18 indicators. At its validation workshop held on August 30, 2013, Indufor found that six indicators were met, two were partially met, three required further clarification of information and five were not met; in other words, about the same pass rate as assessed by RA. However, Indufor then reviewed new evidence submitted by the government of Guyana (GoG) and changed its assessment as follows: "13 of the 16 indicators are met. [One] indicator is not met and 2 indicators are partially met" (Indufor 2013, 1). In other words, Guyana's performance had improved from a 33 percent to 81 percent pass rate in a matter of 3 weeks.

I now examine the four principal indicators related to governance included in the JCN: (1) strategic framework; (2) continuous multi-stakeholder engagement; (3) governance; and (4) the rights of indigenous peoples within REDD+. JCN version 1 stated: "the two Participants will adjust the Indicators of Enabling Activities annually for the subsequent year, based on the detailed REDD-plus governance development plan (RGDP)" (Government of Guyana and Government of Norway 2009). A principal change from JCN version 1 (2009) to JCN version 3 (2012) was a change in how progress was measured—from substantive changes to procedural changes. In practice a substantive change would require an auditor to verify that a named aspect of governance had improved during the assessment period. In contrast a procedural change only requires the auditor to verify the existence of a paper or administrative procedure.

I provide two examples. First, the commitment under the Governance heading in JCN 1, 2 and 3—"A transparent, rules-based, inclusive forest governance, accountability and enforcement system for forest governance in Guyana is being progressively strengthened"—was dropped in JCN 4 (December 2014). That substantive change was replaced by a list of procedural items: "Continue with the process to apply to the Extractive Industries Transparency Initiative (EITI) . . . Advance Guyana's policy of enabling Amerindian communities to 'opt in' to REDD payments." A second example concerns national land use planning, a critical component of any credible REDD scheme. JCN 2 (March 2011) committed that "the development of a national, inter-sectoral system for coordinated land use will continue. The system shall serve to maximize benefits to society and development, while minimizing negative impacts on the environment, from land use." In JCN 3 (December 2012) that commitment was watered down to "Outline in 2013 a GoG (MNRE) programme, with a particular focus on specific efforts to manage degradation from extractive activities where this needs to be done, including, for example: an enhanced miners' environmental knowledge programme." In other words, an initial commitment to an inter-sectoral land use planning process was replaced by commitment to a continuation of top-down outreach activities conducted by the Ministry of Natural Resources and the Environment (MNRE).

JCN Indicator 1 concerns the achievement of an internationally recognized framework for developing a REDD+ program, such as the World Bank's FCPF. It includes evidence that the GRIF was being administered in a transparent manner, with information on all expenditures publicly available on a relevant government of Guyana (GoG) website and through national systems of public disclosure, as well as to the National Assembly.

In 2012, RA noted that disbursement of GRIF funds had been slow up to the time of its audit, but RA did not explain that the reason for the delay was Guyana's non-submission of project proposals (RA 2012). The Guyana government customarily referred to the Norwegian funds as "our earned money," not realizing in the first years that "spends" would be subject to the accounting procedures of the designated partner entities, the IDB and UNDP (Kaieteur News 2010). In 2013, Indufor concluded that Indicator 1 was met, but limited its verification to compliance with IADB (Inter-American Development Bank) procedures (Indufor 2013). Indufor did not mention that only aggregate figures for expenditures were provided on the GRIF website. There is no information on line-itemized REDD expenditures on any GoG website, nor had that information been provided to the National Assembly up to October 2014. Indufor did add that "Parliamentary oversight should be explored as an option to facilitate a smoother transition of responsibility and authority, while at the same time providing opportunity for more open, and hopefully, balanced discussion of the Project" (Indufor 2013, 23). In summary, if Indufor had verified all the goals set out in JCN Indicator 1, it would not have been able to give Guyana a pass grade.

JCN Indicator 2 sought to verify the existence of institutionalized, systematic and transparent processes of multi-stakeholder consultation, involving all potentially affected and interested stakeholders, and evolving over time. A multi-stakeholder steering committee (MSSC) was created for the LCDS in June 2009, chaired by the president of Guyana. Membership was by invitation only. The MSSC never "steered" any REDD process; examination of the 73 monthly minutes of meetings held between June 2009 and November 2014 show no discussion of strategic issues or questions about priorities or budgets or stakeholder consultations. Issues of FPIC, equity and co-benefit with stakeholders were never raised nor addressed. OP did not accept recommendations by IIED (2009) and by LTS in 2010 that there should be at least some parliamentary representation of Opposition parties on the MSSC.[7]

In 2012, RA concluded that JCN Indicator 2 was not met, exemplified by "the recent action of the MSSC led by the representatives of the government ministries and agencies . . . to criticize the action by the combined Opposition parties . . . [which] appears to have created a more partisan, political role for the MSSC" (RA 2012, 6). Two civil society individuals who had refused to sign the prepared letter criticizing the Opposition parties were dismissed from the MSSC in 2012 without explanation.

By the time of the third verification in 2013, Indicator 2 in version 3 of the JCN (December 2012) was revised to audit only compliance with procedures, including:

- Comprehensive minutes of monthly MSSC meetings made publicly available (2a);
- Information and consultation program in place by June 2013 (2b);
- A body for communication, information and consultations by January 2013, operational by the end of 2013 if money available; coordinated information flows related to different parts of LCDS; collaboration with the National Toshaos Council (NTC, a statutory body comprising the elected heads of Amerindian villages); annual plans for stakeholder engagement and awareness, implemented from early 2014; information and consultation routines tailored to Amerindians (2c);
- GRIF and LCDS web pages updated with progress of ongoing processes by January 2013 (2d).

(Government of Guyana and Government of Norway 2012)

In 2013, Indufor verified the subindicators by review of printed minutes of meetings and websites, and through interviews principally with MSSC members. Indufor then pronounced that JCN Indicator 2 was met. Indufor's audit report confirmed that the team did not evaluate the contents of any minutes by verification through discussions with non-governmental stakeholders, as RA had done the previous year (RA 2012, 11–22; Indufor 2013). Indufor included one mild observation: "There is clearly a need to provide summaries of some documents that are both lengthy and technical. More regard must be given to the target audience . . . Questions of access to the internet also arise" (Indufor 2013, 35). None of the latter procedural requirements advanced the quality of governance. The Indufor auditors narrowed their recommendations to suggesting that information and consultation routines were to be improved for getting the government message to the communities.

JCN Indicator 3 concerned the continued development of transparent, rules-based systems for inclusive forest governance, accountability and enforcement. In its 2012 report, RA divided JCN Indicator 3 into three separate procedural indicators that were easy to fulfill: initial structure for independent forest monitoring (IFM) (indicator met); Continuing stakeholder consultation on the European Union Forest Law Enforcement, Governance and Trade (EU-FLEGT) process (indicator met); and Continuing stakeholder consultation on the Extractive Industries Transparency Initiative (EITI) (indicator partially met). In Year 3, Indufor also verified only procedures: application for EITI candidacy by May 2013 (indicator not met); begin formal negotiations with EU for a Voluntary Partnership Agreement (VPA) by the end of 2012 (indicator met); outline in 2013 a program by the Ministry of Natural Resources and the Environment (MNRE) to manage degradation from extractive activities including mining extension service, enhanced dialogue with the relevant sectors (indicator met).

However, in terms of Independent Forest Monitoring (IFM), the indicators selected by the Guyana Forestry Commission (GFC) and included in the JCN were not the comprehensive list set out in the legality scorecard devised by Global Witness for IFM, which the GFC claimed to be following (Global Witness 2005). Instead, verification was mostly limited to paper procedures. The GFC was careful to avoid any investigation of illegal logging, presumably because the main causes of illegal logging are the corruption in the forest sector leading downwards from the apex of government (Bulkan 2014a). A study for USAID of GFC's procedures for verifying legality of logging carried out during the period of the Norway-Guyana MoU showed that the GFC system was ineffective (EFECA 2011). That study was not accepted by the GFC and was not referenced in any of the Norwegian consultancy reports. The GFA Consulting Group reports (2011, 2013, 2014) also showed that from the GFC point of view, IFM meant verifying the existence of a legal framework, not verifying the countrywide, consistent and equitable implementation of that legal framework, nor verifying that each indicator was actually and in every case applicable and applied. GFA noted exemptions negotiated and agreed, but not under what legal procedures those "special conditions" were allowed.

JCN governance Indicator 4 concerned the respect and protection of indigenous peoples and other persons' rights to participation, engagement and decision-making with regards to their own well-being. In its 2012 report, RA found that this indicator was not met. In the following year, Indufor (2013) only monitored procedural goals: presentation of the Amerindian Land Titling (ALT) project to the GRIF steering committee (indicator met); opt-in concept note ready and pilot community for opt-in mechanism selected (partially met); strategy and development of tailored information and consultations for hinterland communities addressed in the outreach program (partially met); and initiating implementation of Community Development Plans (CDPs) through the Amerindian Development Fund (ADF)[8] (indicator met).

By the time of the Indufor verification in 2013 and subsequently, evidence of the government's bad faith and misuse of these projects for political ends was repeatedly reported (APA 2013; Dooley and Griffiths 2014, 76–89). However, Indufor steered clear of any consideration of the content of the projects. Briefly, the ALT ignored the principal request of Amerindians made before Guyana's independence in 1966 that their communal land titles be adjoining partly to prevent the award of mining concessions in the areas between titled villages. Seven Akawaio and Arekuna titled villages had lodged a court case in 1998 to demand GoG compliance with that request so as to try to arrest the devastation of their customary territories by gold miners. That court case is still in process, subject to serial delays, while the ruination of the Akawaio and Arekuna homeland from mining continues (Butt Colson 2013).

Discussion

Norway's REDD+ agreement in general measured progress against procedures, not substance. Five years after the signing of the Norway-Guyana MoU for REDD+ there was little to report in terms of rainforest protection or improved livelihoods or governance. In November 2009, of the 12.9 million ha (Mha) of gazetted state forests, 6.1 Mha were under logging concession and approximately 5 Mha of forest were under partly overlapping mining licences. Although most of the logging and mining concessions are located in and around Amerindian customary lands, there is still no structured process that brings together these separate constituencies for discussion or negotiation.

Only US$378,000 of Amerindian Development Fund monies had reached 22 of the initial 27 Amerindian communities selected by the Ministry of Amerindian Affairs (MoAA), that is, 0.5 percent of the US$69.8 million disbursed by Norway up to 2014. Meanwhile, at least US$1.2 million of that same tranche for the Amerindian Development Fund had been retained for "administration" by UNDP or by MoAA itself. The GRIF Steering Committee approved a further US$10.7 million for the Amerindian Land Titling project in October 2013, most of which was likely to be paid to coastlander (non-Amerindian) and Chinese survey teams for the legally unnecessary demarcation of the natural boundaries of Amerindian titled villages.

None of the transformative changes related to governance outlined in the 2009 Norway-Guyana so-called REDD+ agreement was put into effect. Instead the government of Guyana channeled GRIF funds through state agencies and via patronage politics to selected clients while tightening control over its citizens and state forests. Amerindian leaders, in exchange for token payouts to their communities, delivered signed resolutions of support to the government (OP 2012). The discursive tropes of REDD+ were pressed into this service, then progressively watered down in successive annual amendments to the JCN. The two governments set out the parameters through which the pantomime of REDD+ in Guyana would be enacted, and enlisted international auditors to issue a stamp of approval on the paper procedures that were monitored.

This was the brave new REDD world enabled by the Norway-Guyana agreement and endorsed by international auditors. It is a cautionary tale about how not to draw up a REDD+ contract. Instead of centralizing REDD contracts at the apex of authoritarian governments, there would surely be greater chances of improved governance and protected forest landscapes if REDD contracts were signed with the customary owners and/or users of forests, and if the international auditors were required to also report to those local owners and users as part of their contract terms.

Notes

1. http://www.eu-norway.org/Climate_change/norway_carbon_neutral/.
2. http://www.regjeringen.no/Upload/FIN/cnn/folder.pdf.
3. In 2010 the declared size of the annual budget was US$700 million.
4. Data extracted from reports issued by the Guyana Gold Board and the Guyana Forest Commission's Forest Sector Information Reports.
5. http://hdr.undp.org/en/statistics/.
6. http://www.transparency.org/cpi2012/results.
7. http://www.lcds.gov.gy/index.php/documents.
8. The Amerindian Development Fund (ADF) is a fund under the control of the Office of the President of Guyana. The fund is confusingly composed of distinctly and indistinctly discriminated items. One of the items is the ex gratia "Presidential grants" system, which is used to channel US$900,000 annually to Amerindian communities (Bulkan 2014b). A distinct program, also known as the ADF, is funded under the agreement with Norway. The latter is only for a subset of Amerindian communities selected by the Ministry of Amerindian Affairs.

References

Amerindian Peoples Association (APA). 2013. "Press release: Concerns with Amerindian land titling project under the Guyana/Norway REDD Investment Fund (GRIF)". http://www.forestpeoples.org/topics/redd-and-related-initiatives/news/2013/10/amerindian-peoples-association-apa-press-release-co.
Angelsen, A., D. Boucher, S. Brown, V. Merckx, C. Streck, and D. Zarin. 2011a. *Guidelines for REDD+ Reference Levels: Principles and Recommendations: Prepared for the Government of Norway*. Washington, DC: Meridian Institute.
Angelsen, A., D. Boucher, S. Brown, V. Merckx, C. Streck, and D. Zarin. 2011b. *Modalities for REDD+ Reference Levels: Technical and Procedural Issues: Prepared for the Government of Norway*. Washington, DC: Meridian Institute.
Bade, Heidi. 2012. *Aid in a Rush: A Case Study of the Norway-Guyana REDD+ Partnership*. Master's thesis, University of Oslo: Centre for Development and the Environment.

Boucher, D., P. Elias, J. Faires, and S. Smith. 2014. *Deforestation Success Stories: Tropical Nations Where Forest Protection and Reforestation Policies Have Worked.* Cambridge, MA: Union of Concerned Scientists.

Braswell, B., Hagen, S., Salas, W., Palace, M., Brown, S., Casarim, F., and Harris, N. 2013. Assessing Forest Degradation in Guyana with GeoEye, Quickbird and Landsat. http://www.treesearch.fs.fed.us/pubs/44997.

Bryant, Raymond L. 1997. *The Political Ecology of Forestry in Burma: 1824–1994.* Honolulu: University of Hawaii Press.

Bulkan, Janette. 2014a. Forest grabbing through forest concession practices: The case of Guyana. *Journal of Sustainable Forestry* 33: 407–434.

Bulkan, Janette. 2014b. REDD letter days: Entrenching political racialization and State patronage through the Norway-Guyana REDD-plus agreement. *Social and Economic Studies* 63(3–4): 249–279.

Butt Colson, Audrey. 2013. *Dug Out, Dried Out or Flooded Out? Hydro Power and Mining Threats to the Indigenous Peoples of the Upper Mazaruni District, Guyana.* Survival International. http://assets.survivalinternational.org/documents/1113/book-fpic-oct-2.pdf.

Cadman, T. and T. N. Maraseni. 2011. The governance of climate change: Evaluating the governance quality and legitimacy of the United Nations' REDD-plus programme. *International Journal of Climate Change: Impacts and Responses* 2(3): 103–124.

Dooley, Kate and Tom Griffiths, eds. 2014. *Indigenous Peoples' Rights, Forests and Climate Policies in Guyana: A Special Report.* Moreton-in-Marsh: Amerindian Peoples Association and Forest Peoples Programme.

EFECA. 2011. *Review of Guyana's Legality Assurance System.* Washington, DC: USAID/GTIS.

GFA Consulting Group. 2011. *Guyana—Independent Forest Monitoring Scoping Report.* Hamburg: GFA Consulting Group.

GFA Consulting Group. 2013. Independent Forest Monitoring. Guyana. Summary of First Independent Forest Monitoring. Hamburg.

GFA Consulting Group. 2014. Independent Forest Monitoring. Guyana. Public Summary Report of Second Independent Forest Monitoring. Hamburg.

Global Witness. 2005. *A Guide to Independent Forest Monitoring.* London: Global Witness.

Government Information Agency, Guyana (GINA). 2009. "Gov't to Strengthen Monitoring, Reporting and Verification System." Georgetown: GINA, September 14. http://www.gina.gov.gy/archive/daily/b090914.html.

Government of Guyana and Government of Norway. 2009. Memorandum of Understanding between the Government of the Cooperative Republic of Guyana and the Government of the Kingdom of Norway Regarding Cooperation on Issues Related to the Fight Against Climate Change, the Protection of Biodiversity and the Enhancement of Sustainable Development. http://www.lcds.gov.gy/images/stories/Documents/MOU.pdf and associated Joint Concept Note on REDD+ cooperation between Guyana and Norway. http://www.lcds.gov.gy/images/stories/Documents/Joint%20Concept%20Note.pdf.

Government of Guyana and Government of Norway. 2011. Summary of Joint Concept Note. Version 2. 31 March 2011.

Government of Guyana and Government of Norway. 2012. Joint Concept Note. Version 3. 21 December 2012.

Government of Guyana and Government of Norway. 2014. Joint Concept Note. Version 4. 23 October 2014.

Gruning, Christine and Laura S. Shuford. 2012. *Case study: Guyana REDD-Plus Investment Fund (GRIF): National Climate Finance Institutions Support Programme*. Frankfurt: Frankfurt School of Finance and Management.

Harvard Law School. 2007. *All That Glitters: Gold Mining in Guyana: The Failure of Government Oversight and the Human Rights of Amerindian Communities*. Cambridge, MA: Harvard Law School. http://www.law.harvard.edu/programs/hrp/documents/AllThatGlitters.pdf.

Howden, Daniel and Colin Brown. 2007. "Britain backs Guyana's rainforest plan." *Independent*, December 11. http://environment.independent.co.uk/climate_change/article3241936.ece.

IIED. 2009. *Conceptual Framework on Process for the Multi-Stakeholder Consultations on Guyana's Low Carbon Development Strategy (LCDS)*. Edinburgh: International Institute for Environment and Development. http://www.lcds.gov.gy/images/stories/Documents/conceptual_framework.pdf.

Indufor. 2013. *Independent Assessment of Enabling Activities of the Guyana-Norway REDD+ Partnership: Final Report for Norwegian Ministry of the Environment*. Helsinki, Finland: Indufor.

Inter-American Development Bank (IADB). 2013. "Forest Carbon Partnership Facility Project in Guyana (GY-T1097)." TC document. http://idbdocs.iadb.org/wsdocs/getdocument.aspx?docnum=38259336.

Kaieteur News. 2010. "Norway money . . . Guyana yet to submit projects plan." December 11. http://www.kaieteurnewsonline.com/2010/12/11/norway-money-guyana-yet-to-submit-projects-plan/.

La Rose, Jean. 2013. Guyana: Indigenous Peoples and the Lack of Adequate Consultation on REDD+. *REDD+ Safeguards: More Than Good Intentions? Case Studies from the Accra Caucus*, 5–11. Copenhagen: DANIDA.

Lipton, Eric, Williams, Brooke, and Nicholas Confessor. 2014. "Foreign powers buy influence at think tanks." *New York Times*, September 6. http://www.nytimes.com/2014/09/07/us/politics/foreign-powers-buy-influence-at-think-tanks.html?_r=0.

Lohmann, Larry. 2011. The endless algebra of climate markets. *Capitalism Nature Socialism* 22(4): 93–116.

LTS International, Ecometrica, Indufor Oy, and Chr. Michelsen Institute. 2011. Real-time evaluation of Norway's international climate and forest initiative. Contributions to a Global REDD+ Regime 2007–2010. Evaluation Report 12/2010. Oslo: NORAD.

LTS International, Ecometrica, Indufor Oy, and Chr. Michelsen Institute. 2013. Real-time evaluation of Norway's international climate and forest initiative. Contribution to Measurement, Reporting and Verification. Oslo: NORAD.

Office of the President (OP). 2009. *Transforming Guyana's Economy While Combating Climate Change: A Low-Carbon Development Strategy*. Georgetown: Office of the President.

Office of the President (OP). 2012. *Focus on the LCDS*. Issue 2. Georgetown: Office of the President.

Paterson, M., M. Hoffmann, M. Betsill and S. Bernstein. 2011. The micro foundations of global climate governance: An analysis of the transnational emission trading network. Prepared for the Princeton Conference on Research Frontiers in Comparative and International Environmental Politics, 2–3: 1–24, December. http://www.princeton.edu/~pcglobal/conferences/environment/papers/bbhp.pdf.

Pearson, T. R., S. Brown, and F. M. Casarim. 2014. Carbon emissions from tropical forest degradation caused by logging. *Environmental Research Letters* 9(3): 1–11.

Rainforest Alliance (RA). 2012. Verification of progress related to indicators for the Guyana-Norway REDD+ Agreement. 2nd Verification audit covering the period October 1, 2010–June 30, 2012.

Ram, Chris. 2014. Forestry commission's reports: An affront to decency. http://www.chrisram.net/?p=1456.

REDD-Monitor. 2014. Oil and REDD: Norway pays center for global development to promote REDD in Washington. http://www.redd-monitor.org/2014/09/09/oil-and-redd-norway-pays-center-for-global-development-to-promote-redd-in-washington/.

Solheim, Eric. 2009. "REDD comes with risks but there is no other choice than to try." *Guardian*, October 8. http://www.guardian.co.uk/environment/cif-green/2009/oct/08/redd-norway-brazil-climate-change?intcmp=239.

Stabroek News. 2014. "GFC's $600M payments to NICIL unlawful." January 12. Georgetown.

Strathern, Marilyn, ed. 2000. *Audit Cultures: Anthropological Studies in Accountability, Ethics and the Academy*. London: Routledge.

Thomas, Clive Y. 2009. "Norway and Guyana's rainforest: Why beggars do not choose." *Stabroek News*, December 13. http://www.stabroeknews.com/2009/features/12/13/norway-and-guyana%e2%80%99s-rainforest-why-beggars-do-not-choose/.

Thomas, Clive Y. 2014. "Inequality, poverty and the flaunting opulence of the 'newly-rich.'" *Stabroek News*, September 21. http://www.stabroeknews.com/2014/features/09/21/inequality-poverty-flaunting-opulence-newly-rich/.

Thomas, Clive Y., N. Jourdain, and S. Pasha. 2011. Revisiting the underground economy in Guyana 2001–2008. *Transition* 40: 60–86.

Thompson, M., M. Warburton, and T. Hatley. 1986. *Uncertainty on a Himalayan Scale: An Institutional Theory of Environmental Perception and a Strategic Framework for the Sustainable Development of the Himalaya*. London: Milton Ash.

Trevin, Jorge and Robert Nasi. 2009. *Forest Law Enforcement and Governance and Forest Practices in Guyana*. Bogor: CIFOR—Center for International Forestry Research.

Wilburg, Kiana. 2014. "60 per cent of the economy is run by the underworld." *Kaieteur News*, December 31. http://www.kaieteurnewsonline.com/2014/12/31/60-percent-of-the-economy-is-run-by-the-underworld-dr-clive-thomas/.

6
CORPORATE CARBON FOOTPRINTING AS TECHNO-POLITICAL PRACTICE[1]

Ingmar Lippert

Attempting to tackle climate change with market solutions hinges on the existence of emissions. We know much about the politics of undoing emissions—via offsets (e.g., Böhm and Dabhi 2009). But where do emissions come from? How are they done? Carbon footprinting seems to be the simple answer. Is this merely a technical matter? In this chapter I explore how emissions come into being; carbon accounting emerges as techno-political practice, fraught with non-transparency.

This chapter argues that successful corporate carbon accounting practices efficiently and skillfully ignore significant political implications of the company's practical relation to climate change. 'Successful' in this case signifies what matters for the company to compete well in capitalist markets. By examining voluntary carbon accounting at a financial services corporation, I invite an engagement with how the technicality and politics of carbon interrelate in accounting. I ground my analysis in ethnographic fieldwork across 20 months in the corporate social responsibility (CSR) unit at one of the 50 largest global companies.[2] Over this period, I supported the CSR unit's management of their sustainability data, in exchange for overt and explicit research access to the CSR unit's activities.

Why does carbon accounting matter? Many governments are wedded to the idea that economic growth is the presupposition of any pragmatic policy. Recognizing that the modern economy causes environmental problems, the policy program of ecological modernization suggests greening capitalism as a way out. This policy advocates the establishment of markets in which environmental goods are to be traded. Fitting this general theory, carbon markets allow trading emission offsets; the theory suggests such offsets are effective in reducing emitters' emissions: for example, 5 tons $CO_2e - 1$ ton CO_2e offsets = 4 tons CO_2e. Carbon accounting supposedly provides the foundation to make this theory work:

this class of environmental accounting produces carbon footprints. That is, carbon accounting renders emissions organizationally, economically and politically present.

Examining voluntary carbon accounting is crucial to shedding light on the practical realities of environmental markets. To study carbon markets as neoliberal politics, we need to also understand how carbon accounting works in practice where the nation-state is least involved. Thus, I do not engage with the attempts by governments to command-control accounting; rather, in this chapter I study how carbon accounting is played out in a voluntary setting.

Yet, voluntary carbon accounting does not happen in isolation from society. The company engages with several actors and environmental accounting standards in its accounting; in this study we encounter, for example, the Organisation for Economic Co-operation and Development (OECD) and the World Business Council for Sustainable Development (WBCSD). Precisely because of the company's active relations to other organizations, the findings presented here are applicable beyond the particular case (see Lippert 2014). The practices analyzed here are co-constituting global carbon reports; the practices involve interactions with multiple actors, including the company's peers, standards-setting organizations and brokers, who accept or ignore the company's practices.

Here is an analytical distinction crucial for this chapter: I differentiate between formal realities (in which standards are seen as implementable and effectively implemented) and situated realities (the practical goings-on of humans, their organizations, their technologies—which may involve and relate to, but are never determined by, formal realities). So, situated realities involve formal realities; but formal realities are characterized by an unobtainable imaginary. Formal realities (like a standard) are thought of as capable to determine other realities (e.g., emissions and reports thereof). The following three analytical stories show different types of undesired realities and how they have been marginalized, deleted or ignored within the situated practices of the carbon accounting apparatus. The first story attends to practices that report the company's emissions to global carbon markets via a voluntary emissions disclosure ranking. The second story traces the optimization of the company's accounting apparatus to reduce some of the technocratic tension, allowing us to see the re-alignment of accounting practices to better exclude undesired relations. The third story delves into the globally extended voluntary carbon market (VCM), examining the shifting presence of the VCM Gold Standard for carbon offsets in the company's offsetting discourse and practice, pointing to significant flaws in this market.

An Analytic Distinction: Formal and Situated Realities

This section contrasts two ways of studying organizational practice—one, by attending to formal realities and the other, by scrutinizing situated realities and reality-making. To illustrate this contrast, I draw out a variety of practical social and political concerns in the midst of carbon accounting. I argue we need to raise questions about the who and the what that are included in carbon accounting and that which are excluded, about what counts.

In an analytics that focuses on formal realities, one should ask how the company's carbon accounting is defined. Along with my work contract for the company, I received additional documents formally defining the company. Consider their code of conduct: this document linked the company to the hegemonic discourse of sustainable development and global institutions. 'Through our initiatives for the UN Global Compact programme and the acknowledgement of the OECD Guidelines for multinational corporations we integrate sustainability and social responsibility into our business.' This quote specifies what sustainability and social responsibility are for the company. The company considers its support for the UN framework and its references to the OECD as a practice that works to bring the vague terms sustainability and social responsibility into corporate practices. This statement narrows down what the terms mean, that is, some form of offering support by discursively linking the company to the UN and the OECD.

Such a study of formal reality is quite limited; we are positioned in the hermeneutics of mere statements. More interestingly, on my first day of work, the company's global sustainability manager pointed me to a document that specified the company's environmental management system. Part of this specification was citing the code of conduct; at the same time this management document translated environmental management into carbon accounting; and this carbon accounting was presented as a standardized activity.

1. Formally, according to these specifications, the company conducted environmental management using the International Organization for Standardization standard, ISO 14001. This standard specifies so-called environmental management systems (and critical organization studies problematize such systems as rational myths; Boiral 2007).
2. The company simultaneously claimed its carbon accounting to be standardized according to a voluntary environmental accounting standard, originally emerging in German multinational finance networks under the name VfU.[3] VfU provides, for instance, specific carbon conversion factors. The latter are needed to translate environmental data, for example consumption facts, into carbon emission equivalents (CO_2e). To be at all reliable, these factors need to be clearly documented and stable. Here is the practical messiness that is interwoven with such formal reality. While publicly the company committed to VfU, less visibly it also mentioned the GHG Protocol, developed by the WBCSD, which employs lower conversion factors, effectively reducing the company's emissions on paper.[4]
3. In addition, the company employed the Global Reporting Initiative's methodologies for carbon accounting. These three standards did not completely cohere, prompting accountants to apply 'pragmatic' solutions.

These standards were to define what to measure. Clear definitions were demanded. This demands explicitness from the company about what it included in its measurements and what it excluded. It defined its emissions as caused by office operations (its consumption of electricity and other forms of energy, of water and paper; its business traveling and production of waste). Formally, therefore, all emissions that the company's financial services enabled beyond just office operations did not matter. The company's accounts, for example, neither identified the emissions enabled by the financing of the fossil fuel business nor recognized social problems and conflicts often related to fossil fuels.

To assure shareholders and publics that the company's carbon accounting was rigorous and environmentally legitimate, the company engaged in a partnership with two organizations, each among the biggest in their respective sectors. One partner was one of the world's four biggest accounting firms. The other partner was one of the world's top five environmental conservation NGOs (nongovernmental organizations) (see Jepson 2005). For several million US dollars, the NGO got a seat at the corporate table while their well-known logo reappeared in the company's public relations materials.

Studying formal realities shows that the company discursively engaged with climate change, reported emissions and was committed to various publics. Such engagements were societally translated and shifted through institutions and standards, generating output for rankings and indices that conjured up transparent carbon markets. Studying formal relations as practice yields a grounded understanding of how the company ignores concerns about practical and political realities.

If we do not focus on the formal relations and engagements but on how carbon accountants work, how their accounts take shape and how they are transported and presented, then we move into an analytic space addressing a more ambivalent and situated politics of the technical. To sensitize the reader to this ambiguous form of the political co-constitution of carbon emissions, I offer stories from the margins of what is normally seen as mattering for carbon accounting.

Let's join one of the omnipresent team meetings. We are presented with facts—appearing on slides, projected onto the wall with PowerPoint. Within the CSR department it was considered self-evident that one can separate content from design. This allowed the CSR experts, including the managers of emission facts, to organize their work by using a division of labor. The experts developed the contents for the slides and others would bring the content into shape. This 'bringing into shape' work was outsourced to a company called India Graphics. India Graphics allowed CSR staff positioned in the global North to develop content all day and get cheap southern labor to produce neatly designed slides overnight. Doing carbon accounting now needs to be considered as interwoven with uncertainties about the economic and labor realities that are co-constitutive of emission reports. On one occasion, one of the CSR workers actually questioned whether India Graphics had been checked in terms of 'child labor'. This issue was soon translated into the less threatening and workable category of 'reputational risk'. The risk was managed by asking India Graphics to formally declare that they did not employ child labor.

For such ambivalent economic realities we do not have to shift from the headquarters (HQ) in the North to the South. Consider outdated and useless paper copies. We find them in the trash. Performing neatly the scripted role of a nameless, often smiling and always friendly office cleaner, it was nearly always the female migrant workers who entered the offices of the CSR unit every morning and got rid of the waste that the white-collar workers produced. The production and reproduction of social and economic inequality is not merely found by looking at realities in the South but also in the North's centers. If we engage with the neat offices that produce 'orderly' emissions then we identify the labor relations that sustain the firmness of capitalism.

To sum up, the second analytical category indicates an ambivalent and situated politics: the everyday doing of emissions and its implicated engagements with political-economic realities and imaginaries revolving around carbon. I sketched political concerns (externalities, transparency concerns, outsourcing, gender, migration) that are considered in simplistic accounts of accounting as merely technical matter. Studying situated politics allows rendering present how particular societal realities matter, even though they are not typically assumed to shape carbon.

Differentiating formal relations from situated realities of practical action is helpful to analyze corporate carbon accounting's own logics. In its own theorizing, accounting is a matter of implementing management and accounting standards correctly. To govern emissions and assure stakeholders that emissions are correctly accounted for, the company formally engages with rating agencies, auditors and NGOs. The logic of organizational practice, however, is situated in the midst of societal and political relations.

Thus I identify politics as co-constituting emissions. For any accounting step, we can raise questions about inclusion and exclusion.

Reporting Emissions

Once the company's carbon accountants had brought together all the environmental data—of energy, water and paper consumption, travel and waste services—from across the globally distributed subsidiaries, they were able to assemble that data and extrapolate it, resulting in the global carbon footprint. Elsewhere, I detailed the work processes and politics of assembling carbon data.[5] Here, I retrace how the company's carbon footprint was reported to one of the world's largest carbon ranking organizations. I illustrate the quality of political, tactical and moral decision-making involved in the translation from a spreadsheet into a reporting form for the ranking. I argue that reporting is not a transparent and neutral activity.

Let me introduce Frederik (a pseudonym). He was the company's top carbon accountant. As part of my work, I assisted him in putting together the response to the questionnaire from the ranking. Frederik made clear that copying answers from the company's previous year's response was an apt approach to fill out the current questionnaire, because the previous answer had already been authorized. Based on the prior formulations, he could simply check for mostly quantitative changes. This constituted a strategy of not having to consider new emissions categories. On the categories themselves, he explained that he liked to inscribe the same categories as those in the VfU standard (see the previous section) into the company's response. By this, he explained, he can produce 'coherency'. This is significant for carbon reports: that they cohere with other organizations' reports and standards is not simply given by nature, but rather means that coherence is achieved and brought into existence, serving a strategy not to irritate audiences.

Another type of data practice is related to the precision of numbers. By shaping how and whether numbers appeared, Frederik managed to guide the reader's attention. Here are two interesting moments. He mentioned he liked omitting decimal numbers. For him these constituted nonsense: 'Decimal numbers evoke accuracy. This we do not possess.' Here he related to some numbers in a draft of the response, which were the product of calculations resulting in numbers with decimals. While the decimals may have been technically correct, he decided that the company should not claim to be accurate in its measurements. Particular to this moment, he wanted to steer how the response would be interpreted. If the audience had the impression that the company produces accurate data, then this would set the standard by which its accounting would be measured. And he did not want to allow publics to set such a standard.

On a different occasion he questioned whether the company should offer a number at all. This concerned the company's own quantification of the quality of its environmental data. According to the company's data quality review, much of the environmental data was guessed or produced through some form of calculation (based on contingent assumptions) that resulted in a (low) data quality score number. Frederik did not want to have the number published. Instead, he made clear that all the problems of data quality should be understood as resulting from some individuals' incompetence, while the overall carbon accounting system worked fine.

When I talked with the company's liaison at the ranking agency, I learned the latter were pleased with the data they got from companies; most of the data worked for them. At the same they did not verify whether the data submitted was correct. They accepted companies' data as it came in. The liaison also shared that in cases of mistakes, they are 'quick' to get companies to correct the data. I asked how often corrections took place. He replied that there is lots of checking before companies press 'submit', and about 2–3 percent of the data needed to be corrected.

This section illustrates the relevance of even the smallest judgments and assumptions in carbon accounting. Carbon accounts are not facts of nature, but it is people who produce them. Translating data matters because it constitutes the interpretation and transformation of data into forms, which are accessible to larger audiences. Such practices—tactically conjuring up coherency, managing audiences' impressions of accuracy, invoking naïve trust in companies' data—do partially redistribute responsibilities among the involved actors and audiences and partially make it impossible for actors to responsibly engage with emissions (e.g., if data quality issues are hidden and defined away). And no form or configuration of such translation can be neutral. The human is always part of it. The relevant question is then *how*, rather than *whether*, social, political, economic and organizational tactics shape the emissions.

Optimizing Accounting

Asking how emissions are shaped, I move on to a significant recognition by the company. They noted that their accounting apparatus was not working optimally; it had to be improved. Thus, the company approached optimizing their carbon accounting. Why? The top sustainability managers of the company had gotten the impression that too much of their resources were spent on engaging with technicalities of carbon accounting. In consequence, they initiated a process to optimize data collection, rendering it more 'certain' and 'trimming' it (i.e., streamlining data gathering and analysis processes). I briefly review four tensions the company experienced before I turn to their 'solution'.

The four tensions map onto the ideal(ized) world of carbon accountants: environmental information about consumption of services and resources is readily available, allowing (1) skilled bookkeepers to (2) collect this data and (3) enter it into a central database, where it would be multiplied with appropriate conversion factors and (4) summed up in resulting emissions—and voilà, the carbon footprint would appear.

First, the environmental bookkeeping role was assigned to whoever was somehow seen as available (such as facility managers). This meant that whoever acted as bookkeeper asked the others (e.g., in the financial accounting department or the subsidiary's travel experts) for the invoices on, say, energy consumption or for reports on how many miles or kilometers have been crossed in the prior year.

Second, however, in the practical work of subsidiaries' staff problems abounded. Problems included the absence of data and invoices. Consider countries where offices are rented at a flat rate, thus not documenting consumption as such. Or consider places where waste management is firmly in the hands of an informal economy; there are no invoices. Another class of problems revolved around cases in which bookkeepers could access invoices but could not identify sufficient environmental data (printed onto the invoice). So, financial information was provided but no environmental data. Data collection emerges as data enactment (Lippert 2015). Some of this enactment was standardized; but no standard could ever foresee all eventualities. Finally, there were always people involved in copying data. Copy and paste, in practice, is not straightforward. Sometimes figures are missing, the decimal point is wrongly set, the unit or the category failed.

This, third, resulted in difficulties with submitting data: bookkeepers entered all kinds of 'correct' and 'incorrect' data. People at the company headquarters (HQ) were aware of many of the complexities of sourcing data. Yet, HQ staff could not directly solve the problems. They checked subsidiaries' data; when large deviations (compared to the prior year) were found, HQ staff demanded that the subsidiary staff explain. They developed draft environmental balance sheets, indicating the carbon footprint resulting from the data entered by the subsidiary.

Fourth, these drafts were offered to the subsidiary, making visible the results of the data subsidiary bookkeepers had entered. Thus, data was checked, improved and corrected, often in several rounds, back and forth between HQ and subsidiary. As a result, many circulating versions of footprints existed. Today's footprint result was tomorrow's footprint version's draft. It was not rare that even data entered years back were found to be 'wrong'. Consequently the carbon footprint was shifting and changing. Even baseline data was 'corrected', achieving much more favorable emission reductions: if baseline emissions are increased while current emissions stay stable, then current emissions are relatively less high (Lippert 2013, 227–230).

Eventually, the company assembled a project team to review and optimize carbon accounting. A key artifact of their work was a business process model that reconstructed core intricacies of environmental data work. Their model spanned many pages, thus the model-as-document was unwieldy. They then approached accounting by targeting both the sourcing of data and the processes by which data was processed, working with the process model and streamlining it in two ways. First, they planned to exclude the diverse bookkeepers, to substitute their data sourcing with one clear moment: whenever a financial clerk would enter data on the costs for, say, energy consumption, into the financial accounting system, then the system should prompt them to also enter the corresponding environmental data (e.g., kWh). This understanding of direct data entry allowed the team to presume that loops of data review and corrections would cease. The second difference was that this made them simplify the prospective data flows. In effect, the new data flows and practices were fit into a new business process model that was clearly defined, reducing complexity to two pages. Here, then, was the solution: a streamlined carbon accounting plan.

This description of the solution misses out much of the optimization team's work. A key achievement was to (1) review the details in the old processes and (2) imagine a simplification. The latter achieved identifying data practices they could all agree on as promising to work well. In parallel they identified how some of their key assumptions were unlikely to hold. They recognized that their solution would not work for all the cases where environmental data (for whatever reason) was simply not available. They also noted that the financial clerks could not be expected to care about the environmental data, to check whether the invoices provided realistic environmental data or whether they entered it correctly. Instead, entering environmental data would simply be an additional task for the clerks. This did not promise high-quality environmental data. If we recognize that the team members' voices pointed to the built-in problems of the solution, we can recast the collective silencing and ignorance of these voices as a significant organizational achievement. I propose their achievement as precisely performing good management, data and, hence, environmental governance. The notion of 'good' here points to making the performance of management possible. For if the team members had actually engaged with the complexities, then they would have risked not coming up with the solution, a clear-cut and efficient carbon accounting apparatus.

Golden Offsets

The presentation of the politics, certainties and uncertainties that co-constitute emissions enables us to take a next step and enquire into the textual politics of how the voluntary carbon market reduces emissions. The company was pressed by their NGO partner to only use emission offsets that are certified by the Gold Standard—which is a voluntary standard, described as offering carbon offsets of 'premium value' at a 'premium price' (Bumpus and Liverman 2008, 146).

When I first encountered a Gold Standard credit certificate that the company had bought for offsetting, I got excited. I would leave the HQ office and trace how the company's footprint connected to an offsetting project in the global South. I would study how my company was related to offsets. Alas, this quest failed. And this failure offers an informative moment, resulting in three puzzlements I use to question the workings of voluntary standards and markets.

A first moment of puzzlement occurred when I attended to the certificate's precise formulation. The certificate was formulated to uniquely fit the purpose of offsetting a particular component of corporate emissions—the emissions of several subsidiaries' car fleets. Here is an anonymous, translated transcript of the certificate's key formulation.

> [It] guarantees that all the [emissions of greenhouse gases] caused by the [car fleet] in the year 2008, summing up to 1,550 tonnes CO_2-equivalents, have been offset by means of additional investment in the high quality Gold Standard climate change project *Wind power plant in Maharashtra in India.*
>
> *(Lippert 2013, 526; emphasis in original certificate)*

The closing component of the main clause makes several claims. It says emissions have been offset. It details where the offset took place (in a wind power plant in

Maharashtra) and qualifies the offsetting project as certified by the Gold Standard. My first puzzlement focuses on the means of offsetting. The statement says that the offset was realized 'by means of additional investment'. This (re)specified the claim to assert that the offset was achieved by channeling some investment into the project. The quality or quantity of this investment was not explicated. Taking this formulation seriously means that maybe the monetary investment was limited (even just one cent would qualify), but it would also allow for a non-monetary investment.

Being well puzzled and not having been able to identify any documents that would answer my questions, I started to contact people. I sent inquiries to the carbon broker as well as to the Gold Standard HQ. Neither replied. However, a change occurred. Both broker and the company I studied requalified their claim. They now stopped claiming that the project was a Gold Standard project, instead claiming a certificate 'in accordance' with the standard. Once in 2014 I briefly got hold of a Gold Standard representative and asked him (in person and by email) to update me on this case. A pattern emerged: to date nobody has written to me to explain this case, and none of the organizations has officially replied to me. An informal link between me and another Gold Standard informant, however, clarified that they did not officially know about any such project. Interestingly, the certificate did not contain a carbon registry number. I wonder: what does it mean if VCM players are not ready to explain their methods?

Leaving the field, I pursued some of these questions with a former HQ carbon accountant. She redirected my analysis. I had shared with her the analytical distinction between formal and situated realities. What she told me is simple and straightforward. She explained the puzzlement by calling the project money laundering. According to her, the carbon credits were simply bought from our company's carbon traders. Thus, in her story, the carbon accountants achieved offsetting by buying emission credits from a broker who, respectively, bought the credits from the financial services company itself—constituting a circular relation. I have no means to ascertain the truth of her explanation (I see no reason to doubt her story). What are we left with? The credits are conjured up out of vague 'investments' in an unspecified (if at all existing) wind power project, without a registry number and in a business relation that stifles traceability. The story helps to foreground the significance of engaging with the situated realities of carbon, both of emissions and their alleged offsets. Such situated realities can involve (more or less successfully) staging formal realities.

Opening up the puzzlements in this section links the carbon practices within the company to the textual and economic dynamics of carbon markets. Emission trading is not merely about the projects in the global South. It is also about databases and certificates that are circulated in the North, even if they do not link to actual projects in the South. Performing the VCM may work well enough for players in the North without the necessity for them to stabilize a high-quality link to projects in the South. In the company, emission trading, and thus the

VCM, is practiced as just another set of organizational and textual possibilities. Markets and the political are interwoven.

Conclusion

For a post-carbon economy to work, carbon needs to be 'known'. This does not only refer to carbon offsets but also to emissions themselves. This chapter engaged with the voluntary carbon accounting practices of a major multinational company, troubling how accounting knows emissions. I retraced how the company's carbon accounting apparatus works to produce a reality of climate change in which political realities do not disturb carbon knowledge and, thus, the company. Given that the company was positioned as a major player in the international finance sector, with ties to other major financial players such as the largest accountancy firms, I conceptualize the corporate practices as well aligned to profit making, and thus, capitalism. Three ethnographic stories show how emissions are enacted such that accounting works more smoothly and more efficiently, thereby decreasing the involvement of disturbing realities.

Reading these stories together helps to understand how capitalism performs carbon accounting while not engaging with related political problems at any level. With the case 'reporting emissions', I showed carbon information is produced to increase the appearance of coherent and good data practices while uncertainties are hidden. I propose that the enactment of such data helps the market to perform more smoothly, while disconnecting the market from the entities it supposedly represents. The case 'optimizing accounting' shows how the accounting apparatus was reorganized to better exclude distributed actors who would care about environments and data practices. At the same time the optimization results in systematically not engaging with data gaps and quality problems. These disconnects help the company to stage its carbon accounting as clear-cut, rendering carbon footprinting more efficient. The case 'golden offsets' shows how the presence of voluntary standards cannot ensure high quality emission realities.

Carbon accounting is key to enabling carbon markets. To study carbon markets it is appropriate to scrutinize their socio-technical constitution, that is, technocratic emission realities. This chapter finds corporate practices are (re) organized to render carbon accounting less disturbing, thus aligning accounts of environmental impact to perform well in capitalism.

Notes

1. I thank the informants in the company I studied, the editors of this book, and Lydia Stiebitz and Rachel Douglas-Jones for critiques on drafts of the chapter.
2. This chapter partially re-analyzes results of a study published elsewhere (Lippert 2013). The core of this analysis is positioned at an intersection of ethnomethodology and actor-network theory (Lippert 2014) and includes a systematic reconstruction of situated,

material-semiotic data practices at corporate carbon accountants' workplaces. Ethnographic observation resulted in field notes of about 300 pages that guided an inductive qualitative analysis. The material re-analyzed in this chapter sketches the relevant observations for the chapter's argument. Complementary analyses have been conducted, for example, problematizing categorization and classification practices (Lippert 2012) and reconceptualizing environments as datascapes (Lippert 2015). The identity of the company is confidential.
3. VfU is the German abbreviation for the Association for Environmental Management and Sustainability in Financial Institutions.
4. To illustrate, consider a subsidiary reporting that they conducted business travel with long-haul flights, crossing a distance of 365,387 km. With the WBCSD's factor, this distance is translated into about 65.8 tCO_2e; with VfU's factor, about 119.2 tCO_2e would have resulted. For a more detailed account, see Lippert (2013, 88–107).
5. See note 2.

References

Böhm, Steffen and Siddhartha Dabhi, eds. 2009. *Upsetting the Offset: The Political Economy of Carbon Markets*. London: MayFly.

Boiral, Olivier. 2007. Corporate Greening through ISO 14001: A Rational Myth? *Organization Science* 18, 1: 127–146. doi:10.1287/orsc.1060.0224.

Bumpus, Adam G. and Diana M. Liverman. 2008. Accumulation by Decarbonization and the Governance of Carbon Offsets. *Economic Geography* 84, 2: 127–155. doi:10.1111/j.1944-8287.2008.tb00401.x.

Jepson, Paul. 2005. Governance and Accountability of Environmental NGOs. *Environmental Science & Policy* 8, 5: 515–524. doi:10.1016/j.envsci.2005.06.006.

Lippert, Ingmar. 2012. Carbon Classified? Unpacking Heterogeneous Relations Inscribed into Corporate Carbon Emissions. *Ephemera* 12, 1/2: 138–161. http://www.ephemeraweb.org/journal/12-1/12-1lippert.pdf.

Lippert, Ingmar. 2013. *Enacting Environments: An Ethnography of the Digitalisation and Naturalisation of Emissions*. PhD thesis, Augsburg University.

Lippert, Ingmar. 2014. Studying Reconfigurations of Discourse: Tracing the Stability and Materiality of 'Sustainability/Carbon'. *Journal for Discourse Studies* 2, 1: 32–54.

Lippert, Ingmar. 2015. Environment as Datascape: Enacting Emission Realities in Corporate Carbon Accounting. *Geoforum* 66: 126–135. doi:10.1016/j.geoforum.2014.09.009.

7
REGULATING FAIRNESS IN THE DESIGN OF CALIFORNIA'S CAP-AND-TRADE MARKET

Patrick Bigger

On January 1, 2013, California launched a cap-and-trade carbon market as the centerpiece of the most significant climate change regulations in the United States. The creation of this market is the result of 7 years of legislative maneuvering and regulatory negotiation among myriad stakeholders. The bill that authorized the creation of market-based climate policy, AB32, the California Global Warming Solutions Act of 2006, offered significant discretion[1] to regulators on how to achieve California's emissions reduction goal of returning to 1990 levels by 2020. What has resulted is a state-based emissions trading system with limited use of offsets that covers about 85 percent of the state's documented emissions (ARB 2011). The market is part of a regulatory suite that blends direct regulations, including performance standards and clean energy quotas, with expansive market-based governance mechanisms. While the basic principles of carbon market design are fairly straightforward, the nuances in rule-making are multitudinous, the negotiations over those nuances protracted, and the number and types of interlocutors trying to influence all of the moving parts of the system are vast.

This chapter seeks to elucidate one of the key ways that stakeholders have tried to shape market rules to their liking, specifically through arguments about 'fair treatment' in design elements large and small that almost invariably call for the costs to be imposed on industry and consumers to be minimized, turning matters of environmental governances into matters of cost containment. The number and types of decisions that must be made in order to implement any regulatory environmental market are substantial, but they are multiplied and magnified in carbon markets because of the more abstract qualities of the material being governed, the potential impacts on the productive economy that emits greenhouse gases, and the political visibility and stakes of regulating climate change gases

(Tietenberg 2006; Bumpus 2011; Boykoff and Yulsman 2013). These decisions range from the broad, such as how to distribute pollution permits among different industries, to the specific, such as accounting for the ways that power import contracts are written and for the emissions embodied in that power.

In California, regulated industry, industrial business groups, financial concerns, market-enthusiast environmental NGOs (nongovernmental organizations), academics, and sundry others have argued about market design elements using nearly every rhetorical strategy imaginable, from cajoling, to threatening, to reasoning, and sometimes even pleading. These strategies are deployed using the language of economic efficiency, political expediency, or technological feasibility. But one commonality across all of these strategies, and the languages used to express them, is the presence of sentiments relating to just and fair treatment under the regulation. In the course of rule-making, 'fairness' has come to be defined within a narrow register, reflecting one pole of thought on how carbon pricing can achieve desired outcomes. This is chiefly through cost containment for regulated entities rather than through establishment of a substantial price on carbon that would induce changes in the composition of polluting industries in California.

The US government recently updated its social cost of carbon to US$37 a ton, which refers to the total social and environmental benefits that accrue from avoiding the emission of one ton of CO_2, or alternatively, the damage done by releasing that ton of CO_2 (IWGSCC 2013). Some models indicate that the price needed to induce structural change to a low-carbon economy, sufficient to avoid catastrophic impacts of climate change, is around US$200 per ton in the United States (Hope 2013). Throughout consultation on the creation of the cap-and-trade market, regulated industries in California have repeatedly demonstrated that a price of that magnitude is unacceptable, and regulatory decisions indicate concurrence on the part of the state (ARB 2010b). This points to a potential contradiction in the development of carbon markets, where price is supposed to engender widespread socio-technological change, yet the price signals allowed by political compromises in the regulatory design process may not be sufficient to achieve transition to a 'green' economy.

In California, stakeholders' arguments about fair costs in the formulation of the program resulted in a rhetorical inversion: What started as a market with the potential to make 'polluters pay,' embodying the economic notion that a substantial carbon cost would drive down emissions of both climate change gases and co-pollutants, was transformed to a market where 'pay to pollute' became the operating principle. Environmental justice organizations have long feared such an outcome, believing that emissions will not be avoided and that polluters can eschew responsibilities to impacted communities through accounting tactics and the outsourcing of reductions with offsets.

The chapter begins with a brief discussion of the methodological approach taken to creating and analyzing the data on which my arguments are based. It then describes how 'fairness' was deployed in the rule-making process for the California cap-and-trade program, and how the concerns of environmental justice advocates came to be sidelined through the regulatory decision to conduct a specific kind of carbon pricing. I will consider how industrial advocates were able to make ideas about fairness largely a matter of cost reductions, in order to influence specific aspects of the regulation. The chapter will conclude by demonstrating the success of these arguments in shaping market rules, resulting in a situation where many more allowances than needed to raise the price of carbon to a behavior-changing level were given away. Along with macroeconomic factors, the widespread use of offsets and the relaxation of rules around electricity imports have blunted the price signal that is supposed to be the driving force for changes to the underlying economy.[2]

Methods and Data

The analysis in this chapter is based on 14 months of fieldwork in Sacramento, California, the capital of California and home to the California Air Resources Board (ARB). California has both a tradition of cutting-edge environmental policy-making and strong public transparency laws, which make it an ideal location to examine how innovative environmental policy is implemented (ARB 2010a, CGC §11120–11132). It is based on textual analysis of presentations by regulators, draft regulatory text, and publicly available comments on these regulatory documents by stakeholders from 2008 to 2013, on topics ranging from new offset protocols to public information disclosure. I used Nvivo qualitative data analysis software to code for an initial set of terms, use the terms to find related ideas, and then recode the texts using both sets of terms relating to just, fair, consistent, or reasonable treatment, what Crabtree and Miller term the 'immersion/crystallization' method (1999). For each of these data points I coded two kinds of information: (1) the type of organization making the argument; and (2) the way that ideas about fairness were being constituted.

Twenty-four kinds of organizational actors were recorded deploying arguments about fairness.[3] The groups that most often made arguments related to fairness were compliance entities, the polluting businesses that are required to participate in the cap-and-trade market. This category includes investor and publicly owned utilities (39 and 31 comments, respectively), oil refiners (22), and power traders and importers (22). The second largest set of compliance actors who made claims about fair treatment and application of the carbon market were industry and business lobbying groups. Among these, the most prevalent were electric power industry lobbying groups, particularly the Western Power Trading Forum (51 in total) and Joint Power Authorities (25). Oil and petrochemical refining groups often made these kinds of arguments (17), while environmental groups also made sustained efforts to influence market formation using arguments of fairness, both on their own and in coalitions (56). The remaining organizational actors express the breadth of those trying to influence the deployment of cap-and-trade in California, ranging from borax mine owners, insulation manufacturers, and academics, to actors specific to carbon markets, such as the International Emissions Trading Association and offset developers.

In coding for notions of 'just' treatment under the regulation, 27 different ideas about fairness were identified. It is perhaps unsurprising, given the diversity of participants, that arguments about specific aspects of the program, and the desirability of the program itself, are varied, often complicated, and in some cases, contradictory. Indeed, as noted by a coalition of environmental groups regarding how to allocate emissions permits, "most options could be called 'fair' from the perspective of some party" (Coalition 2009), but this statement could be applied to nearly any topic under consideration. Statements about fair treatment attempting to influence the regulation were often about matters of spatial governance (77), enforcement and administrative burden on regulated entities (43), perceived inequity either between industrial sectors or between entities in the same industrial sector (70 and 71, respectively), perceived reasonableness of compliance costs (53), and used as an overarching theme to structure entire regulatory worldviews (31). Finally, many arguments about fairness pertained explicitly to the idea that just outcomes were best achieved through the unfettered operation of markets (83), despite the many other arguments that this had too much potential to create unreasonable costs to polluters. This is a key challenge for regulators in market construction, who must balance the logic of neoclassical economics underlying the financialization of nature with the realities of market failures and potential price volatility. The situation regulators face is that markets created by government can lose political support, costing the regulator the opportunity to regulate greenhouse gases at all.

Legislating Fairness

California is one of the few jurisdictions that has codified ideas about environmental justice in state law, where it is defined as "the fair treatment of people of all races, cultures, and incomes with respect to the development, adoption, implementation, and enforcement of environmental laws and policies" (CGC §65040.12). However, not all claims to fairness are equally influential. Many claims about fairness at the beginning of AB32 rule-making were concerned about the use of market mechanisms at all. Environmental justice (EJ) groups actively mobilized against using markets, drawing on voluminous literature on the potential for market mechanisms, and particularly the use of offsets, to create or exacerbate the unjust distribution of the negative consequences of pollution (see for example Chan 2009; Pastor et al. 2010; Lohman 2011). These ideas about justice, and specific cases of environmental injustice related to the spatial distribution of greenhouse gases, impacts of pollutants that often occur alongside greenhouse gases like fine particulate matter, and unequal impacts of climatic change in California, did not exactly fall on deaf ears. Regulators I spoke with take environmental concerns seriously; after all, the mission of the Air Resources Board is primarily protection of public health. Further, the text of AB32 specifically directed ARB to convene an Environmental Justice Advisory Committee (EJAC) to alert ARB to potential unjust outcomes impacting economically disadvantaged communities and communities of color (§38591). Other regulatory action taken by ARB is designed to combat criteria pollutants,[4] though not at the pace that impacted communities and their advocates would like (EJAC 2014).

In terms of actual regulatory outcomes from the rule-making process for cap-and-trade, the claims of EJ advocates failed to make much of a material difference. Offsets can be used to fulfill about half of polluters' emissions reductions (Haya n.d.); and while the reduction of co-pollutants through GHG emissions reductions is seen by regulators as a positive outcome, it is not one of the key goals of the market.[5] Indeed, the only area where environmental justice advocates made headway was in achieving legislative direction that at least 25 percent of revenues generated through the auction of emissions allowances must be directed to reducing emissions in state-designated 'environmental justice communities' (SB535). Even then, the way that these funds can be spent is still subject to other juridical conceptualizations of fairness, stemming from state case law about the use of revenues generated by fees (*Sinclair Paint v. State of California*). Most EJ groups were (and still are, EJAC 2014) opposed to the implementation of any form of market environmental governance that would allow the perpetuation of patently unfair distribution of environmental externalities (EJAC 2008; Barboza 2014). Nevertheless, many stopped trying to influence the regulatory process when it became clear that cap-and-trade would go forward and include the use of offsets. This is borne out by the suspension of the Environmental Justice Advisory Committee that AB32 required. The advisory committee was eventually reconstituted, largely with different members in in 2013 after a 3-year hiatus. But in the meantime EJ advocates reoriented their strategy to litigation and legislative lobbying efforts to redress their grievances[6] (Perkins, forthcoming). The absence of these groups gave others more space to discuss different ideas about fairness.

Other kinds of appeals to fairness have indeed proven more successful at moving the regulatory needle. These claims are made by regulated entities and their industry associations, often dealing with programmatic minutiae that fail to generate much excitement beyond the handful of people in the room considering a particular rule change. The ways that appeals to fairness are framed can help us to better understand what kinds of claims are effective at creating incremental changes that then add up to major programmatic characteristics, ultimately substantially altering function and, in effect, defining what the market is for. In the case examined here, the arguments framed California's contribution to climate change as an industrial and technological problem, one that could be solved through economic regulation predicated on neoclassical economic efficiency rather than the large-scale social transformation that the challenge of climate change requires (Klein 2014).

This was demonstrated as early as 2009, during arguments about how to allocate emissions permits. ARB gave specific instructions to a group of economists hired to conduct analyses to look for allowance allocation options that would be simple, fair, cost-effective, and environmentally effective. Fairness became a design criteria primarily of interest to regulated entities, especially because, as noted by a coalition environmental groups, definitions of 'fair' and 'environmentally effective' were much less well developed than those mandating simplicity and cost containment (Coalition 2010). This gave interpretive license to those who would claim the mantle of fair treatment, even while the imperative to keep costs down was explicitly delineated. Over the course of initial implementation, programmatic goals were evacuated of their transformative potential and took on the Swyngedouwian imperative of finding ways "to change so that nothing really has to change" (2010, n.p.).

The topics of discussions invoking fairness discourse are as varied as the arguments themselves, but here I will analyze three key decisions influenced by claims about fairness that are emblematic of the rule-making trajectory toward cost minimization. The first is broad, and revolves around how to distribute emissions permits to regulated industries. The second is specific, and deals with the relaxation of rules pertaining to a phenomenon called 'resource shuffling,' which will be explored later in this chapter. The third illustrates the ability of regulators to push back on claims to fairness made by industry through limitations imposed on the expansion of the offset program.

Matter of Fairness 1: Allocation

In California, once emissions levels have been determined, a cap set, and regulated industries identified, the ARB had to decide how to allocate emissions permits. Research participants told me nearly universally that this was the most divisive and important decision in market design. The options for allocation included distribution of permits for free to regulated industries; conduct of an auction through which regulated entities purchase their pollution permits; or some combination of the two.

It is easy to see why regulated industry would fight against auctions. Based on the decision to compel all polluters with emissions of greater than 25,000 tons of CO_2[7] per year to participate in the market, auctioning could create substantial costs for the biggest polluters like oil refineries and electric utilities, and be potentially burdensome for the smallest polluters, including food processors or universities that operate on slim economic margins. Full auctioning was advocated by environmental justice organizations and most environmental advocacy groups on the grounds of 'polluter pays,' the idea that emissions permits represent the distribution of privatized slices of a global atmospheric commons. In this view, the proceeds of that privatization ought to go to the state, ultimately to be redistributed to the public to generate a powerful incentive for polluters to reduce their emissions. Auctioning was also advocated, primarily by academic commenters, on the grounds that it would be more akin to a 'true' market through the generation of a price signal representative of the physical cost of abatement[8] (EAAC 2010). Empirical analysis has also shown that substantial free allocation can undermine other program goals, as was the case in the European Union carbon market that led to delayed emissions reductions and prices under one euro per ton (McAllister 2009).

The decision on auctioning versus free allocation was further complicated by numerous existing regulations and restrictions exemplifying the tension between the economic theory that supports market environmental governance on one hand, and the practical implementation of market devices on the other. Given that markets are supposed to achieve environmental objectives through a price signal for both industrial emissions producers and consumers, free allocation would presumably mute this signal, leading to continued 'irrational' behavior as far as polluting the atmosphere was concerned. However, AB32 specifically directs ARB to minimize 'leakage,' which occurs when "a reduction in emissions of greenhouse gases within the state . . . is offset by an increase in emissions of greenhouse gases outside the state" (AB32, 4). This situation arises when producers move their operations out of the regulated territory in order to avoid the added pollution cost. Arguments about free allocation and its relationship to leakage generated voluminous comments, often pertaining directly to fair treatment of producers, the relationship between producers and consumers, and to the economic theory underlying the program.

Arguments about spatial equity and leakage minimization also centered on connecting California's market to other jurisdictions' carbon markets. California's market was originally conceived as part of a regional carbon market, the Western Climate Initiative (WCI), that was to cover most of the western United States and Canada (WCI 2008). Following the ascension of Republican legislatures across the western states in 2010, most of those jurisdictions withdrew their support, leaving California to 'go it alone,' a situation perceived by the industry as fundamentally unfair because of the product market distortions caused by carbon pricing, particularly when climatic benefits of abatement accrue globally.[9] The withdrawal of other WCI jurisdictions has given rise to increasingly urgent commentary about leakage potential. The AB32 Implementation Group, an industry group, urged ARB to "immediately amend the regulation so that all industries will be allocated 100% of allowances thereby eliminating the auction. This is reasonable given that there is no strategy to address leakage and job loss created by an auction" (AB32 IG 2012, 1).

To attempt to allay these fears and comply with the AB32's leakage minimization requirement, ARB developed convoluted formulas to determine how much pressure from imports each industry was under, and then assigned free permits based on their risk of being undercut by industry in jurisdictions without carbon constraints. The initial regulations provided significant assistance to a majority of covered industries, to the tune of around 90 percent free allocation in the first compliance period that ran through 2014. However, lingering concerns about the health of the California economy, coupled with sustained lobbying on the part of regulated entities, resulted in across-the-board increases in allocations to trade-exposed industries, ranging from breweries to natural gas suppliers. This reduced the number of allowances auctioned by more than a quarter in 2015, and eliminated the need for auction participation for some industries through 2020, the entire design life of the market (Doan 2013). Under current regulations, ARB is projected to give away over 700 million CO_2 worth of allowances over the life of the market just to utilities, which is roughly the equivalent of the climate pollution generated by Germany in 1 year (UNFCCC 2010; Cullenward and Weiskopf 2013).

Matter of Fairness 2: Resource Shuffling

Another outcome of the rule-making process that dramatically changed the California carbon market program was the relaxation of prohibitions around a practice called resource shuffling. This occurs when a company that imports power from outside the state changes its power-sourcing behavior to favor lower carbon sources. This allows it to decrease compliance obligations within the state, but may not actually engender any material reductions in emissions if that power is then sold elsewhere in unregulated markets. For example, a California utility could contract to buy hydro power from Oregon rather than coal-fired power from Arizona, but that coal power would still be used elsewhere, resulting in no net emissions reductions. This is a matter of concern in California; while the in-state energy mix is relatively clean, the state imports about 30 percent of its power, almost 75 percent of which is coal-fired generation (Cullenward and Weiskopf 2013 cited in Rossi and Smith 2014). A provision in the initial regulatory text forbade resource shuffling as any "plan, scheme or artifice to receive credit for emissions reductions that have not occurred, involving the delivery of electricity to the California grid" (§95802(a)(250)).

Power generators, importers, and their business allies argued vociferously that a blanket prohibition infringed on their ability to conduct business as they otherwise might, to comply with other regulations that encouraged the winding down of power import contracts from out-of-state coal generation, and could ultimately destabilize the power grid in California. As argued by PowerEx, the energy exporting branch of B.C. Hydro:

> The definition of 'Resource Shuffling' ... is too vague and too subjective to provide the regulated community with adequate certainty as to what ARB will consider to be legitimate electricity imports and which it would deem to constitute illegal 'resource shuffling.' There are two critical problems with the definition. First, the term 'any plan, scheme, or artifice' is

inherently subjective and requires *ex post facto* determinations of intent. What a member of the regulated community may genuinely believe to be a plan to pursue normal market incentives may be viewed by another as an illegal 'plan, scheme, or artifice.'

(Beveridge and Diamond 2012, 3)

As a result of these and allied arguments, ARB ultimately adopted 13 'safe harbor' provisions while nominally preserving the prohibition against resource shuffling. However, as demonstrated by Cullenward and Weiskopf (2013), these safe harbor provisions are likely to result in significant leakage, damaging the environmental efficacy of the program both directly through increased emissions, and indirectly: lowered demand on the part of utilities and power importers for allowances will serve to further depress prices, muting price incentives. Safe harbor provisions could result in leakage of between 108 and 187 million MtCO$_2$e, which corresponds to *up to 197 percent of emissions reductions* mandated by AB32 (Cullenward and Weiskopf 2013, see Kama 2014 for a contrasting case in the EU Emissions Trading System).

Matter of Fairness 3: Expansion of Offsets

As noted earlier, offsets can be used to satisfy up to 8 percent of an entity's compliance obligation. What this means over time—as the cap is lowered—is that roughly half of the mandated aggregate emissions reductions could come from sources outside of the cap, because the overall cap is scheduled to decline by about 16 percent over the course of the program (Haya, n.d.). This raises concerns about polluting industries simply buying their way out of making serious pollution reductions, and the attendant health impacts of the co-pollutants created in industrial processes. The 8 percent limit doubles the original limitation proposed in the draft regulations issued by ARB in 2009, and was a move to make compliance entities more willing to accept the program as a whole (CEI 2010). This decision came on the heels of lobbying by groups like the California Chamber of Commerce (2010), which argued that:

> A well-designed program is one that includes a broad use of offsets. Policies that increase the likelihood of an inadequate supply of offsets and the inability to link to other cap-and-trade programs will greatly decrease the potential for a cost effective California program . . . CalChamber is supportive of increasing the offset availability to at least 8% of total emissions under the cap, again emphasizing the importance of offsets as an effective cost containment mechanism.

Doubling offsets allowed under the cap has increased the theoretical maximum of offsets that can that be used for compliance to 25.8 million CO$_2$ in the first year of the program, which expanded commensurately in 2015 as more economic sectors were brought into the regulation.

There are a number of things that offsets are supposed to do for the market. Primary among them is to serve as a cost-containment mechanism, by providing an option to regulated entities that is less costly than obtaining permits to pollute in the open market or than actually reducing emissions (ARB 2010b). The inclusion of offsets in the California program is predicated on the conceit that all emissions reductions, regardless of how or where they happen, are entirely fungible or equivalent in ecological terms, an already questionable proposition and point of contention for environmental justice advocates (McAfee 2012).

This supposition of universal fungibility has engendered continuous pushback on limits to offsets by regulated industry and the developers of offsets, even after the limits were doubled. They concluded that if all emissions reductions are equal, and criteria pollution is best ameliorated through targeted regulation, then there should be no limit to the amount of offsets they can use (AB32 IG 2009). Other quirks in California's offset system include buyer liability in the case that ARB finds an offset project to have not fulfilled its promised emissions reductions according to standardized methodologies, a limited number of approved offset methodologies, and the restriction of offset creation to the contiguous United States.[10] These quirks create theoretical limits, but thus far they have not, and may never, be reached (Thomson Reuters 2013). While EJ advocates were (and continue to be) against the inclusion of offsets in the program,[11] environmental groups were in support of limited (though still fairly expansive) inclusion, while regulated entities, their industrial associations, and the offsets industry have continually pushed for unlimited usage. This is one area where advocates for stringency have had success, despite widespread and sustained pressure on regulators to do otherwise. This demonstrates that, among other things, it is inappropriate to understand ARB as 'the handmaiden of capital,' acting solely on behalf of regulated entities, but rather as a regulator acting within increasingly path-dependent policy options, bounded by the master narrative of cost control.

Regarding buyer liability, the Western Power Trading Forum (2010, 3) argued that "such a buyer liability rule would be patently unfair to covered entities, who would effectively be penalized for the misconduct of others." In regards to limitations on the variety of offset protocols eligible to generate credits, The Carbon Offset Providers Coalition (2010, 6) argued,

> standards-based offset methodologies provide the most efficient, objective and easily calculated means for determining whether a given offset project should be approved or not. That said, ARB also should provide means by which new offset project types can be approved on a case-by-case basis.

Each of these arguments was rejected by regulators, though they continue to explore new protocols.

One of the key arguments in this vein is that if emissions reductions from all sources are truly fungible, then it is unfair to limit the amount of offsets compliance entities can use (and developers can develop, and brokers can broker) on both ecological and economic grounds. Despite analyses of the market that demonstrate moderate to severe oversupply of allowances through at least 2019, compliance entities and offset developers continue to make dire predictions about price spikes that can only be mitigated through the dramatic expansion of the offset program (Thomson Reuters 2013). Compliance entities' dissatisfaction with the limited availability of offsets is illustrated by the similarity of comments about the fairness of restriction offsets that read nearly identically from 2008 to present.

However, things have changed from an EJ perspective as negotiations for inclusion of global South offsets continue, new offset protocols are approved, and contracts for offset provisions are increasingly being standardized in a way that reverts invalidation liability to the producer rather than the purchaser, increasing the likelihood that polluters will pay their way out of reducing emissions as much as possible. The inclusion of REDD+ offsets has been simmering, as regulators have indicated their interest in developing protocols and have signed memoranda of understanding with the states of Chiapas, Mexico, and Acre, Brazil (see Blanchard and Vila, this volume). These lay the groundwork for future collaboration on the deployment of REDD+ projects if the spatial restriction is lifted (ROW 2013). ARB has also recently approved a compliance offset protocol that will generate credits for flaring methane from inactive coal mines, which detractors claim will unduly profit the coal industry, effectively the inverse of what climate policy through price mechanisms is supposed to accomplish (SELC 2013).

Conclusion

Generous free allocation of pollution permits, stubbornly low prices, and persistent questions about carbon markets' abilities to alleviate environmental and social injustices seem to be ubiquitous in the construction of cap-and-trade markets globally. By examining the ways that competing notions of fair treatment under California's Global Warming Solutions Act were deployed by diverse stakeholders, this chapter demonstrates some of the mechanisms by which 'pay to pollute' comes to stand in for the more transformative formulation of 'polluter pays.' California has taken important steps to begin reigning in greenhouse gas emissions, but the focus on maintaining low costs for regulated industry has likely limited the regulation's ability to create a more just emissions landscape. A high carbon price could reduce climate pollution and co-pollutants associated with industrial production in the state. When environmental justice groups abandoned attempts to influence market design after it became clear that carbon trading would become the centerpiece of California's climate change strategy, industry and its allies maintained significant pressure on regulators.

Polluters couched their arguments in the language of fairness, to advocate for policies that would perform to their benefit. In doing so, the door was left open to the possibility that California polluters would be well supplied with options to comply with the law by way of low carbon prices and abundant offsets. These options reduce compliance costs for covered entities and blunt the transformative potential of attaching a price to carbon emissions. This does not necessarily mean that the market will not achieve its quantitative goals for relatively modest emissions goals; indeed, it appears that California will quite likely meet its 2020 target (California.Carbon.info 2014). But this success may well be predicated on falling emissions from changing macro-economic factors, and pollution will simply be paid for as a cost of doing business by polluters. Changes to the regulation also open the possibility for 'paper reductions' in the form of shifting contracts for imported electricity from dirty sources to cleaner ones that result in no net emissions reductions, and an expansive use of non-local offsets that fail to address the causes of long-standing unequal health impacts of air pollution in California. The changes were achieved in part through the appropriation of the language of fairness and equity by the very polluters the regulation was designed to constrain. It seems that transformative carbon reduction impacts will more likely be achieved via direct regulation, such as renewable energy standards and other policies that are described as merely 'complementary' to the carbon market.

Notes

1. AB32 is only 13 pages long. The federal Waxman-Markey climate change bill 1 was over 1,400 pages long and contained nearly every design element conceivable, whereas AB32 did not even mandate the creation of market mechanisms, much less programmatic specifics.
2. The European Union Emissions Trading System (EU ETS) is the current example par excellence of what happens when a carbon price drops precipitously. Its failures to deliver emissions reductions are well documented (Carbon Trade Watch 2012). California has avoided some of the pitfalls experienced by Europe by preventing windfalls profits to the largest polluters despite the number of free permits allocated to them.
3. This coding exercise excluded comments from Environmental Justice groups, of which there were only three. Environmental Justice organizations stopped trying to influence many aspects of the regulatory process in 2010 for reasons explained in the next section.
4. Criteria pollutants are six toxins regulated by the US Environmental Protection Agency that have significant negative impacts on human health, including fine particulate matter and lead (USEPA 2012). Criteria air pollution has fallen statewide since the early 1980s, though some areas of the state continue to have the worst air quality in the United States, and those areas map quite neatly onto poor communities of color (CalEnviroScreen 2014).
5. In an early victory for EJ advocates, the text of AB32 does prohibit the increase of criteria pollutants in 'environmental justice communities' as a result of regulatory compliance (Perkins forthcoming). Industrial and academic interlocutors opined that regulating criteria pollution through cap-and-trade would be ineffective because it would distort market performance, resulting in increased prices and potentially

uneven compliance costs depending on differences in criteria emissions from different kinds of processes, (Analysis Group 2009).
6. Unsuccessful for the most part, as cap-and-trade survived EJ groups' lawsuit, as well as lawsuits from the right brought by business associations seeking to enjoin any attempts to mitigate GHGs.
7. The measurement used in regulatory documents is tCO_2e, or tonnes of carbon dioxide equivalent. This is used in order to equate six different greenhouse gases in terms of their potential to warm the atmosphere (see MacKenzie 2008).
8. The question of what a price signal in environmental markets actually represents is convoluted, as is any discussion involving value, state, and nature (Robertson and Wainwright 2013).
9. This is the 'free-rider' problem that Pigouivan (1920) policies are designed to address in the first place, and points to other challenges for any attempt to regulate GHGs absent coordinated global effort.
10. Offsets will be allowed in Quebec following linkage of the California and Quebec carbon markets in mid-2014.
11. EJ advocates have taken the legislative lobbying approach to limiting offsets spatially and quantitatively though the introduction of a series of bills, though these have been unsuccessful so far.

References

AB32 Implementation Group. 2012. RE: AB 32 Implementation Group's Comments Regarding the May 24, 2012 "Public Consultation on Investment of Cap-and-Trade Auction Proceeds." June 22, 2012. Public comments to the California Air Resources Board.

Analysis Group. 2009. RE: The Scope and Options for the Economic and Allocation Advisory Committee. June 29, 2009. Public comments to California Air Resources Board.

Assembly Bill 32 (AB32) Implementation Group. 2009. Public Comments on Economic and Allocation Advisory Committee. August 6, 2009. Public comments to the California Air Resources Board.

Barboza, Tony. 2014. L.A., Central Valley have worst air quality, American Lung Assn. Says. *Los Angeles Times*, April 29, 2014. http://www.latimes.com/science/la-me-0430-air-pollution-20140430-story.html (accessed May 4, 2014).

Beveridge and Diamond. 2012. Re: Comments of PowerEx Corp. on the Stakeholder Workshop re Cap-and-Trade Compliance Requirements for First Deliverers of Electricity. May 11, 2012. Public comments to the California Air Resources Board.

Boykoff, Maxwell, and Tom Yulsman. 2013. Political economy, media, and climate change: Sinews of modern life. *Wiley Interdisciplinary Reviews: Climate Change*, 4(5): 359–371.

Bumpus, Adam. 2011. The matter of carbon: Understanding the materiality of tCO_2e in carbon offsets. *Antipode*, 43(3): 612–638.

CalEnviroScreen. 2014. CalEnviroScreen 2.0 Draft and Webinar. http://oehha.ca.gov/ej/ces2.html (accessed May 7, 2014).

California Air Resources Board. 2010a. History of Air Resources Board. http://www.arb.ca.gov/knowzone/history.htm (accessed May 7, 2014).

California Air Resources Board. 2010b. Appendix E: Setting the Program Emissions Cap. http://www.arb.ca.gov/regact/2010/capandtrade10/capv3appe.pdf (accessed May 12, 2014).

California Air Resources Board. 2011. Overview of ARB Emissions Trading Program. http://www.arb.ca.gov/newsrel/2011/cap_trade_overview.pdf (accessed March 24, 2014).
CaliforniaCarbon.info. 2014. State of the California Cap-and-Trade Market 2013/2014. http://californiacarbon.info/2014/03/21/sotm-1314/ (accessed April 24, 2014).
California Chamber of Commerce. 2010. CalChamber Comments on June 22, 2010 Cap-and-Trade Public Workshop Cost Containment & Offsets. July 12, 2010. Public Comments to the California Air Resources Board.
California Environmental Insider. 2010. Special Report: California's Cap-and-Trade Program. November 10, 2010. http://www.ceitoday.com/pdfs/CapandTrade.pdf (accessed May 11, 2014).
California Government Code. Environmental Justice. Article 3 section 65040.12. http://www.leginfo.ca.gov/pub/99-00/bill/sen/sb_0101-0150/sb_115_bill_19991010_chaptered.html (accessed May 8, 2014).
California Government Code. Public Meetings. Article 9, Sections 11120–11132. http://www.leginfo.ca.gov/cgi-bin/displaycode?section=gov&group=11001-12000&file=11120-11132 (accessed May 8, 2014).
Carbon Offset Providers Coalition. 2010. Re: Preliminary Draft Regulation for a California Cap-and-Trade Program. June 7, 2010. Public Comments to the California Air Resources Board.
Carbon Trade Watch. 2012. Green Is the Color of Money: The EU ETS Failure as a Model for the "Green Economy". http://www.carbontradewatch.org/downloads/publications/EU-ETS_Report-web.pdf (accessed May 7, 2014).
Chan, Michelle. 2009. *Subprime Carbon? Rethinking the World's Largest New Derivatives Market*. Washington, DC: Friends of the Earth.
Coalition Letter to the EAAC. 2009. RE: Suggestions for the EAAC. November 9, 2009. http://www.climatechange.ca.gov/eaac/comments/2009-11-09_Natural_Resources_Defense_Council_Coalition.pdf (accessed May 7, 2014).
Crabtree, Benjamin and William Miller (eds.). 1999. *Doing Qualitative Research* (2nd edition). London: Sage.
Cullenward, Danny and David Weiskopf. 2013. Resource Shuffling and the California Carbon Market. *Environmental and Natural Resources Law & Policy Program Working Paper*.
Doan, Lynn. 2013. California 'Freebies' Drive Carbon to 2013 Low: Energy Markets. *Bloomberg News*. http://www.bloomberg.com/news/2013-08-21/california-freebies-drive-carbon-to-2013-low-energy-markets.html (accessed March 24, 2014).
Economic and Allocation Advisory Committee. 2010. Allocating Emissions Allowances Under a California Cap-and-Trade Program: Recommendations to the California Air Resources Board and California Environmental Protection Agency. http://www.climatechange.ca.gov/eaac/documents/eaac_reports/2010-03-22_EAAC_Allocation_Report_Final.pdf (accessed November 10, 2013).
Environmental Justice Advisory Committee. 2008. Environmental Justice Advisory Committee Recommendations on Proposed AB32 Scoping Plan. http://www.arb.ca.gov/cc/ejac/proposedplan-ejaccommentsfinaldec10.pdf (accessed May 7, 2014).
Environmental Justice Advisory Committee (California Air Resources Board). 2014. Final Recommendations Environmental Justice Advisory Committee on Proposed AB32 Scoping Plan. http://www.arb.ca.gov/cc/ejac/meetings/041014/final_ejac_recommendations.pdf (accessed May 13, 2014).

Haya, Barbara. n.d. California's Carbon Offsets Program: The Offsets Limit Explained. http://bhaya.berkeley.edu/docs/QuantityofAB32offsetscredits.xlsx (accessed March 21, 2014).

Hope, Chris. 2013. "How high should climate change taxes be?" In Fourquet, Roger (ed.), *Handbook on Energy and Climate Change*, 404–414. Northampton, MA: Edward Elgar.

Interagency Working Group on Social Cost of Carbon (IWGSCC). 2013. Technical Update of the Social Cost of Carbon for Regulatory Impact Analysis. US Government. http://go.nature.com/vzpkkb (accessed March 3, 2014).

Kama, Karg. 2014. On the borders of the market: EU emissions trading, energy security, and the techno-politics of 'carbon leakage'. *Geoforum*, 51: 202–212.

Klein, Naomi. 2014. *This Changes Everything: Capitalism vs. the Climate*. New York: Simon and Schuster.

Lohman, Larry. 2011. The endless algebra of carbon markets. *Capitalism, Nature, Socialism*, 22(4): 93–116.

MacKenzie, Donald. 2008. Making things the same: Emission rights and the politics of carbon markets' accounting. *Organizations and Society*, 34(3–4): 440–455.

McAfee, Kathleen. 2012. The contradictory logic of global ecosystem services markets. *Development and Change*, 43(1): 105–131.

McAllister, Lesley. 2009. The over-allocation problem in cap-and-trade: Moving toward stringency. *Columbia Journal of Environmental Law*, 32(4): 395–445.

Pastor, Manuel, Rachel Morello-Frosch, James Sadd, and Justin Scoggins. 2010. *Minding the Climate Gap: What's at Stake if California's Climate Law Isn't Done Right and Right Away*. Los Angeles, CA: USC Program for Environmental and Regional Equity.

Perkins, Tracy. 2014/forthcoming. "Climate Justice Advocacy in the Global North: A Case Study from California." Environmental Justice Research: Contemporary Issues and Emerging Topics III, American Association of Geographers Annual Meeting, Los Angeles, CA, April 13.

Pigou, Arthur. 1920/2002. *The Economics of Welfare*. New Brunswick, NJ: Transaction.

REDD Offset Workgroup. 2013. California, Acre, and Chiapas: Partnering to Reduce Emissions from Tropical Deforestation. http://greentechleadership.org/documents/2013/07/row-final-report-executive-summary.pdf (accessed May 13, 2014).

Robertson, Morgan and Joel Wainwright. 2013. The value of nature to the state. *Annals of the Association of American Geographers*, 103(3): 890–905.

Rossi, Jim and Andrew James Dearing Smith. 2014. Electric power resource 'shuffling' and subnational carbon regulation: Looking upstream for a solution. *Forthcoming San Diego Journal of Climate & Energy Law*. http://ssrn.com/abstract=2397464 (accessed May 13, 2014).

Stanford Environmental Law Clinic. 2013. Mine Methane Capture Compliance Offset Protocol. http://switchboard.nrdc.org/blogs/pmiller/2013-07-01%20Comment%20to%20ARB%20on%20MMC%20Offset%20Protocol%20from%20researchers%20at%20Stanford%20University%20%282%29.pdf (accessed May 12, 2014).

Swyngedouw, Eric. 2010. Desalinating the Seas: Fixing the Spanish Water Conundrum or . . . 'How to Change so that Nothing Really Has to Change'. Keynote address at Dimension of Political Ecology: Conference on Nature/Society. Lexington, KY.

Thomson Reuters Point Carbon. 2013. Carbon Market Analysis: New California Emissions Model and Revised WCI Price Forecast. September 4.

Tietenberg, Tom. 2006. *Emissions Trading: Principles and Practice*. Washington, DC: Resources for the Future.

United Nations Framework Convention on Climate Change. 2010. National Inventory Submissions 2010. http://unfccc.int/national_reports/annex_i_ghg_inventories/national_inventories_submissions/items/5270.php (accessed May 3, 2014).

United States Environmental Protection Agency. 2012. National Ambient Air Quality Standards (NAAQS). http://www.epa.gov/air/criteria.html (accessed March 24, 2014).

Western Climate Initiative. 2008. Design Recommendations for the WCI Regional Cap-and-Trade Program. http://www.ecy.wa.gov/climatechange/WCIdocs/092308WCI_ReportsSec1_DesignRecs.pdf (accessed March 24, 2014).

Western Power Trading Forum. 2010. Comments of the Western Power Trading Forum on the California Air Resources Board's Preliminary Draft Regulation for a California Cap and Trade Program. January 13, 2010. Public comments to the California Air Resources Board.

SECTION III
National and Subnational Framings

8

CARBON, CARBON EVERYWHERE

How Climate Change Is Transforming Conservation in Costa Rica

Robert Fletcher

Introduction

The growing urgency of the global campaign to address anthropogenic climate change is transforming environmental governance throughout the world in ways that have yet to be rigorously conceptualized. As climate change has, in the past decade, become what While and colleagues (2009, 2) call "the new 'master discourse' in environmental governance" globally, it has changed the way other significant environmental issues are understood and addressed. Increasingly, both funding and programming for environmental protection are focused on carbon capture and emissions reductions to mitigate climate impacts, forcing environmentalists to frame other issues in terms of their contribution to climate action. This has, ironically, created new conflicts among environmental priorities, for "discourses of climate change can be mobilized politically to justify social and technical fixes for states that environmentalists find unacceptable" (While et al. 2009, 10). At the same time, climate action is spurring novel social conflicts, in that "climate change discourse has a powerful moral imperative, especially when it intersects with questions of national security" (While et al. 2009, 10), and can therefore be used to override the interests of particular groups when issues of national (and/or global) security are deemed to be at stake.

In this chapter, I describe how the growing preoccupation with climate change is transforming environmental governance in Costa Rica, long considered one of the world's most important sites for experimentation in innovative forms of environmental management (Boza et al. 1995; Evans 1999). In this sense, an analysis of environmental governance in Costa Rica may hold important lessons for the rest of the world as well, helping not only to forecast what is likely to occur in other places but also offering suggestions for how to address these consequences most effectively. Currently, the growing focus on climate change is precipitating a number of interrelated environmental and social impacts, as I elaborate upon in this chapter. While for heuristic purposes I address these dynamics separately, they are of course intimately conjoined in complex socio-ecological systems (Ostrom and Cox 2011). Environmental impacts include: a reevaluation of conservation hotspots and appropriate mechanisms for preserving them; the resurgence of hydroelectric power as an ostensibly clean energy source; conflict between biodiversity preservation and development for green energy production; and the undermining of ecotourism as one of the main sources of conservation funding. Socially, preoccupation with climate change is increasing the marginalization of groups in possession of resources deemed important for carbon mitigation; it is also increasing inequality as the growing revenue from climate-related action becomes concentrated in the hands of those best positioned to capture it.

I begin my analysis by describing how climate change has increasingly captured center stage in the global environmental movement and the implications of this for other issues of environmental and social concern. Following this, I show how all of this is impacting Costa Rica, where the government recently garnered international praise by declaring its intention to become the world's first carbon neutral nation by 2021 (see Fletcher 2013). I conclude by highlighting the consequences of this analysis for the future of environmental governance, both within Costa Rica and elsewhere.

This analysis is grounded in political ecology, an interdisciplinary field comprising anthropologists, geographers, sociologists and others that emphasizes the interrelationship between political-economic structures and environmental management as well as among governance processes at multiple scales (see e.g., Biersack and Greenberg 2006). It is based on 6 years of ethnographic research investigating the development of environmental governance in Costa Rica, documented in a series of previous publications upon which the present analysis draws. In order to explore interconnections among processes at different scales, my research has transcended the traditional anthropological emphasis on long-term residence in a relatively small, bounded area (see Stocking 1983) to employ what anthropologists increasingly call "multi-sited ethnography" (Marcus 1995; Hannerz 2003): participant observation and semi-structured interviews with policy-makers and local residents in various locations throughout the country. To extrapolate beyond the Costa Rica context, the analysis also draws on a secondary literature analyzing environmental governance elsewhere.

A Changing Climate for Conservation

In the past, preservation of the world's imperiled biodiversity was the main focus of the global conservation movement—a trend Adams (2004) calls working "against extinction." The chief mechanism for achieving this was of course the protected area (PA), epitomized by the national park. Originating in its modern form with the development of the US National Park System in the later nineteenth century, this model was subsequently exported to Africa via colonialism and then the rest of the world though a global campaign spearheaded by the International Union for the Conservation of Nature (IUCN). The intention was to compel all governments to designate a substantial portion (ideally at least 11 percent) of their territory (both terrestrial and marine)[1] as formally protected area (see Igoe 2004; Brockington et al. 2008).

Protected areas were originally governed in terms of what Brockington (2002) calls the "fortress" model: state-centered command-and-control protection, entailing formal designation of strictly defined borders and imposition of sanctions for their violation. Yet this model was widely challenged in the 1980s and 1990s due to its widespread social impacts: PA creation commonly caused the eviction and impoverishment of large numbers of local inhabitants (Brockington 2002; Igoe 2004; Dowie 2009). As a result, conservation managers increasingly adopted a community-based conservation (CBC) approach intended to redress these problems by incorporating local people into PA planning and administration (Borgerhoff Mulder and Coppolillo 2005). The mechanism through which CBC was most commonly implemented was the integrated conservation and development project (ICDP), designed to generate income from the biodiversity preserved in PAs and thereby redress the poverty PAs often created while simultaneously providing encouragement (incentive) for local people to support conservation (West 2006).

The main financing mechanism for achieving this was ecotourism, defined by the International Ecotourism Society as "responsible travel to natural areas that conserves the environment and improves the well-being of local people" (Honey 2008, 6). Partly as a consequence of this, for the past several decades, ecotourism has been the fastest growing segment of a global tourism industry that rivals oil production for the status of world's largest industry (UNWTO 2015).

The growing focus on climate change is changing all of this. As the new "master concept" of environmental governance (While et al. 2009, 2), climate change now competes with biodiversity as the central organizing principle of conservation efforts in many places. Whereas previously most conservation efforts had been advanced in the name of preserving biodiversity (for instance, combating deforestation), increasingly such efforts are now framed additionally, if not primarily, in terms of their contribution to mitigating and/or adapting to climate change (e.g., Reid and Swiderska 2008)—a dynamic that Jinnah and Muñoz Cabré (2011) label "climate change bandwagoning." Biodiversity preservation has thus shifted from a worthy conservation goal in its own right to an instrument, in many cases, in the campaign against climate change.

This newfound preoccupation with climate change has a number of important implications for the practice of conservation that are only beginning to be documented and conceptualized. First and foremost, the focus on climate change is fast altering the distribution of conservation funding, both programmatically and geographically. Nongovernmental organization (NGO) agendas have always been strongly shaped by funders' interests (Levine 2002), and as these interests shift, so do NGOs' priorities. As one of my former graduate students observes, "donor organizations are quickly changing regional and thematic foci away from Latin America and sustainable development in favor of Asia and climate change mitigation projects" due to this latter region's greater contribution to greenhouse gas (GHG) emissions (Ani Zamgochian, August 17, 2009).

In addition, climate change is altering the very nature of PAs' relationship with the biodiversity they are intended to protect. Species migrate in response to climatic change, yet the PAs that preserve them are fixed in place (UNEP 2009). Hence, the ecological impacts of climate change are likely to have increasing implications for the future of PAs, already called into question in many places by dynamics of human displacement and their status as essentially unsustainable forest islands (Borgerhoff Mulder and Coppolillo 2005).

Added to this is the fact that concerns over climate change are compromising ecotourism's future as an effective conservation strategy. While in theory at least, ecotourism has significant potential to assist in biodiversity preservation (its efficacy in actual practice is debated; see e.g., West and Carrier 2004), its standing vis-à-vis climate change is more problematic. After all, one of the principle criticisms of ecotourism's ostensible contribution to environmental efforts points to the activity's contribution to climate change through emissions from long-haul air transport, upon which the ecotourism industry fundamentally depends (Carrier and Macleod 2005). If biodiversity conservation is ecotourism's aim, these emissions can be considered externalities to some degree. If the focus of conservation is climate change itself, on the other hand, ecotourism's emissions must be included in calculations of the activity's overall benefit, which as a result is greatly diminished.

In response to this recognition, recent years have witnessed increasing attention to, and discussions concerning mitigation of, the tourism industry's carbon footprint (see e.g., Gössling and Peeters 2007). Proposals advanced thus far focus primarily on offset programs and efficiency improvements in transport technology. In October 2009, however, NGO Responsible Travel, one of the first ecotourism promoters to endorse offsets, reversed its position, stating, "We believe that the travel industry's priority must be to reduce carbon emissions, rather than to offset."[2] This decision followed a similar pronouncement by Friends of the Earth, which contends, "Carbon offsets distract tourists from the need to reduce their emissions. They create a 'medieval pardon' for us to carry on behaving in the same way (or worse)."[3] If such recommendations for reducing emissions through decreased travel are widely acted upon, the tourism industry's growth (and thus its potential for expanding conservation work), nearly continuous for the past half-century, may be significantly reduced.

Climate change is altering the calculus of conservation in other ways as well. One of the most significant of these concerns controversy over hydroelectric dams. Long a staple of the global development industry, large dams in particular became subject to a growing resistance movement due to their environmental and social impacts, resulting in the release of a scathing critique by the World Commission on Dams (WCD 2000) and a dramatic reduction in the incidence of dam building worldwide (McCully 2001). Now, however, this trend is reversing, in large part due to the growing global concern over climate change (Imhof and Lanza 2010). Within climate change discourse, hydroelectric power is widely considered a clean source of renewable energy that produces minimal greenhouse gas emissions relative to other electricity-generating mechanisms (e.g., coal, oil). As a result, the growing focus on mitigating climate change provides a new incentive for those interested in reducing their carbon footprint to invest in hydropower production. On the other hand, while hydropower generation is thus intended to influence climate change, research suggests that future climate variation will impact hydropower generation as well, due to new fluctuations in precipitation levels affecting river flows—and thus generation potential—that climate changes will likely engender (see, e.g., McCully 2001, xxxvii–iii). The incentive to invest in hydro power is enhanced by the Kyoto Protocol's Clean Development Mechanism (CDM), which includes investment in hydroelectric dams among the various projects for which offset credits can be granted (see Fletcher 2010).

Yet the environmental benefits of hydro power are far from unequivocal. In fact, a growing body of research asserts that large dams, particularly in tropical areas, actually produce substantial GHG emissions from the methane produced by submerged vegetation released during outflow or drawdown (see Mäkinen and Khan 2010). In addition to increased emissions from deforestation, substantial biodiversity is often lost when reservoirs are filled (Goldman 2004; Fletcher 2010). In cases such as this, then, different environmental goals are drawn into conflict with one another (Fletcher 2010), forcing policy-makers to decide which will take priority.

Similar tradeoffs occur with respect to other issues. For instance, plantation forestry can be justified in relation to climate change on the basis that young trees sequester more carbon during their initial growth than mature trees in an unmanaged forest, which has already captured most of the carbon it is able to contain, despite the fact that biodiversity in forest plantations is far less than in most unmanaged forests. Even if plantation trees are harvested for timber, replanting means that the plantation contributes to a net gain in carbon capture over time, as long as harvested trees are turned into relatively durable products. This implies that if climate change is the central focus of conservation efforts, when considering reforestation strategies, plantation cultivation can be considered more beneficial than "natural" regeneration that delivers far more biodiversity.

The Social Costs of Carbon Control

In addition to these various environmental implications, growing focus on climate change has important social impacts. This is particularly apparent with respect to hydroelectric power. Large dams, in general, have always exacted substantial human costs; conservative estimates of the total number of people displaced by dams in the twentieth century alone range from 40 million to 80 million (WCD 2000). While the problems associated with this displacement were an important part of the anti-dam movement and resulting WCD report, the newfound framing of dams as a climate-friendly form of energy production is changing this association, allowing issues of ostensible national energy security to prevail over the interests of groups living in areas slated for flooding by dam construction. This is particularly true for indigenous peoples, whose 300-year history of oppression on the part of settler populations is compounded in this process (Finley-Brook and Thomas 2010, 2011). This has already led to numerous conflicts worldwide (see, e.g., Barrionuevo 2011; Finley-Brook and Thomas 2011).

Human rights issues have also been raised with respect to other carbon offset projects in southern communities (see Checker 2009), in particular the REDD+ (Reduced Emissions through Deforestation and forest Degradation) mechanism developed via recent UNFCCC negotiations. REDD+ seeks to build on the global carbon market to develop offset projects by stopping deforestation before it starts, and thereby claiming the emissions subsequently avoided as the basis for sequestration payments. Announcement of the REDD+ mechanism was met by protest from groups around the world, particularly indigenous peoples concerned that it would exacerbate the type of negative social impacts created by similar initiatives in the past (see e.g., Newswire 2011). And in a number of places, initial implementation of the mechanism is indeed bearing out such concerns. In Kenya, for instance, the No REDD in Africa Network (NRAN) recently condemned forced eviction of the Sengwer people from the Cherangany Hills in association with a project supported by the World Bank's REDD readiness program, claiming that the people "are now facing complete annihilation under the guise of 'conservation' under REDD."[4]

Focus on Costa Rica

Climate change has become an increasing focus of attention in Costa Rica since former President Oscar Arias captured international headlines by declaring the nation's intention to become the world's first carbon neutral country by 2021 (see Fletcher 2013). In pursuit of this ambitious goal, the Ministry of Environment and Energy (MINAE) drafted a National Climate Change Strategy (NCCS), calling for comprehensive mitigation and adaptation measures across all sectors of society (Dobles 2008). In response to such initiatives, Climate Action Tracker praised Costa Rica as a "role model" for the world (cited in Hermwville 2011, 10).

Even prior to this, however, Costa Rica had been on the forefront of carbon mitigation efforts for some time. In 1997, with assistance from the World Bank and USAID, the government launched an innovative national payment for environmental services (PES) program, intended to link to the global carbon market mandated by the Kyoto Protocol established in the same year. An initial offset payment of $2 million from Norway helped to jumpstart the program, which has expanded rapidly ever since and now encompasses more than 700,000 ha nationwide (Daniels et al. 2010). In addition to this state-sponsored initiative, numerous private carbon offset programs throughout the country link to the voluntary carbon offset (VCO) market (see Bumpus and Liverman 2008) that has developed parallel to the Kyoto structures.

The NCCS, through which the carbon neutrality campaign is to be pursued, continues this effort to link domestic climate action to global carbon markets. The principle means of financing the plan is to build on the PES program to access these markets. As the strategy explains, "carbon markets are the opportunity to establish links between climate change and the competitiveness of national strategies," while "appropriate financial instruments and carbon markets provide effective incentives for developing countries" (Dobles 2008, 20). In conjunction with this, the national carbon neutrality initiative is intended as a branding mechanism to leverage "C-Neutral" products and services through "the sustained creation of value for target customers in the market or segment of interests, which proves to be superior to the value offered by the competition" (Dobles 2008, 14).

This growing focus on climate change has impacted conservation in Costa Rica in numerous ways. Over the past several years, several of the prominent international NGOs (e.g., The Nature Conservancy, Conservation International, World Wildlife Fund) long instrumental in Costa Rican conservation, have downsized their activities considerably, in part due to the shifts in global funding priorities noted earlier. As NGOs chase climate change funding to Asia and beyond, their interest in and ability to continue their conservation work in Costa Rica is diminished—an ironic turn of events given the nation's reputed leadership in such work in the past. This irony is compounded by the fact that it is precisely Costa Rica's success in reducing deforestation rates in the past that is diminishing the country's attractiveness to conservationists in the present—a dynamic mirrored in the structure of the REDD+ mechanism, which in its focus on reducing deforestation perversely tends to reward those nations least successful at combating forest clearing thus far.

Climate change is also challenging Costa Rica's celebrated PA system, which currently covers more than a quarter of the nation's territory (Evans 1999). As former park service director Alvaro Ugalde explained in a recent interview, "The future of the parks in Costa Rica is highly dependent on climate change effects now . . . Changes in water patterns and temperatures are already forcing species to move around" (May 5, 2010). Hence, he asserted, the capacity of protected areas to function as effective mechanisms for conservation hangs on the success of carbon mitigation measures throughout the country to minimize climate impacts.

Concern for climate change is affecting Costa Rican conservation in other ways as well. While ecotourism has been the main economic support mechanism for conservation in the country over the past 30 years (Honey 2008), climate change is calling this relationship into question for the reasons described earlier. This issue is complicated by Costa Rica's carbon neutrality plan. What is taken into account in neutrality calculations, of course, always depends as well upon how entities "define their system boundaries" (Gössling 2009, 21). The Kyoto Protocol, for example, does not require any specific parties to assume responsibility for emissions from international air transport, while the UN World Tourism Organization's (UNWTO) more recent Davos Declaration, intended to address the tourism industry's climate implications (see UNWTO 2007), demands "that destinations would only be responsible for emissions released during the tourists' stay" (Gössling 2009, 21). Yet revenue generated through international tourism arrivals constitutes one of Costa Rica's most important revenue streams (Honey 2008). Hence, arguably, international air transport should be included in Costa Rica's carbon assessment if it is to constitute a meaningful measure, yet this would make achieving genuine neutrality exceedingly difficult.

A similar dynamic arises with respect to energy production. Costa Rica is widely viewed as "a world leader in renewable energy use" (NRDC 2007, 1), due to the fact that most of its domestic energy (currently around 93 percent) is produced without fossil fuels, the majority of this (about 80 percent of the total) generated by hydroelectric dams (ICE 2009). The substantial GHG emissions actually resulting from hydro production, however, are neither measured nor taken into account in calculating Costa Rica's carbon balance. Given that the nation's electricity demand is projected to increase 6 percent per year for the foreseeable future, and that the majority of this is to be met with increased hydroelectricity production, including such emissions in neutrality calculations (especially if air travel emissions are also included) would make it virtually impossible for the country to achieve its neutrality goals.

Social Costs Redux

In general, Costa Rica's human rights record is exemplary relative to most other countries in the region (Brysk 2009). One of the main exceptions to this concerns the treatment of indigenous peoples, who have generally been regarded as second-class citizens and relegated to the margins of society (Evans 1999). In the past, as in many other places, dam building has created conflicts between national development priorities and the interests of indigenous peoples. In the south-central part of the country, a controversial proposal to build a mega-dam for electricity generation inspired resistance on the part of Boruca people facing displacement by the project; their efforts eventually killed the initiative entirely after more than 30 years of discussion (see Carls and Haffar 2010).

Recently, however, a new, somewhat smaller dam has been proposed in the same area. Called the El Diquís project, at 170 m in height and with a projected generating capacity of 631 MW, it still promises to be the largest dam in Central

America.[5] The project is the cornerstone of Costa Rica's future energy generation plan, which calls for expansion of "renewable energy sources [that] provide the double function of reducing dependency on petroleum and permitting clean and sustainable development" (ICE 2009, 21). The national electricity commission (ICE) claims that of the approximately thousand people who will likely be displaced by the project, all have been subject to a thorough consultation process and—significantly—that none of the displaced will be "indigenous" per se (ICE 2010). Representatives of the Terraba people, however, have opposed the project, complaining that they will in fact be adversely affected by it. Such claims are supported by a report published by the Human Rights Clinic of the University of Texas School of Law in July 2010, which asserted:

> Costa Rica has failed to respect and protect the human rights of its indigenous peoples in the areas of information, property, representation and effective participation in decisions surrounding the [project] . . . Its national electricity authority, ICE, has not obtained the effective participation of the Teribe [sic] peoples as required under international law.
> (UTSLHRC 2010, 4)

An additional issue concerns the distribution of benefits from climate-related projects. This is particularly apparent with respect to the PES program. The first priority of the program since its inception has been to expand conservation; use of payments as a form of income generation in the interest of poverty alleviation has always been a secondary concern (Pagiola 2008). This, combined with issues of efficiency in managing the program, has meant that

> a significant proportion of payments tend to go to areas with lower opportunity costs, relatively large farms and private companies. The PES programme's lukewarm approach to dealing with poverty has limited its impact in this field, for while many poor landowners theoretically meet the criteria to participate in the programme, it is actually the wealthier landowners who tend to benefit most from its policies.
> (Porras 2010, 1)

Indeed, the share of PES revenue captured by private firms, as opposed to the individual landholders envisioned as the ideal payment recipients, "has increased steadily since the beginning of the programme, rising from about 30 per cent of all contracts (and 20 per cent of annual funds distributed) in 1997 to almost 50 per cent of contracts and funding by 2008" (Porras 2010, 13). Such firms include the Miami-based C and M Investment Group Ltda. and international forestry corporation Heartwood Timberlands LLC. Lansing (2013), moreover, demonstrates that the program's controversial reforestation (through tree plantation) modality effectively serves as an indirect subsidy to industrial agriculture interests, in that most of the wood produced is turned into pallets for use in agricultural export. Recent policy innovations have sought to address these issues, by for instance, prioritizing areas of the country with a low Human Development Index (HDI) and including collectively held land or that without secure legal tenure, but the extent to which such measures will succeed in countering this trend toward revenue capture remains unclear.

Conclusion

Climate change is a serious issue, with far-reaching implications for socio-ecological systems worldwide. Yet its importance should not obscure the presence of other equally serious social and environmental issues that also require policy-makers' attention—not merely as subcomponents of climate change policy, but as worthy foci of planning and funding in their own right. At the moment, the overwhelming focus on climate change is forcing both development and environmental organizations in many places to join the bandwagon, regardless of which issues they wish to prioritize. As described in the preceding analysis, this can create conflicts among contradictory aims, in ways that have yet to be rigorously documented. This chapter therefore calls for more investigation of such dynamics in other contexts so that emergent conflicts can be identified and addressed as the global climate governance regime unfolds.

This analysis also highlights significant challenges associated with employing carbon markets in support of progressive climate action. That is because efforts to harness the economic value of conserved resources become subject to the same capitalist dynamics as other, more conventional industries, creating powerful incentives to externalize both ecological and social costs of production in the interest of increased profit, while also encouraging the consolidation of such profits in increasingly fewer hands. This, as I have described in this chapter, is precisely what is happening with the rise of carbon control in Costa Rica, and it is clear that such dynamics—what Fairhead et al. (2012) call "green grabbing"—are only likely to accelerate in the future if reliance on relatively unregulated capitalist markets is the dominant approach to carbon control strategies.

Problems arise, in this case, whether or not sufficient revenue is actually generated from conserved resources to make them profitable. If revenue is not sufficient, projects will be unable to compete with forms of alternative land use (agriculture, logging, cattle raising, etc.), the opportunity costs of which are intended to be countered by market-based mechanisms (Fletcher 2012). If projects are able to generate significant profit, on the other hand, elites are incentivized to capture these profits and increase them through externalization of production costs—and the greater this potential profit, the greater the incentive.

Consequently, the attempt to internalize socio-ecological externalities, a central tenet of market-based carbon control strategies, commonly ends up creating new externalities. Lohmann contends that complexities of the "performative equations" involved in calculating equivalencies between carbon emitted and sequestered via carbon markets means "that, with respect to the climate crisis in particular, internalizing externalities through commodity formation, however profitable the result, constantly gives rise to fresh externalities that are so overwhelming that, from an environmental point of view, they invalidate the project" (2014, 178). In this chapter, I have described similar dynamics in the development of carbon control policies in Costa Rica, showing how efforts to mitigate climate change end up creating environmental externalities in terms of the other ecological issues marginalized and/or intensified by these policies, while at the same time obscuring aspects of the projects themselves that detract from their overall ecological benefit.

In addition, however, I have demonstrated a parallel process with respect to social externalities. Market-based carbon control is employed as a mechanism for income generation and to thereby internalize the social problems (e.g., income inequities and forms of marginalization) created by conventional capitalist production. These, in turn, create new forms of inequality, due to the same drive to consolidate profit that these mechanisms are intended to combat. It is hard to imagine how this dynamic can be countered without heavy-handed intervention by the state or some other entity to appropriate and redistribute resources to prevent excessive accumulation (as the Costa Rican state is indeed starting to do with PES).

In acknowledgement of such dynamics, it is clear that effectively addressing the climate crisis in a manner that encompasses concerns for both social and environmental justice will require thinking beyond the growing hegemony of market mechanisms in climate governance, to develop novel forms of policy and practice. A variety of potential models for such efforts have been advanced in recent years (see Dressler et al. 2014). How these will be translated from theory into practice is a vital question for the future.

Notes

1. Recently amended to 17% of terrestrial area and 10% of marine via the Convention on Biological Diversity's 2010 Aichi Targets.
2. http://www.responsibletravel.com/Copy/Copy902116.htm (accessed November 28, 2010).
3. Ibid.
4. http://www.justconservation.org/forced-relocation-of-sengwer-people-proves-urgency-of-canceling-redd (accessed March 10, 2014).
5. http://www.grupoice.com/esp/ele/infraest/proyect/icelec/proyecto_diquis_icelec.htm (accessed March 23, 2010).

References

Adams, William. 2004. *Against extinction: The story of conservation*. New York: Routledge.
Barrionuevo, Alexei. 2011. Brazil rejects panel's request to stop dam. *New York Times*, April 5. http://www.nytimes.com/2011/04/06/world/americas/06brazil.html?emc=eta1&_r=1& (accessed March 8, 2014).

Biersack, Aletta and James B. Greenberg. 2006. *Reimagining political ecology*. Durham: Duke University Press.

Borgerhoff Mulder, Monique and Peter Coppolillo. 2005. *Conservation*. Princeton: Princeton University Press.

Boza, Mario, Diane Jukofsky and Chris Wille. 1995. Costa Rica is a laboratory, not ecotopia. *Conservation Biology* 9, 3: 684–685.

Brockington, Dan. 2002. *Fortress conservation: The preservation of the Mkomazi game reserve, Tanzania*. Oxford: James Currey.

Brockington, Dan, Rosaleen Duffy and Jim Igoe. 2008. *Nature unbound: Conservation, capitalism and the future of protected areas*. London: Earthscan.

Brysk, A. 2009. The little country that could: Costa Rica. In *Global good Samaritans: Human rights as foreign policy*, 95–118. Oxford: Oxford University Press.

Bumpus, Adam and Diane Liverman. 2008. Accumulation by decarbonisation and the governance of carbon offsets. *Economic Geography* 84: 127–156.

Carls, Jürgen and Warren R. Haffar. 2010. *Conflict resolution of the Boruca hydro-energy project*. New York: Continuum.

Carrier, James G. and Donald V.L. MacLeod. 2005. Bursting the bubble: The sociocultural context of ecotourism. *Journal of the Royal Anthropological Institute* 11: 315–334.

Checker, Melissa. 2009. Double jeopardy: Carbon offsets and human rights abuses. *Counterpunch*, September 9. http://www.counterpunch.org/2009/09/09/double-jeopardy-carbon-offsets-and-human-rights-abuses/ (accessed March 10, 2014).

Daniels, Amy E., Kenneth Bagstad, Valerie Esposito, Azur Moulaert and Carlos Manuel Rodriguez. 2010. Understanding the impacts of Costa Rica's PES: Are we asking all the right questions? *Ecological Economics* 69: 2116–2126.

Dobles, Roberto. 2008. *Summary of the national climate change strategy*. San José, Costa Rica: Ministry of Environment and Energy.

Dowie, Mark. 2009. *Conservation refugees: The hundred-year conflict between global conservation and native peoples*. Boston, MA: MIT Press.

Dressler, Wolfram, Bram Büscher and Robert Fletcher. 2014. Conclusion: The limits of Nature™ Inc. and the search for vital alternatives. In *Nature™ Inc.: Environmental conservation in the neoliberal age*, Bram Büscher, Wolfram Dressler and Robert Fletcher eds., 246–253. Tucson: University of Arizona Press.

Evans, Sterling. 1999. *The green republic: A conservation history of Costa Rica*. Austin: University of Texas Press.

Fairhead, James, Melissa Leach and Ian Scoones. 2012. Green grabbing: A new appropriation of nature? *Journal of Peasant Studies* 39, 2: 237–261.

Finley-Brook, Mary and Curtis Thomas. 2010. From malignant neglect to extreme intervention: Treatment of displaced indigenous populations in two large hydro projects in Panama. *Water Alternatives* 3, 2: 269–290.

Finley-Brook, Mary and Curtis Thomas. 2011. Renewable energy and human rights violations: Illustrative cases from indigenous territories in Panama. *Annals of the Association of American Geographers* 101, 4: 863–872.

Fletcher, Robert. 2010. When environmental issues collide: Climate change and the shifting political ecology of hydroelectric power. *Peace & Conflict Review* 5, 1: 14–30.

Fletcher, Robert. 2012. Using the master's tools? Neoliberal conservation and the evasion of inequality. *Development and Change* 43, 1: 295–317.

Fletcher, Robert. 2013. Making "peace with nature": Costa Rica's campaign for climate neutrality. In *Climate change governance in the developing world*, David Held, Charles Roger and Eva-Maria Nag eds., 155–173. London: Polity Press.

Goldman, Michael. 2004. Constructing an environmental state: Eco-governmentality and other transnational practices of a 'green' World Bank. *Social Problems* 48, 4: 499–523.

Gössling, Stephen. 2009. Carbon neutral destinations: A conceptual analysis. *Journal of Sustainable Tourism* 17, 1: 17–37.

Gössling, Stephen and Paul Peeters. 2007. "It does not harm the environment!": An analysis of industry discourses on tourism, air travel and the environment. *Journal of Sustainable Tourism* 15, 4: 402–417.

Hannerz, Ulf. 2003. Being there . . . and there . . . and there! Reflections on multi-site ethnography. *Ethnography* 4, 2: 201–216.

Heindrichs, Thomas. 1997. *Innovative financing instruments in the forestry and nature conservation sector of Costa Rica*. Eschborn, Germany: Deutsche Gesellschaft für Technische Zusammenarbeit (GTZ) GmbH.

Hermwille, Lukas. 2011. *The race to low-carbon economies has started: Developing countries leading low-carbon development (briefing summary)*. Berlin: Germanwatch.

Honey, Martha. 2008. *Ecotourism and sustainable development: Who owns paradise?* 2nd ed. Washington, DC: Island Press.

Igoe, Jim. 2004. *Conservation and globalization: A study of national parks and indigenous communities from East Africa to South Dakota*. Belmont, CA: Wadsworth/Thompson.

Imhof, Aviva and Guy R. Lanza. 2010. Greenwashing hydropower. *Worldwatch*, January/February. http://www.worldwatch.org/node/6344 (accessed September 10, 2011).

Instituto Costarricense de Electricidad (ICE). 2009. *Plan de expansión de la generación eléctrica 2010–2021*. San José, Costa Rica: ICE.

Instituto Costarricense de Electricidad (ICE). 2010. *Proyecto Hidroeléctrico El Diquís*. San José, Costa Rica: ICE.

Jinnah, Sikina and Miguel Muñoz Cabré, eds. 2011. *Global Environmental Politics* 11, 3, special issue on "Climate Change Bandwagoning."

Lansing, David. 2013. Understanding linkages between ecosystem service payments, forest plantations, and export agriculture. *Geoforum* 47: 103–112.

Levine, Arielle. 2002. Convergence or convenience? International conservation NGOs and development assistance in Tanzania. *World Development* 30, 6: 1043–1055.

Lohmann, Larry. 2014. Performative equations and neoliberal commodification: The case of climate. In *Nature™ Inc.: Environmental conservation in the neoliberal age*, Bram Büscher, Wolfram Dressler and Robert Fletcher eds., 158–180. Tucson: University of Arizona Press.

Mäkinen, Kirsi and Shahbaz Khan. 2010. Policy considerations for greenhouse gas emissions from freshwater reservoirs. *Water Alternatives* 3, 2: 91–105.

Marcus, George E. 1995. Ethnography in/of the world-system: The emergence of multi-sited ethnography. *Annual Review of Anthropology* 24: 95–117.

McCully, Patrick. 2001. *Silenced rivers: The politics and ecology of large dams*. 2nd ed. London: Zed Books.

Natural Resources Defense Council (NRDC). 2007. *Costa Rica: Setting the pace for reducing global warming pollution and phasing out oil*. Washington, DC: Natural Resources Defense Council.

Newswire. 2011. Declaration of the indigenous peoples of the world at COP 17. *Newswire*, December 7. http://intercontinentalcry.org/newswire/declaration-of-the-indigenous-peoples-of-the-world-at-cop-17/ (accessed April 26, 2012).

Ostrom, Elinor and Michael Cox. 2011. Moving beyond panaceas: A multi-tiered diagnostic approach for social-ecological analysis. *Environmental Conservation* 37, 4: 451–463.

Pagiola, Stefano. 2008. Payments for environmental services in Costa Rica. *Ecological Economics* 65: 712–724.

Porras, Ina. 2010. *Fair and green? Social impacts of payments for environmental services in Costa Rica.* London: International Institute for Environment and Development (IIED).

Reid, Hannah and Krystyna Swiderska. 2008. "Biodiversity, Climate Change, and Poverty: Exploring the Links." IIED Briefing, February.

Stocking, George W., Jr. 1983. The ethnographer's magic: Fieldwork in British anthropology from Tyler to Malinowski. In *Observers observed,* 12–59. Madison: University of Wisconsin Press.

United Nations Environment Program (UNEP). 2009. *Vital forest graphics.* Nairobi: UNEP.

United Nations World Tourism Organization (UNWTO). 2007. *Davos Declaration: Climate change and tourism; Responding to global challenges.* http://www.unwto.org/pdf/pr071046.pdf (accessed April 26, 2012).

United Nations World Tourism Organization (UNWTO). 2015. *Tourism highlights 2014.* Madrid: UNWTO.

University of Texas School of Law Human Rights Clinic (UTSLHRC). 2010. *Swimming against the current: The Teribes peoples and the El Diquís hydroelectric project in Costa Rica.* Austin: University of Texas.

West, P. 2006. *Conservation is our government now: The politics of ecology in Papua New Guinea.* Durham: Duke University Press.

West, Paige and James C. Carrier. 2004. Ecotourism and authenticity: Getting away from it all? *Current Anthropology* 45, 4: 483–498.

While, Aidan, Andrew E. G. Jonas and David Gibbs. 2009. From sustainable development to carbon control: Eco-state restructuring and the politics of urban and regional development. *Transactions of the Institute of British Geographers* 35, 1: 1–19.

World Commission on Dams (WCD). 2000. *Dams and development: A new framework for decision-making.* London: Earthscan.

9

CUSTOMARY LANDOWNERS, LOGGING COMPANIES AND CONSERVATIONISTS IN A DECENTRALIZED STATE

The Case of REDD+ and PES in Papua New Guinea

David Lipset and Bridget Henning

Papua New Guinea: A New Postcolonial State in the Southwest Pacific

Set on a rugged, high island geography in the southwest Pacific, the sovereign state of Papua New Guinea (PNG) governs the eastern half of the island of New Guinea (Figure 9.1). Its people, known as Melanesians, arrived over land bridges from the north in several waves of prehistoric migration some 60,000 years ago and then by outrigger canoe as recently as 4,000 years ago. They lived in domestic, kinship-based groups who practiced various forms of subsistence production while secret warrior societies defended their territories against neighboring tribes. Although the inhabitants began to have intermittent contact with European explorers as early as the sixteenth century, Germany and Great Britain did not colonize the eastern half of the island until the late 1800s. After the outbreak of WWI, their two colonies became an Australian protectorate until PNG was granted independence in 1975 under the leadership of the nation's "founding father," Sir Michael Somare. Today, PNG is a vital, relatively stable, parliamentary democracy with a Westminster-style system of government and 20 provinces, each with its own government, and each with district- and local-level leadership.

In addition to English, Tokpisin and Motu, its nearly 7 million people speak approximately 20 percent of the total number of languages in the world (800). In other words, the nation has great cultural diversity. It is also blessed with enormous economic resources. Over two-thirds of Papua New Guinea is covered by the third largest intact tract of rainforest, after the Amazon and Congo Basins, making it a target of REDD activities (FAO 2011). In addition, Papua

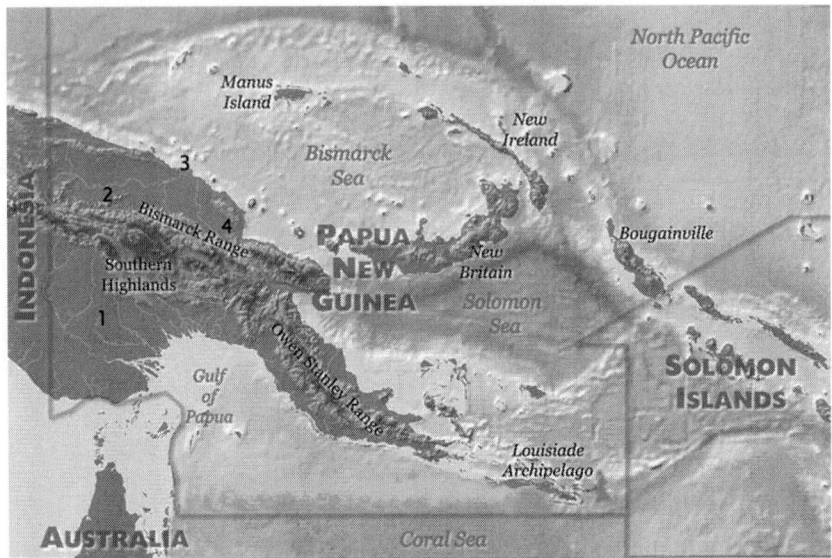

FIGURE 9.1 Map of Papua New Guinea indicating locations of case studies further discussed: (1) Kamula Doso Forest Area, (2) April-Salomei Forest Area, (3) Murik Lakes, and (4) Wanang Conservation Area.

New Guinea is a noteworthy case study of REDD for historical reasons, as it was one of the founding members of the Coalition for Rainforest Nations. In 2005, Costa Rica and PNG first proposed the notion of REDD to the eleventh Conference of the Parties agenda (Somare 2005). This initiative led to the decision of the UN Framework Convention on Climate Change to support the future inclusion of REDD in approaches to emissions reductions. As a result, PNG became a priority country for REDD development for the UN and the World Bank.

However, foreign industrial logging interests have largely operated in PNG without state control since the 1990s, threatening the forests central to REDD (Lattas 2011). According to recent estimates, current rates of degradation and deforestation compare to those found in the Amazon basin. Some predict that 83 percent of commercially accessible forest in PNG will be cleared or degraded by 2021 (Shearman et al. 2009). Commonly cited figures hold that anywhere from 70 percent to 90 percent of logging in PNG may be illegal with unsustainable logging practices, political corruption, human rights abuses and an absence of sustained benefits for local communities (World Bank 2006; ODI 2007; Greenpeace 2008).

In addition, the combination of population growth and swidden horticulture contributes to deforestation (Shearman et al. 2008). People live in rural, kinship-based communities where a domestic economy still supports about 80 percent of the population through semi-subsistence hunting, horticulture and fishing.

Although the population is largely subsistent, PNG as a nation has pursued a strategy of a *rentier* state, relying on revenue from export-driven growth. During the 1980s and 1990s, natural resource industries, including logging, grew to account for 90 percent of the government's revenue. Extraction of large-scale, non-renewable minerals, as well as oil and gas, also grew. Agro-forestry projects involving the four major export crops—coffee, cocoa, copra and palm oil—are also increasing rapidly.

Around 77 percent of the world's forest is owned by governments, and state ownership of the land predominates in the majority of rainforest nations (White and Martin 2002);[1] however, PNG is unique among REDD+ countries in that the vast majority of land in the country (97 percent) is legally held by kinship-based groups. As part of the government's commitment to decentralization, customary land tenure was enshrined in the country's constitution. Although successive governments have deviated from the spirit and letter of the constitution, kinship-based groups continue to retain rights of access, use, management and exclusion of their customary land, which are passed on to their next generation (Anderson and Lee 2010).[2] However, landowners do not possess alienation rights and may not sell their land (Sillitoe 2000). Instead, they may form a separate entity known as an Incorporated Land Group (ILG) under which agreements with the state to lease land or grant specific rights are permitted. Not all landowners, nor all land, is registered in an ILG and many view registration with suspicion. Although ILGs are officially registered with the government, their property is not surveyed and ILGs offer limited assurance of exclusive rights, as customary boundaries are often undefined and open to renegotiation between kinship-based groups. ILGs are often grouped together as a landowner company, which despite legislation removing their ability to deal in timber rights, frequently continue to act in such a capacity, negotiating deals between landowners and industrial logging companies (Filer 2012, 603).

In a centrifugal political-legal environment, in which customary landowners must provide informed consent for any developments on their property (Ase 2011), governing, planning and decision-making requires consultation with them (Laurance et al. 2011). Potential developers of logging projects, agricultural plantations or mines negotiate agreements with the state, which in turn brokers agreements with the identified ILGs of the forest area of interest. In many cases, however, informed consent of the customary landowners has not been secured or due process has not been followed (Babon and Gowae 2013). If land use planning is a first step in the implementation of REDD+, effective social mapping, boundary demarcation, ILG formation and participatory decision-making at the landowner level are vital.

At the national level, the electoral system is highly and hotly contested and features large numbers of candidates. Regionally based voting blocs and constituencies exert undue power over representatives. Incumbency bestows little by way of political tenure, as the turnover rate is about 50 percent per election. In the 40 years since independence, there have been 13 coalition governments and 7 different prime ministers. Political parties are not united by ideology as much as by regional affiliation and the goal of maintaining power through patronage. Until the past few years, regular no-confidence votes cut short the rule of governments to 2.5 years. Ministers and senior officials play a large role in policy administration, and frequent turnover has made policy commitment difficult to maintain. Notwithstanding pressures from local-level political systems (Allen and Hasnain 2010), democracy has held its own in this postcolonial state. It is the case, however, that the public service has struggled. Despite many well-educated staff, political interference has resulted in nepotism and the politicization of appointments (May 2009; Gelu 2010). Like many postcolonial states in the global South, providing basic services and infrastructure has tested the government. In 2008, PNG ranked 149 out of 179 in the Human Development Index, which includes educational attainment, life expectancy and child mortality rates (AusAID 2010). This is also partly a result of widespread corruption and other law and order problems. The limited government presence at the local level, as well as inconsistent market penetration, has resulted in persisting long-term exchange relations embedded in ancestor-based cosmologies and non-state forms of leadership (Strathern 1988; Jebens 2004).

PNG, the Origins of REDD+ and Policy Efforts

PNG climate change policy has passed through four distinct phases in its development: international REDD agenda setting, an illegitimate gold rush by so-called carbon-cowboys, national strategy development, and national strategy implementation (Howes 2009; Babon 2011; Filer and Wood 2012). Prior to the development of REDD, PNG had no existing strategy on payments for ecosystem services. The first phase in REDD policy lasted 2 years and culminated in 2007 at the Bali climate change conference, when PNG government representatives played a big role in advancing the international REDD agenda through the UNFCCC.

In the second phase from 2007 to 2009, foreign investors began negotiating voluntary REDD projects with local landowners amid a lack of national policy or guidelines, creating speculation and inflating landowner expectations (Melick 2010). Meanwhile foreign aid agencies lent some support for the development of a national policy framework. In 2008, PNG co-signed the PNG-Australia Forest Carbon Partnership, applied to participate in the World Bank's Forest Carbon Partnership Facility, and established an Office of Climate Change and Environmental Sustainability and Carbon Trading (Somare 2009). This office was tasked with developing a climate change policy and managing carbon trading. It was housed within the prime minister's office and was placed under the leadership of Kevin Conrad, a former adviser to the prime minister. Its legality was questioned and allegations of corruption and financial mismanagement were lodged against the executive director in 2009 (Lang 2009b; Wynn and Creagh 2009). Questions about nepotism arose and the legitimacy of PNG's climate change institution was challenged both domestically and internationally (Lang 2009c).

Although PNG was a global leader in the development of REDD+, the domestic policy vacuum failed to provide a clear framework for REDD development. Too much initial emphasis was placed on carbon accounting and valuation at the expense of community engagement. The rush by some businesses, nongovernmental organizations (NGOs), researchers and various political interests to establish pilot projects and develop carbon markets took place without the understanding of the forest-based communities. Unregulated carbon deals and preemptive trading occurred. Carbon-trading scandals within the PNG Office for Climate Change led to the suspension of senior officials (Filer and Wood 2012).

Hinting at the prospects of fast money, middlemen raced to secure carbon rights from villagers. A gold-rush mentality ensued. Rumors circulated that villagers were paying for worthless carbon certificates from carbon cowboys who were roaming about the countryside buying carbon rights from remote communities, making deals that were neither legal nor authorized by informed consent. Preferential allocation of carbon deals, conflicts of interest within the government and forward trading of unverified carbon rights for nonexistent projects were all reported in the national and international media.[3]

In 2010, an editorial in the prestigious journal *Conservation Biology* concluded that the actual number of forest-carbon projects in PNG was "hard to determine" (Melick 2010, 360) because of contentious land tenure disputes. At that time, although 90 projects, said to cover over 5 million ha, had been "signed," no single one that was underway on the ground could be documented. Several "demonstration" projects were planned. The Australian government had provided A$200,000 to four NGOs to develop them (The Nature Conservancy, the Wildlife Conservation Society, Conservation International and Live and Learn). In addition, a local NGO, called FORCERT, began working at the community level to develop a forest carbon inventory methodology as part of a trial payment for environmental services (PES) project.[4]

Carbon rights remained ambiguous. The state's Office of Climate Change indicated a preference for ownership/control of carbon at the national level yet there was no legislation to support this; various academics argued that carbon/carbon sequestration rights were held by customary landowners (Ase 2011; Bingeding 2009).

During the third phase of REDD development from 2009 to 2011, the government tried to end voluntary carbon schemes between foreign investors and local landowners, choosing instead to support a process of policy development that sought financial support from foreign aid agencies. After corruption concerns arose, PNG's National Executive Council abolished the Office of Climate Change and established a new governmental unit to coordinate its action. A National Climate Change Committee, composed of secretaries of all key governmental departments, was established to serve as the new decision-making body to oversee climate change policy. An Office of Climate Change and Development was then established. A Technical Working Group on REDD+ was also convened, with diverse membership that included the state's Forest Authority and Department of Conservation; the Australian consulting firm McKinsey and Co.; several international and local NGOs such as The Nature Conservancy, World Wildlife Fund, and PNG Eco-Forestry Forum; donors such as UN-REDD, the EU and AusAID (the Australian Agency for International Development); and industry associations such as the PNG Forest Industry Association.

In the current phase starting around 2011, national government representatives, international and local NGOs, and foreign aid agencies have tried to engage in implementing the new REDD action plan (Filer 2011). The PNG government released "Climate-Compatible Development Strategy" in March 2010, with the help of McKinsey and Co.; the goal of the plan is to achieve carbon neutrality by 2050 while maintaining annual economic growth at 7 percent (Lang 2010). The strategy did not seek a complete halt to commercial logging, but only a reduction that would be supplemented by improved secondary forest management, which was intended to provide opportunities for REDD+. The new policy faced implementation delays amid criticism that it lacked national ownership, due to the Australian authorship by McKinsey and Co. and very limited national consultation.

In 2011, the Policy Board of the UN Collaborative Programme on Reducing Emissions from Deforestation and Forest Degradation in Developing Countries (UN-REDD) approved the National Programme Document, which set out how the government of PNG proposed to achieve a state of REDD+ readiness by 2014. This approval triggered the release of US$6.4 million to establish a monitoring, reporting, and verification (MRV) system in preparation for individual projects to receive compensation funds from the international community. The Office of Climate Change and Development was to have taken charge of administering forest conservation and sustainable forest management. However in the midst of a change of government that took place in 2012, questions were raised whether this office possessed the leadership to do so (Filer 2012).

The PNG Forest Authority, for its part, proposed a model of carbon rights similar to timber rights, whereby carbon would be owned by landowners but managed by the state on their behalf. In the absence of legislation, the government asserted that no legal basis for establishment of forest carbon sequestration schemes in PNG existed. In the meantime, however, two projects, described in the next section, were initiated in different parts of the country, outcomes of which illustrate the on-the-ground obstacles REDD+ projects have to overcome in PNG.

Initial REDD+ Projects

In 2003, negotiations began between customary landowners, the government and the largest logging enterprise operating in PNG, the Malaysian company Rimbunan Hijau, for rights to a forest area called Kamula Doso (790,000 ha) in the Middle Fly District of the Western Province (Figure 9.1). After 2 years, negotiations stalled (Filer and Wood 2012). By the end of 2007, a pair of rival Australian entrepreneurs started talks with the directors of Tumu Timbers Development Ltd., a landowner company representing 52 local ILGs, about alternative uses of the Kamula Doso area. An initial notion was that the logging company would build a road to the district capital in exchange for partnership with local ILGs. As land surveys were undertaken that same year, some "rogue directors" of Tumu Timbers began to pursue carbon trade in preference to partnering with the logging company (GoPNG 2011, 94). In a proposal, the Tumu Timbers landowners' board stated that it was converting "the Forest Management agreement to a Carbon Project and recognize . . . rights to carbon sequester in the project area" (TTD 2010, 10). By late 2008, the head of the Office of Climate Change granted rights to 1 million tons of carbon credits from the Kamula Doso area to one of the Australian entrepreneurs, Nupan Trading (Filer and Wood 2012). In the wake of media scrutiny (Lang 2009a), he demurred that the Kamula Doso "carbon credit" agreement did not represent a "real deal," but merely a mockup document stolen out of his desk drawer (Gridneff 2009a). In 2009, the government granted Tumu Timbers a 99-year license for carbon trading—which was immediately contested.

The Kamula Doso landowners split into rival factions that were competing for control of Tumu Timbers. Despite a government announcement that it opposed voluntary carbon schemes, a REDD proposal submitted to the Climate, Community and Biodiversity Alliance in early 2010[5] claimed to represent as many as 52 ILGs with rights to the Kamula Doso forests. In the course of subsequent comments, 50 letters appeared, purportedly signed by the chairmen of the land-holding groups. The writers subsequently declared that they had been made shareholders of Tumu Timbers without their consent, that their signatures had been forged, and that they had been forced into signing documents they could neither read much less understand. By early 2011, the deal for carbon trade was off and landowners announced that they preferred to lease their lands to agroforestry projects and sustainable logging projects (Laepa 2011). Needless to say, the leases that resulted did not necessarily include "most of the customary landowners in the area" (Filer and Wood 2012, 10) or result in any action at all, for that matter. The outcome of the Kamula Doso initiative showed that customary landowners in PNG would not yield the use of their land despite the unlawful and dishonest efforts by foreigners to do so (Filer and Wood 2012, 11).

Results of a second REDD-related project were much the same. The April-Salomei Forest Management Area, located in an isolated region of the East Sepik Province, is one of the largest tracts of undisturbed lowland rainforest in PNG (Figure 9.1). In 2008, PNG's Office of Climate Change entered into a partnership with two carbon brokers, Earth Sky Ltd. and Climate Assist PNG, to finance development of an "avoided deforestation" REDD+ project within the boundaries of the April-Salomei Forest Management Area by selling US$500 million forest carbon offsets (Lang 2009d; Leggett and Lovell 2011, 122). The executive director of the Office also wrote that the two companies were "prepared to put in A$10 million to assist the establishment" of the Climate Change Office.[6] The project had the support of Michael Somare, the prime minister, as well as of 163 local landowning groups (ILGs). Stated goals were an annual reduction of 1 million tons of CO_2 emissions, to protect and conserve the April-Salomei forest area and its species, to develop sustainable livelihoods of local communities and to harvest sustainable logs from native species forest plantations. A Swiss company was contracted to complete the project design, and in late 2008 the government recognized the rights of a landowner company to "produce and sell carbon credits in the . . . April-Salomei" (OCCES 2009).

Research in 2009 concluded that despite efforts by developers to interact with them, landowners nevertheless expressed "a considerable lack of awareness and understanding" of the goals of the project (Leggett 2012, 124). A commonly held belief was that the project would produce "money from the sky." That is, either a cargo mentality prevailed according to which money would magically materialize without having to work for it, or possibly the view was that they would be paid "money *for* the sky," where carbon payments were compensation for the purity of the local air (Filer 1997). Perhaps both ideas were in play. In addition to the contested understandings of what REDD+ meant, Hunstein Range Holdings Ltd., the landowning company that claimed to be the sole authorized representative body of the communities in the project area, was "strongly and repeatedly rejected" by the landowners who had "never" given the company "permission to act as their representatives" (Leggett and Lovell 2011, 127).

Landowners also raised concerns about how the benefits would be distributed. The project seemed to lack safeguards to protect their revenue. It also favored educated elites and the authority of a top-down, board of directors of Hunstein Range Holdings Ltd. that was to serve as gatekeeper for rural villagers, who in turn would only be allowed access to revenue on a project-by-project basis. Under the initial design from 2009, landowners and their livelihoods would have been subject to a project that they had no hand in designing, would play no part in managing, and did not understand (Leggett and Lovell 2011). Yet by March 2014, the project had met the requirements of the Verified Carbon Standard per the Pacific Forest Alliance.

Blue Carbon and PES

The synoptic history of REDD+ in PNG outlined earlier, as well as the accounts of the Kamula Doso and the April-Salomei projects, depict a REDD+ landscape made up of limited governmental authority, an opportunistic private sector as well as puzzled, suspicious and defiant rural communities. We now turn to brief discussions of two more rural societies, in which the authors, Lipset and Henning, have themselves done research. In the case Lipset observed, debate at the community level was just starting out, while in the other, studied by Henning, a relationship with a conservation group had been going on for more than a decade. Taken together, they add further evidence of the tests REDD+ projects face in PNG that we have already encountered, but also suggest ways to deal with them.

Blue Carbon in the Murik Lakes

A mangrove is a tree, a forest and an ecosystem, all with specialized adaptations to brackish water. Mangrove forests supply several valuable ecosystem goods and services, such as water quality control, fisheries production, nursery habitats and storm protection (Wells et al. 2006; Albert et al. 2012). Mangroves are also efficient carbon dioxide sinks, which, along with seagrasses and other marine reservoirs, are collectively known as "blue carbon." As blue carbon comprises more than one-half of the biological carbon sinks on earth, the conservation and restoration of blue carbon habitats has belatedly been recognized by the UN to play an important role in climate change mitigation (Nellemann et al. 2009).

Recent assessments indicate that mangroves are among the most carbon-rich forests in the tropics (Mumby et al. 2004; Donato et al. 2011), and in PNG carbon brokers have expanded their sights to include them. The Murik Lakes provide a case in point. They are a large, intertidal system of brackish lagoons and mangrove forests (120 square miles) located just west of the mouth of the Sepik River (Figure 9.1). The Murik Lakes are a restricted commons in which Murik people claim rights to fish, harvest shellfish, gather firewood and cut bush materials for building houses (Figure 9.2). In 2012, Lipset encountered talk about carbon among the Murik, the villagers with whom he has conducted long-term research since the early 1980s (see Lipset 1997).

As mentioned earlier in this chapter, rural communities in PNG are administered by democratically elected officials, who link villagers to the national and global moment. At a weekly meeting Lipset attended in a Murik village, an elected councilor of the local level government announced that residents should immediately start to "license families" in advance of the imminent arrival of carbon investors. "A lot of money will be coming, but we will only get very little of it," he predicted rather caustically, unless kin registered themselves into ILGs. A university graduate, who was then on leave from work in the national capital, added that companies were competing to represent carbon trade in PNG and were even emailing him. The Murik Lakes "have no rival in PNG as far as the size of our mangroves. We will make big money." In the ethnographic record, of course, rural Melanesians are famous for a great many things, sacred and profane, but breathless optimism and unfounded expectations about what modernity might have to offer is certainly one of them (Schwartz 1962).

What did villagers understand about carbon payments? Carbon trade, so one man proposed, was for "wind . . . Dirty smoke . . . comes from machines or

FIGURE 9.2 Two brothers out for a turn in their canoe at low tide, 2012.

Credit: David Lipset.

engines . . . is destroying everything. They will be paying us for [our] air." He, as well as others, advised Lipset to go talk to a Murik man called Erik Komang, who was then living in Wewak, the provincial capital. A university graduate, and a sister's son of the prime minister, Komang headed a company called Pacific Carbon Trade Ltd. that was intending to register kin groups and represent them to foreign investors it was enlisting (Gridneff 2009b). However to the very same extent that villagers held up Erik Komang as knowledgeable about carbon trade, he was regarded with no less wariness. That is to say, many people did not view him as a neutral entrepreneur who could be trusted. Rather than an opportunity to make money, they saw him as a fellow kinsman who only had his own lineage rather than tribal interests as a whole at heart.

Lipset found Erik Komang staying with kin in a peri-urban encampment where Murik people live in the provincial capital. Pacific Carbon Trade Ltd. was one of 14 such companies in PNG, he told him, but it was the only one registered in the East Sepik Province. "In the East Sepik," he boasted, "it is me alone." What, Lipset asked, would stakeholders be obliged to do in return for carbon payments? Not fish, harvest shellfish, or collect firewood from a demarked area? In response, Komang talked vaguely about hectares of rainforest, which seemed tangential to mangrove ecology and went on to list the several stages by which the size and value of the Murik mangrove would be measured.

The first step was to register stakeholders into ILGs, a process well-known for its cost and for the frustration it inevitably creates (Jorgensen 2007, 67).

Identifying property and proprietary units from an overlapping jumble of rival lineages was not going to be easy. Aligning customary marine estates with corresponding statutory declarations might well turn out to be impossible. His preference was to duck the problem. "I would like to distribute ... money village by village. The clan system [is] ... too complicated." In any case, the whole process was not really yet ready to move forward. Families had not yet registered as ILGs and the government had issued no regulations for a carbon market.

Elsewhere in PNG, communities have been pressured to inventory and record themselves with the state for economic development (Weiner and Gaskin 2007). The problem they have faced is having to create an exclusive list, e.g., a legible kind of group, not only because of indefinite lineage boundaries, but also because of adoption, out-migration and intertribal marriage in contemporary society. No less than a complex social structure, and a shifting, intertidal environment, the Murik Lakes are also subject to *cathexis*, to an investment, not just of jural rights, but of emotional attachments. They are associated with and arouse deep-seated, irresolvable jealousies (Lipset 2014).

Indeed, in the Murik Lakes the promise of compensation payments in return for looking after a plot of mangrove forest exposed the main fissure in their commons. Potential beneficiaries feared that they might not profit equally (or at all) from carbon. Their anxiety did not arise from the invention of new social categories that never existed in the past (see Ernst 1999, 88). It rather derived from the continued *absence* of supra-community interests and identity in Murik culture, an absence that may be traced back to pre-colonial times, and which continues to inspire little by way of fair-mindedness today. In the contexts that rural Murik think about the tangible boundaries of corporate groups that own estates, they think of family, lineage-based ceremonial exchanges and community ethnohistory, but not tribe, much less the nation-state and its legal bureaucracy. Eric Komang, equipped with his Memoranda of Agreement, was not understood or expected to advocate for the society as a whole in an evenhanded way, only for a subset of relationships within it.

The Wanang Conservation Project

Clearly, the prospect of payments for blue carbon aroused the kind of anticipation and suspicion among the Murik that recalls rivalries and doubts we have already encountered in the brief history of REDD+ in PNG. The Wanang Conservation Project in Madang Province provides us with a different perspective (Figure 9.1). Not itself a REDD+ project, but one that is based in payments for environmental services (PES) or direct payments for conservation. It arose from a conservation agreement among nine landowning clans (consisting of 10–30 people each) of Wanang village, who chose to prohibit logging on 10,000 ha of their forested land in 2000 and eventually received financial support. European and American biologists with the New Guinea Binatang Research Center (BRC) began to conduct research in the Wanang forest in 2001 and developed a long-term research plot.[7] Biologists provided landowners with employment opportunities on their research teams. Starting in 2009, in return for the intact research site, villagers received annual payments to each clan and village-wide benefits, such as a school, with funding provided by research grants, corporate sponsors, and conservation NGOs. Biologists trained villagers to assist long-term monitoring of forest dynamics (comparable to the kind of work needed in REDD+ monitoring). The project grew and changed over time, but it persists, not only as the first project of its kind in the country to pay landowners for conservation or ecosystem services but also the longest.[8]

As we have argued, REDD+ in PNG struggles with the issue of community support. Thus it is remarkable that Wanang landowners more or less initiated the project themselves.[9] Customary institutions and clan structure were used in the legally binding conservation deed. Filip Damen, a local "big man," championed it and managed ceremonial exchange relationships between villagers and BRC at milestones and to resolve disputes (Figure 9.3). Although both biologists and landowners credit Filip Damen for holding the project together, misgivings nevertheless do surface about his use of money and power.

Compared to REDD projects, the Wanang forest area is relatively small; in part, its scale has made the project viable. Individual landowners have relationships

FIGURE 9.3 Wanang villagers use customary exchange to resolve a conflict with BRC, 2012.

Credit: Bridget Henning.

with all other participants within it. By contrast, intervillage competition has limited alliances outside Wanang. Wel, a neighboring village backed by logging interests, had violent confrontations with Wanang during the early stages of the project. The Malaysian logging company, Rimbunan Hijau, was rumored to have funded them to oppose the project. Indeed, 11 clans initially expressed interest in joining with conservation, but two withdrew to allow logging on their land.

REDD+ advocates emphasize the need for employing mechanisms of conflict resolution, customary boundaries being continually subject to renegotiation in PNG (Paka and Ondopa n.d.). The Wanang Conservation Project has been able to deal with conflicts and property disputes by holding lengthy meetings, in which everyone may voice concerns and work out ancestral histories of the land. Formal mechanisms of social control of this kind require a measure of mutual trust. After the conservation project began, for example, a family returned to the village claiming land rights because of a grandfather who had been born in Wanang but had left as a young man during World War II. Meetings were held within the village until the claim was resolved through the retelling of ancestral histories, which re-established relationships and land claims. The resolution was affirmed by ceremonial exchange.

While the use of customary institutions fortified the project within the community, it should be pointed out that their use does not ensure political equality or equitable benefit distribution at the local level, both major concerns of REDD+ developers. In Wanang, big man leadership and engagement with biologists situate certain men to receive greater benefits and exert wider influence than ordinary villagers. Rival men objected to such privileges at one point, which led to efforts to invent more project management positions to be more inclusive, yet this accomplished little change in terms of the exercise of power. Annual payments are made to clans, and distribution is the prerogative of their individual big men; money is either divided up fairly, or simply kept by the leader. Despite perceived material inequities, villagers rarely complain to BRC because what leaders do is regarded as a matter of custom and not subject to outside mediations.

Complaints they do express to BRC concern the overall level of benefits. As we have seen, Wanang landowners, no less than elsewhere in PNG, expect huge windfalls of money. Although they appreciate receiving village-wide benefits, such as the school, they complain that as individuals they have been forgotten. Landowner expectations are not based on comparisons with logging, or opportunity costs. Rather, they seem to derive from an assumption that there should be equality with the exchange partner, who is, in this case, BRC. Landowners see their relative wealth and become unhappy that BRC seems to have become wealthier and are able, for example, to add vehicles to their fleet, while villagers see their own level of wealth as unchanged.

REDD+ project proposals in PNG have claimed that education and awareness will correct these kinds of exaggerated expectations (Hooper et al. 2011). But in the PNG context, we suggest that they may be the result of cultural understandings rather than ignorance. Some people in Wanang continue to believe that the accumulation of wealth may be facilitated by exchange with ghosts. Indeed, a number of Wanang landowners understood the conservation project to be part of long-term exchanges with deceased ancestors that will grant access to the abundant wealth of the afterlife.

Another concern we have seen in the REDD+ literature in PNG is that landowners do not give free, prior and informed consent, as villagers do not understand the premises underlying the nature of contracts (Leggett 2012). Although a good faith effort has been made in Wanang to hold meetings to establish rules and agreements, understandings of the project vary within the community. Women are less engaged in conservation due to their domestic duties, but a number of them expressed fears that their access to the forest had become illegal and could result in their men being jailed. Furthermore, a contract to restrict forest use in exchange for the construction of a school was seen as but one transaction in a long-term exchange relationship that would involve future obligations. Accordingly, both BRC staff and landowners have been continually frustrated by each other, yet the project has succeeded because both groups have remained committed. Had BRC staff insisted on following the terms and conditions of the contract in a strict and rigid way, we believe that their relationship with villagers would not have withstood a 12-year test of time.

Conclusion: The Future of REDD+ in PNG

Despite presenting what must be seen as an ideal legal framework for REDD+ implementation, such as constitutionally secure land rights and a process for landowner registration, many analysts who support the concept of REDD+ have concluded that PNG is not ready for it (Greenpeace 2010; Hunt et al. 2011). A number of issues, in their view, must first be resolved. According to Brockhaus and Angelsen (2012, 16–17), achieving effective and fair REDD+ outcomes anywhere is likely to require a *transformational change* in national policy, which is to say, "a shift . . . that leads to policy formulation and implementation away from . . . approaches that directly or indirectly support deforestation and forest degradation." Indeed, challenging powerful logging interests is one challenge, from which PNG is not exempt.

Among other things, the four cases in this chapter demonstrate that the design of a national REDD+ strategy must be rooted in the (however shifting) consent of landowners (Filer 2012).[10] But arriving at a shared understanding of an agreement has proved difficult for several reasons, not least of which are the implicit obligations Melanesians often take for granted. Moreover, although gaining local consent is necessary, it is not sufficient to ensure equitable REDD+ implementation, as elite interventions and jealousies in the April-Salomei, Murik Lakes, and Wanang Conservation cases illustrated.

Despite a general understanding that all parties have legitimate rights to claim benefits commensurate with their contribution to REDD+, conflict management between landowners and conservation interests, as well as among landowners themselves, has been a persistent problem. Long-term participation will require mediation and conflict management at multiple levels especially in cases where numerous ILGs are dealt with together. Here, the flexibility shown by both landowners and BRC in the case of the Wanang Conservation Project is exemplary.

Perhaps then, prospects for REDD+ projects in PNG need not be anticipated in entirely negative terms. A potential surely exists to find and provide incentives that mobilize support for the sustainable management of PNG's forests while also being suitable to the local cultural context. Support for REDD+ persists, and, in developing a decentralized, governmental structure that can integrate the needs and perspectives of landowners, together with multiple stakeholders, the state has shown some progress over the past decade. Indeed the quick transition from carbon cowboy deals to strategic national planning with stakeholder consultation does seem to indicate that strides are indeed being made toward fair and equitable REDD+. But realistically, the best we can anticipate about future prospects for REDD+ PNG is that they must be viewed with uncertainty.[11]

Notes

1. Cotula and Mayers (2009) argue that the stated level of government control of and access to forest resources often bears little resemblance to reality, and that in fact local control of forests is increasing. Furthermore, forest-dependent communities are not powerless actors in the management and development of their resources: their actions can dictate the long-term success or failure of carbon forestry projects (see also Oestreicher et al. 2009).
2. In May 2010, the PNG Parliament amended the Environment Act of 2000 to restrict landowners' rights should a project be deemed in the "national interest."
3. See http://www.economist.com/node/13724646 (accessed April 28, 2014).
4. Donors included UN-REDD, AusAID (the Australian Agency for International Development), Japan International Cooperation Agency (JICA) and the EU.
5. The Climate, Community & Biodiversity Alliance is an NGO founded in 2003 with the goal of promoting land management that mitigates climate change, reduce poverty and conserve biodiversity. Its members include Conservation International, The Nature Conservancy, Rainforest Alliance and Wildlife Conservation Society. Its donors include the Blue Moon Fund, the Kraft Fund, BP, Hyundai, Intel, the Rockefeller Foundation, S.C. Johnson and Weyerhaeuser, among others. See http://www.climate-standards.org/about-ccba/ (accessed April 28, 2014).
6. http://www.redd-monitor.org/2009/06/07/anatomy-of-a-deal-the-april-salome-redd-project-in-papua-new-guinea/ (accessed May 1, 2014).
7. The New Guinea Binatang Research Center (BRC) on the coast of Madang, was devised after the Christensen Research Institute (CRI) closed in 1999. Shortly thereafter, Vojtech Novotny and other researchers who had been working at CRI established BRC to house their research. Novotny continues to direct BRC, which has expanded from the central station to three additional satellite field stations and over 25 full-time staff. BRC has been supported by grants from the US National Science

Foundation (NSF), the Darwin Initiative for the Survival of Species, the United Kingdom, the Czech Academy of Sciences, and the Czech Science Foundation.
BRC continues to conduct biological research but also trains students and paraecologists. BRC conservation efforts focus on protecting their research areas.
8. Support for the Wanang Conservation Project has been pieced together from several donors. Originally and throughout the project, research grants from the NSF and the Czech Academy of Sciences have been used for employment and compensation to landowners for research sites. The World Wildlife Fund gave money to support mapping of the boundary. Swire and Steamships contributed funds for the research station, forest plot research, and community compensation. Seacology paid for the construction of the permanent school building. The Christensen Fund also contributed to general expenses and for a landowner exchange with another conservation area.
9. With the assistance of the Bismarck-Ramu Group, a local NGO that works to help organize initiatives of this sort at the local level.
10. Several local NGOs have become active in promoting customary rights to land and natural resources. These include the PNG Eco-Forestry Forum, the Foundation for People and Community Development, the Bismarck-Ramu Group and the Forest Management and Product Certification Service.
11. We wish to acknowledge Wanang and the Murik Lake villagers for their hospitality, time, and cooperation in helping us to understand Melanesia and REDD+ in PNG.

References

Albert, J. A., K. Warren-Rhodes, A. J. Schwarz, and N. D. Duke. 2012. *Mangrove ecosystem services & payments for blue carbon in Solomon Islands.* Solomon Islands: WorldFish Center. AAS-2-12-06.

Allen, Matthew and Zahid Hasnain. 2010. Power, pork and political patronage: Decentralization and the politicization of the development budget in Papua New Guinea. *Commonwealth Journal of Local Governance* 6: 7–31.

Anderson, Tim and Gary Lee, eds. 2010. *In defence of Melanesian customary land.* Sydney: AID/WATCH.

Anonymous. 2008. Forest carbon partnership facility, readiness plan idea note: External review (PNG). http://www.forestcarbonpartnership.org/fcp (accessed April 21, 2014).

Ase, Damien. 2011. Meaningful landowner participation in REDD projects: What do we mean by this? Paper presented at REDD+, PES and Benefit Sharing Workshop, February 17–18, 2011, Institute of National Affairs/Institute of Global Environmental Strategies, Central Province, Papua New Guinea.

Australian Agency for International Development (AusAID). 2010. Forest carbon initiative concept development grants. http://www.ausaid.gov.au/keyaid/forest-carbon.cfm (accessed April 4, 2010).

Babon, Andrea. 2011. Snapshot of REDD+ in Papua New Guinea. Info brief 40. http://www.cifor.org (accessed May 4, 2014).

Babon, Andrea and Gae Yansom Gowae. 2013. *The context of REDD+ in Papua New Guinea: Drivers, agents and institutions.* Occasional Paper 89. Bogor, Indonesia: Center for International Forestry Research.

Bingeding, Nalau. 2009. *Carbon trade: Do we know what we are doing?* Spotlight with NRI 3, 5: 1–4. Papua New Guinea: National Research Institute.

Brockhaus, Maria and Arlid Angelsen. 2012. Seeing REDD+ through 4Is: A political economy framework. In *Analysing REDD+: Challenges and choices,* Arlid Angelsen,

Maria Brockhaus, William Sunderlin, and Louis Verchot eds., 15–30. Bogor, Indonesia: Center for International Forestry Research.

Callick, Rowan. 2009. The rush is on for "sky money": The struggle begins to control the immense wealth certain to flow from carbon trading. *Weekend Australian*, September 5.

Cotula, Lorenzo and James Mayers. 2009. *Tenure in REDD: Start point or afterthought?* Natural Resources Issues No. 15. London: International Institute for Environment and Development.

Donato, Daniel C., J. Boone Kauffman, Daniel Murdiyarso, Sofyan Kurnianto, Melanie Stidham, and Markku Kanninen. 2011. Mangroves among the most carbon-rich forests in the tropics. *Nature Geoscience* 4: 293–297.

Ernst, Thomas M. 1999. Land, stories and resources: Discourse and entification in Onabasulu modernity. *American Anthropologist* 101: 88–97.

Filer, Colin. 1997. Compensation, rent and power in Papua New Guinea. In *Compensation for resource development in Papua New Guinea*, Susan Toft ed. Port Moresby, PNG: Law Reform Commission (Monograph 6).

Filer, Colin. 2011. REDD+ at the crossroads in Papua New Guinea. *East Asia Forum*. http://www.eastasiaforum.org/2011/07/23/redd-plus-at-the-crossroads-in-papua-new-guinea/ (accessed April 22, 2014).

Filer, Colin. 2012. Why green grabs don't work in Papua New Guinea. *Journal of Peasant Studies* 39, 2: 599–617.

Filer, Colin and Michael Wood. 2012. The creation and dissolution of private property in forest carbon: A case study from Papua New Guinea. *Human Ecology* 40 (5): 665–677. doi:10.1007/s10745-012-9531-2.

Food and Agriculture Organization (FAO). 2011. *State of the world's forests*. Rome: Food and Agriculture Organization of the United Nations.

Gelu, Alphonse. 2010. Politics and governance. In *Papua New Guinea's development performance 1975–2008*, Thomas Webster and Linda Duncan, eds. Boroko, PNG: National Research Institute.

GoPNG (Government of Papua New Guinea). 2011. *Commission of inquiry into SABL: Transcript of proceedings, 22 November 2011*. Waigani, PNG: Department of Prime Minister and NEC.

Greenpeace. 2008. *Preserving paradise: The value of protecting Papua New Guinea's forests for climate*. Amsterdam: Greenpeace International. http://www.greenpeace.org/austria/Global/austria/dokumente/Reports/wald_preserving-paradise-PNG_2008.pdf (accessed May 5, 2014).

Greenpeace. 2010. *Papua New Guinea: Not ready for [REDD]*. Sydney, Australia: Greenpeace Australia Pacific. http://www.greenpeace.org/ (accessed April 30, 2014).

Gridneff, Ilya. 2009a. Carbon conmen selling the sky. *Sydney Morning Herald*, June 13.

Gridneff, Ilya. 2009b. PNG's PM nephew "pushing carbon deals". *Age*, July 3.

Hooper, S., C. Kaluwin, J. Duguman, G. Gowae, A. Asmann, E. Kwa, O. Gideon, and S. Saulei. 2011. *Project design document: April Salumei, East Sepik*. Papua New Guinea: Climate Community and Biodiversity Standards, prepared by Rainforest Project Management.

Howes, Stephan. 2009. Cheap but not easy: The reduction of greenhouse gas emissions from deforestation and forest degradation in Papua New Guinea. *Pacific Economic Bulletin* 24, 1: 13.

Hunt, Colin A., Damien Ase, David S. Cassells, Colin Filer, Peter Hitchcock, Frances Hurahura, Kenn Mondiai, Thomas Paka, Quentin Reilly, Ross Sinclair, and Simon

Saulei. 2011. Asking the question: Is Papua New Guinea REDD-ready? http://www.colinhunt.com.au/index.php/download_file/view/59/139/ (accessed April 30, 2014).

Jebens, Holger. 2004. Talking about cargo cults in Koimumu (West New Britain Province, Papua New Guinea). In *Cargo, cult, and culture critique*, Holger Jebens ed., 1–14. Honolulu: University of Hawaii Press.

Jorgensen, Dan. 2007. Clan-finding, clan-making and the politics of identity in a Papua New Guinea mining project. In *Customary land tenure and registration in Australia and Papua New Guinea: Anthropological perspectives*, James Wiener and Katie Glaskin, eds. Asia-Pacific Environment Monograph 3. Canberra: ANU ePress. http://www.google.com/url?sa=t&rct=j&q=&esrc=s&source=web&cd=1&ved=0CCgQFjAA&url=http%3A%2F%2Fwww.oapen.org%2Fdownload%3Ftype%3Ddocument%26docid%3D458933&ei=La1nU-yHHu2n8QGxn4HwDA&usg=AFQjCNFkJfsvfGHRkTcBQLvYRzdsWF-70g&sig2=fF0xQlkNPSZPVxt4_xYKbQ&bvm=bv.65788261,d.b2U/ (accessed May 4, 2014).

Laepa, Caldron. 2011. Carbon Trade Deal Terminated. *PNG Post-Courier*, February 7.

Lang, Chris. 2009a. PNG update: Yasause suspended, dodgy carbon credits and carbon ripoffs. REDD-Monitor, July 2. http://www.redd-monitor.org/2009/07/02/png-update-yasause-suspended-dodgy-carbon-credits-and-carbon-ripoffs/ (accessed May 5, 2014).

Lang, Chris. 2009b. Kevin Conrad on REDD, irregularities and carbon cowboys in PNG. REDD-Monitor, July 9. http://www.redd-monitor.org/2009/07/09/kevin-conrad-on-redd-irregularities-and-carbon-cowboys-in-png/ (accessed May 5, 2014).

Lang, Chris. 2009c. More questions than answers on carbon trading in PNG. REDD-Monitor, September 11. http://www.redd-monitor.org/2009/09/11/more-questions-than-answers-on-carbon-trading-in-png/ (accessed May 5, 2014).

Lang, Chris 2009d. Anatomy of a deal: The April Salome REDD project in Papua New Guinea. REDD-Monitor, June 7, 2009. http://www.redd-monitor.org/2009/06/07/anatomy-of-a-deal-the-april-salome-redd-project-in-papua-new-guinea/#more-2159 (accessed September 23, 2014).

Lang, Chris. 2010. McKinsey's REDD plans in Papua New Guinea: Nice work if you can get it. REDD-Monitor, October 7, 2010. http://www.redd-monitor.org/2010/10/07/mckinseys-redd-plans-in-papua-new-guinea-nice-work-if-you-can-get-it/ (accessed May 4, 2014).

Lattas, Andrew. 2011. Logging, violence and pleasure: Neoliberalism, civil society and corporate governance in West New Britain. *Oceania* 81, 1: 88–107.

Laurance, William F., Titus Kakul, Rodney J. Keenan, Jeffery Sayer, Simon Passingan, Gopalassamy R. Clements, Felipe Villegas, and Navjot S. Sodhi. 2011. Predatory corporations, failing governance, and the fate of forests in Papua New Guinea. *Conservation Letters* 4: 95–100.

Leggett, Matt. 2012. The status of REDD+ in PNG. USAID/ASIA. http://www.leafasia.org/library/status-redd-papua-new-guinea/ (accessed May 5, 2014).

Leggett, Matthew and Heather Lovell. 2011. Community perceptions of REDD+: A case study from Papua New Guinea. *Climate Policy* 12, 1: 115–134.

Lipset, David. 1997. *Mangrove man: Dialogics of culture in the Sepik estuary*. Cambridge: Cambridge University Press.

Lipset, David. 2014. Place in the Anthropocene: A mangrove lagoon in Papua New Guinea in the time of rising sea-levels. *HAU* 4, 3: 215–243.

May, Ronald James, ed. 2009. Policy making and implementation: Studies from Papua New Guinea. In *Studies in State and Society in the Pacific* 5. Canberra, Australia: State, Society and Governance in Melanesia Program in association with the National Research Institute, Papua New Guinea.

Melick, David. 2010. Credibility of REDD and experiences from Papua New Guinea. *Conservation Biology* 24, 2: 359–361.

Mumby, Peter J., Alasdair J. Edwards, Ernesto Arias-Gonzalez, Kenyon C. Lindeman, Paul G. Blackwell, Angela Gall, Malgosia I. Gorczynska, Alastair R. Harborne, Claire L. Pescod, Henk Renken, Colette C. C. Wabnitz, and Ghislane Llewellyn. 2004. Mangroves enhance the biomass of coral reef fish communities in the Caribbean. *Nature* 427: 533–536.

Nellemann, Christian, Emily Corcoran, Carlos M. Duarte, Luis Valdes, Cassandra De Young, Luciano Fonseca, and Gabriel Grimsditch, eds. 2009. Blue carbon: A rapid response assessment. United Nations Environment Programme, GRID-Arendal. http://www.grida.no/publications/rr/blue-carbon/ (accessed April 29, 2014).

Oestreicher, Jordan S., Karina Benessaiah, Maria C. Ruiz-Jaen, Sean Sloan, Kate Turner, Johanne Pelletier, Bruno Guay, Kathryn E. Clark, Dominique G. Roche, Manfred Meiners, and Catherine Potvin. 2009. Avoiding deforestation in Panamanian protected areas: An analysis of protection effectiveness and implications for reducing emissions from deforestation and forest degradation. *Global Environmental Change* 19, 2: 279–291.

Office of Climate Change and Environmental Sustainability (OCCES). 2009. http://climatepng.org/index/php?option=com_content&view=article&id=4:0cces-clarifies-mediaissues&catid=6:press-releases/ (accessed May 4, 2014).

Overseas Development Institute (ODI). 2007. *Issues and opportunities for the forest sector in Papua New Guinea*. London: Overseas Development Institute.

Paka, Thomas and Justin Ondopa. n.d. The need for avenues for redress and complaints for indigenous peoples and local communities under REDD+—the PNG case. Ecoforestry Forum. https://www.google.com/url?sa=t&rct=j&q=&esrc=s&source=web&cd=1&ved=0CCgQFjAA&url=https%3A%2F%2Fseors.unfccc.int%2Fseors%2Fattachments%2Fget_attachment%3Fcode%3D5HFFZO8W32QYEJ2G4NVKLZ99V4FGBJ1H&ei=9DZqU7mrIOfH8AHjkYDQDw&usg=AFQjCNGOpOPWnsdk—iNplVXEsloGGMsDw&sig2=P7kHRRLC_TG_YLvb-Gj08Q&bvm=bv.66111022,d.b2U (accessed May 7, 2014).

Schwartz, Theodore. 1962. The Paliau movement in the Admiralty Islands, 1946–1954. *Anthropological Papers of the American Museum of Natural History* 49, 2.

Shearman, Phil, Julian Ash, Brendan Mackey, Jane Bryan, and Barbara Lokes. 2009. Forest conversion and degradation in Papua New Guinea, 1972–2002. *Biotropica* 41, 3: 379–390.

Shearman, Phil, Jane Bryan, Julian Ash, Peter Hunnam, Brendan Mackey, and Barbara Lokes. 2008. *The state of the forests of Papua New Guinea. Mapping the extent and condition of forest cover and measuring the drivers of forest change in the period 1972–2002*. Port Moresby, PNG: University of Papua New Guinea.

Sillitoe, Paul. 2000. *Social change in Melanesia: Development and history*. Cambridge: Cambridge University Press.

Somare, Michael. 2005. Statement by Sir Michael T. Somare at the global roundtable on climate change. Columbia University, New York. http://www.rainforestcoalition.org/SirMichaelSomareGROCCSpeech-FINAL.pdf (accessed March 8, 2012).

Somare, Michael. 2009. Parliamentary statement: Carbon trading and office of climate change and environmental sustainability. *National*, August 5.
Strathern, Marilyn. 1988. *The gender of the gift: Problems with women and problems with society in Melanesia*. Berkeley: University of California Press.
Tumu Timbers Development Limited (TTD). 2010. *Kamula Doso improved forest management carbon project*. Arlington, VA: Draft Report to Climate, Community & Biodiversity Alliance.
Weiner, James and Katie Gaskin. 2007. *Customary land tenure and registration in Australia and Papua New Guinea: Anthropological perspectives*, James F. Weiner and Katie Glaskin, eds. Asia-Pacific Environment Monograph 3. Canberra: ANU ePress. http://www.google.com/url?sa=t&rct=j&q=&esrc=s&source=web&cd=1&ved=0CCgQFjAA&url=http%3A%2F%2Fwww.oapen.org%2Fdownload%3Ftype%3Ddocument%26docid%3D458933&ei=La1nU-yHHu2n8QGxn4HwDA&usg=AFQjCNFkJfsvfGHRkTcBQLvYRzdsWF-70g&sig2=fF0xQlkNPSZPVxt4_xYKbQ&bvm=bv.65788261,d.b2U/ (accessed May 4, 2014).
Wells, Sue, Corinna Ravilous, and Emily Corcoran. 2006. *In the front line: Shoreline protection and other ecosystem services from Mangroves and Coral Reefs*. Cambridge: United Nations Environment Programme World Conservation Monitoring Centre.
White, Andy and Alejandra Martin. 2002. *Who owns the world's forests? Forest tenure and public forests in transition*. Washington, DC: Center for International Environmental Law.
World Bank. 2006. *Strengthening forest law enforcement and governance: Addressing a systemic constraint to sustainable development*. Washington, DC: Environment and Agriculture and Rural Development Departments.
Wynn, Gerard and Sunanda Creagh. 2009. Forest carbon market already shows cracks. *Reuters*, June 4.

10

INTERROGATING PUBLIC DEBATES OVER JURISDICTIONAL REDD+ IN CALIFORNIA'S GLOBAL WARMING SOLUTIONS ACT

Implications for Social Equity

Libby Blanchard and Bhaskar Vira

Introduction

As no binding global commitment has yet been developed to respond to the threat of climate change once the Kyoto Protocol expires, state governments have begun to develop regional, transnational, and subnational climate change mitigation strategies.[1] One such subnational strategy is California's Global Warming Solutions Act (AB 32), passed in 2006. This was the first enforceable statewide legislation in the United States to cap and reduce GHG emissions from major industries. It requires California to reduce its level of GHG emissions to 1990 levels by 2020, and 80 percent reduction from 1990 levels by 2050. California has since implemented a number of regulations and a cap-and-trade system to achieve this goal.

California is also at the vanguard of linking this initiative with other subnational climate change mitigation strategies. On January 1, 2014, California formally linked its cap-and-trade program with that of Quebec, Canada, to deliver regional emission reductions between the two states at lower costs by creating a larger carbon market; they will soon link with Ontario.[2] In addition, California is considering linking "jurisdictional REDD+" forest carbon credits into its program with Chiapas, Mexico, and Acre, Brazil. Jurisdictional REDD+ would allow capped entities in California to buy forest carbon offset credits from projects nested in, monitored and verified in aggregate GHG emissions reductions at a jurisdictional (state- or province-wide) level (Nepstad et al. 2013), in partnerships that are disparate from national and international climate agreements. If California does ultimately incorporate these novel jurisdictional REDD+ carbon offset credits into its program, it could be the first official compliance market to use REDD+ credits. At present, it is the only jurisdiction in the world that is formally considering REDD+ offsets as part of its GHG compliance system.[3]

While California's Air Resources Board (ARB) and governor were expected by some to make a decision on jurisdictional REDD+ by the end of 2013, as of this writing in November 2015, jurisdictional REDD+ has not yet been approved for California's cap-and-trade system. When the first update to the AB 32 Scoping Plan was published by ARB in May 2014, it stated only that it will "continue to consider international sector-based offsets from programs designed to Reduce Emissions from Deforestation and Forest Degradation (REDD)"; it did not cite a specific timeline for a future rulemaking (California.Carbon.info 2014, 88). This delay could be attributed to the vibrant public debate in California over the potential inclusion of these offsets in the state's cap-and-trade initiative.

This chapter examines the contours of this debate, with specific focus on the social impact narratives developed by actors participating in the debate. It then considers the extent to which groups who will be impacted by the proposed policy have voice or are represented within the debate, and what legitimacy this representation holds. Finally, the chapter considers the implications of participation and representation, including how the privilege of participation or voice frames the public discourse on particular social and environmental impacts of jurisdictional REDD+. Ensuring equity across multiple actors and scales of implementation has driven debates over REDD+ at international and national levels (Peskett et al. 2008). Insights from this subnational debate may be useful for understanding other scales of climate governance at which REDD+ is being proposed and debated, potentially contributing to more effective, equitable, and inclusive climate mitigation strategies.

Methodology

We collected documentary evidence of the debate over California's proposed use of jurisdictional REDD+ in the public sphere, including open letters, opinion articles, blogs, reports, web pages, and online videos created by different actors representing various positions within this debate between November 2010 and December 2013 via a systematic review, following a methodology outlined in *Guidelines for Systematic Review in Conservation and Environmental Management—Version 3.1* (CEBC 2009). These parameters represent the period of time beginning when California developed and signed their Memorandum of Understanding with Chiapas, Mexico, and Acre, Brazil, in November 2010, and the end of 2013, when ARB was expected to make a decision on whether or not to incorporate REDD credits into California's cap-and-trade program.

This documentary evidence was then examined with a focus on the narratives developed by actors of the perceived social impact of jurisdictional REDD+, using environmental narrative analysis. An environmental narrative is defined as "a simplified explanation of cause and effect relationships that assigns roles to different actors who are implicated (or not) in an environmental problem" (Beymer-Farris and Bassett 2012, 334). Such narratives are stories, with premises and conclusions, revolving "around a sequence of events or positions in which something happens or from which something follows" (Roe 1991, 288). Roe (1991) argues that such narratives "have the objective of getting their hearers to believe or do something" (p. 288). Further, the knowledge production and problem framing behind environmental narratives is dependent on the privilege of participation (Forsyth and Walker 2008; Scoones 2009). Environmental narrative analysis is helpful for recognizing and making explicit the politics of environmental knowledge (Boyd 2009; Forsyth and Walker 2008). Thus, environmental narratives, and the actors who create them, influence which challenges or problems are framed and what policies and responses are prioritized (Forsyth 2003; O'Brien et al. 2007).

Social Impact Narratives of Jurisdictional REDD+

As with all REDD+ debates, this debate exists, in part, because REDD+ as a concept is arguably a boundary object—a concept that is "plastic, open to interpretation by different actors and valuable to each for different reasons" (McDermott et al. 2012, 64). This is especially the case in regards to jurisdictional REDD+, as it is a new and novel REDD+ design still at its conceptual stage: its operational mechanisms for benefit sharing with stakeholder groups have not been fully designed and stakeholders do not yet know what the precise terms of their engagement might be.

The proposed use of jurisdictional REDD+ in California resulted in the emergence of a broad array of actors who have deployed narratives on the social impacts of such a strategy (See Table 10.1). This group included individuals acting in their personal capacity; biodiversity conservation organizations; social and environmental justice organizations; certification bodies; special interest groups and cause coalitions; and private foundations. We collected 71 public documents from actors engaged in this debate: 25 in support of California's proposal for jurisdictional REDD+ and 46 against it. The majority of the actors engaged in this debate were based in California or were from the global North.

These actors created new and novel actor coalitions. One large actor coalition consisted of those who signed Code REDD's (a carbon offsets advocacy group based in California) open letter of support for jurisdictional REDD+. This group included conservation organizations such as the Wildlife Conservation Society and for-profit companies, including the Walt Disney Company and the Pacific Gas and Electric Company.[4] This trend of biodiversity conservation organizations partnering with the private sector used to be unheard of (MacDonald 2010); and, indeed, the relationship was frequently far more adversarial, but is now becoming more common along with the emergence of narratives about the green economy and the associated "neoliberalization" of conservation (Büscher et al. 2012; MacDonald 2010; Pirard 2012).

Actors and coalitions have promoted various narratives about the impact of jurisdictional REDD+ on particular groups of people. Those who have argued in favor of California's proposed use of jurisdictional REDD+ (interestingly, 24 of 25 documents that argued in favor were created by actors or organizations based in the global North) see the strategy as an economically efficient way for the state to reach its emissions reduction goals, thus benefiting the state's businesses and citizens. Such actors embody a market-liberal worldview toward REDD+ that has been seen in other REDD+ debates (Hiraldo and Tanner 2011) and often employ win-win narratives. For example, The Nature Conservancy, in a press release from March 26, 2013, after a workshop on safeguard policies for jurisdictional REDD+, applauded government officials, indigenous leaders, and international policy experts for their work to "tackle climate change, promote sustainable development in local communities and conserve tropical forests."[5] Likewise, the Code REDD letter of support argues that REDD+ would provide both forest conservation and economic growth. They write that "REDD+ is a proven, cost-effective tool" that will result in "further growth in sustainable industries both domestically and abroad."[6] In a July 22, 2013, opinion piece in *Forbes Magazine* by David Rothschild of the Skoll Foundation, and Karin Burns of Code REDD, there is a similar claim that REDD+ is "a viable climate solution . . . empowering local communities, protecting wildlife and slowing deforestation."[7] Other actors that employed win-win narratives included the Environmental Defense Fund (Keohane 2013) and the Forests 4 Climate Network.[8]

On the other side of the debate, actors and coalitions have argued against the use of jurisdictional REDD+ on the basis of perceived adverse social (and environmental) consequences that may result from its implementation. Many of their social equity arguments echo those in national and international discourses on climate change mitigation strategies (e.g., Agarwal and Narain 1991), including arguments for inter- and intra-generational equity and the ethical importance of the global North in taking historical responsibility for emissions reductions (e.g., Oilwatch International's letter on California's proposed REDD+[9]). These actors also discuss the potential negative social impacts of REDD+ on people in the developing world, including arguments that REDD+ perpetuates preexisting inequalities between the global North and South (Picq 2013) and could result in land grabs (e.g., Friends of the Earth[10] and many others[11]) and human rights abuses.[12]

Further, critics of jurisdictional REDD+ in California discuss concerns over the legal limits of social safeguards that have been developed in any interstate arrangement, which "ultimately rest in the domain of domestic law" (Osborne 2013). Dr. Tracey Osborne writes—as part of a series of essays produced by the University of Arizona's Public Political Ecology Lab focusing on REDD+ in California's cap-and-trade scheme—that California has no legal ability to enforce how REDD+ is implemented in partner jurisdictions and whether safeguards are enforced. This is one of the major challenges with implementing and enforcing REDD+ or affiliated social and environmental safeguards across any international border.

One important and widely overlooked narrative in REDD+ debates generally that takes precedent in the California REDD+ debate is the focus on negative social impacts of REDD+ to Californians themselves, especially those who live or work near large California emissions sources. Alan Ramo and Janet Redman write in a *Los Angeles Times* op-ed that international offsets force California residents, especially those in proximity to point source emitters, to "endure the excess pollution of the company buying the credit," thus suffering increased exposure to co-pollutants and greater health and environmental risks.[13] This concern is echoed by Oilwatch International; Activist San Diego in collaboration with over 30 like-minded actors in a July 10, 2012, open letter to California's Governor Brown;[14] a coalition of environmental conservation and social justice groups, including Amazon Watch, Asian Pacific Environmental Network and others in a May 3, 2013, open letter; and by Friends of the Earth. Friends of the Earth write:

> California's use of carbon trading means that as individual plants trade away their obligations to reduce emissions, communities already burdened with dangerously high concentrations of carbon co-pollutant from power plants, refineries, and other industrial sources would likely see few benefits of reduced toxic smog and particulate matter that would accompany local greenhouse gas emissions reductions.[15]

Further, Oilwatch International points out that those in California who live closest to the state's largest emitters are disproportionately low-income communities and people of color, and that their being exposed to increased co-pollutants from such industries buying carbon offsets is an act of "environmental racism."[16] This focus on the negative impacts of offsets on people in the developed world (in this case, California) by actors and actor coalitions in this debate is a new but important narrative in REDD+ debates, bringing up important additional social issues that are important to consider with any REDD+ (or offset) strategy at any scale.

Participation and Representation within the Debate

Those who have the privilege to shape and deploy environmental narratives in this debate play a powerful role in promoting perceived cause and effect relationships between jurisdictional REDD+ and its social impact in the public sphere, with resultant implications for the policy's popularity and even its potential adoption. Therefore, it is important to consider not only which actors have had the privilege of participating in the debate, but also which actors, some of whom will likely be impacted by jurisdictional REDD+, are absent from the debate. Finally, it is important to consider how actors who are absent or do not have direct voice have been represented in the debate (if at all), and the legitimacy of such representation.

TABLE 10.1 Actors and organizations' positions in the REDD+ debate in California

Actors and Organizations' Positions in the REDD+ Debate in California

Due to space limitations, this table summarizes the narratives and many of the primary actors in the California debate. A representation of the full complexity of coalitions, politicians, and actors outside of the state of California is described in the text.

Actors Publicly for the Use of REDD+ in California	Actors Publicly against the Use of REDD+ in California
Narratives: Generally market-liberal, with win-win narratives about the impact of REDD+ on economy, environment, and social equity.	**Narratives:** Generally focused on inter- and intra-generational equity, with narratives about historical responsibility and/or that REDD+ may perpetuate preexisting social inequality.
Individuals (with organizational affiliations)	
Individuals, speaking in their individual capacities, but associated with organizations Code REDD, Skoll Foundation, and The Nature Conservancy.	Individual researchers and faculty from various universities, individuals associated with the Institute for Policy Studies, and Friends of the Earth.
Biodiversity Conservation Organizations	
The Nature Conservancy; and the Environmental Defense Fund.	Friends of the Earth and Greenpeace.
Social and Environmental Justice Organizations	
	Oilwatch International, Activist San Diego, Amazon Watch, Asian Pacific Environmental Network, California Environmental Justice Alliance, No REDD in Africa Network, Indigenous Environmental Network, Communities for a Better Environment, Global Justice Ecology Project.
Certification Bodies, Special Interest Groups and Cause Coalitions and Private Foundations	
Code REDD, Forests 4 Climate Network, Green Technology Leadership Group, Earth Innovation Institute, Skoll Foundation, Verified Carbon Standard, and Wildlife Works.	Americans Against Offsets, Global Alliance of Indigenous Peoples, and Local Communities on Climate Change against REDD and for Life.
Businesses	
The Walt Disney Company and Pacific Gas and Electric Company.	

Individuals from highly vulnerable or marginalized groups, such as the poorest of the poor (including landless people and migrants in Chiapas and Acre), who likely have the most to gain or lose from REDD+ (Strassburg et al. 2012), have not had direct voice and are largely absent from this debate. Even organized landholders and civil society groups from Acre and Chiapas have had limited involvement in the debate. For example, there were no documents or direct involvement in the debate from the Chiapas-based civil society coalition REDDeldía (translated as "REDD-ellion," as in "rebellion"), even though their views against jurisdictional REDD+ were represented in the *Yes!* magazine article, "Should Chiapas Farmers Suffer for California's Carbon?"[17] There are many impacted stakeholders who are absent from the public conversation simply because they are not aware of the proposed jurisdictional REDD+ initiative or the development of policy recommendations regarding it, while others believe they are being actively excluded. Cecilia Simon, a climate change consultant from Mexico, wrote in her draft recommendations to the REDD Offsets Working Group (ROW)—an ad hoc technical expert working group that developed technical and policy recommendations for how to implement jurisdictional REDD+ that were presented to ARB—that

> I think it will be wise to involve more stakeholders in the next steps of the process, especially from other jurisdictions other than California. In Mexico the process is not well known and I think it will benefit from other points of view.
>
> *(Simon 2013, p. 1)*

This concern was echoed by a coalition of environmental justice organizations from Brazil, who stated in an open letter to Governor Brown:

> In addition to our opposition of this proposal, we also challenge the legitimacy of the "consultation" process underway in Chiapas with regard to this matter, due to the lack of effective participation by the communities in Acre and Chiapas who depend on the forests to maintain their way of life and will be directly affected by this REDD+ proposal.[18]

Representation, like voice, is mediated by power relations, where the power being exercised is that of the privileged actor taking part in the debate, not that of the represented groups. This reality lends itself to the question of what legitimacy representation has when it does occur. While representation can be beneficial by creating some visibility for specific marginalized groups who do not have direct voice, it could also simplify, limit, or inaccurately convey the concerns of those being represented. Thus, the engagement of a particular actor, or representation of a particular stakeholder within a narrative (e.g., an indigenous leader or an indigenous group), may be taken (incorrectly) to imply that *all* indigenous groups or leaders take the same stance toward the perceived social impact narrative presented. For example, when The Nature Conservancy, the Skoll Foundation, or Code REDD contend that REDD+ benefits "local communities,"[19] or when the Environmental Defense Fund writes that jurisdictional REDD+ benefits "indigenous peoples,"[20] this is suggestive, though not accurately so, of the representation of all people within such broad categories and conceals the heterogeneity of different impacted actors and stakeholders within such groups. These narrative oversimplifications belie the growing body of research that shows that REDD+ could benefit, or harm, particular groups of people even within populations—such as the poorest farmers and women—depending on the particular context in which project activities arise (Corbera et al. 2007; McDermott and Schreckenberg 2009). Representation does not adequately replace a direct voice in the public sphere, and is not a fully equitable solution for those impacted by jurisdictional REDD+.

To a similar end, focusing on particular groups may also conceal the impact of jurisdictional REDD+ on other impacted groups. When a narrative focuses on a specific group, for example on those in a different geographic location, such as Chiapas or Acre, this may draw the focus of the debate away from other marginalized groups, for example, the lower income people in California who live near California's largest polluting entities who also may be impacted by the strategy. This has been seen in virtually all REDD+ debates until the emergence of this particular debate in California, where concerns of the poor in the developed world are now being raised as an important, additional social equity issue.

The privilege of voice and representation to shape and perpetuate particular narratives, has and will continue to have significant implications on California's cap-and-trade program and whether the state will move forward with a jurisdictional REDD+ initiative. Due to the emerging narratives about the negative impacts of REDD+ and other offsets on low-income and disadvantaged communities in California, State Senator Ricardo Lara introduced SB605 in February 2013, a bill that targeted emissions reductions and intended to prohibit the use of out-of-state offsets. The original reasoning for this bill was for AB32 to focus on emissions reductions and to capture the economic benefit of any offsets within the state.[21] While clean air, public health, and community advocacy groups such as the California Environmental Justice Alliance and Physicians for Social Responsibility Los Angeles supported the bill, the Environmental Defense Fund[22] and The Nature Conservancy only supported the bill if amended[23] to allow for out-of-state and international offsets.[24] Given opposition and political tensions, the bill did not pass in its original form and was eventually transformed to direct ARB to create and implement a strategy to reduce emissions of short-lived climate pollutants, defined as "an agent that has a relatively short lifetime in the atmosphere, from a few days to a few decades, and a warming influence on the climate that is more potent than that of carbon dioxide,"[25] with no mention of offsets. Thus, powerful actors who were able to shape social, environmental, and economic narratives over the importance of offsets successfully affected (and will continue to affect) this particular bill and potentially the ultimate design of California's cap-and-trade program.

Other bills addressing social equity and climate change emissions in California have been more successful. Due to concerns that low-income and disadvantaged communities in California face health disparities due to poor air quality in their communities, SB535 was introduced by State Senator De León in September 2012. This bill calls for 25 percent of the available monies in the Greenhouse Gas Reduction Fund—a fund from the auction of sale of allowances as part of the market-based compliance mechanism—to go to "projects that provide benefit to disadvantaged communities" and 10 percent to "projects located within disadvantaged communities."[26] The governor approved this bill in late September 2012.

The public debate in California over jurisdictional REDD+ contributed to the outcome of these two bills. Further, many individuals on both sides of the debate believe that the public debate itself has caused ARB to take a cautious approach with moving forward with approving jurisdictional REDD+, due to perceived social impacts both in the developed and developing world.

Privileged actors frame this public debate, the narratives about social equity within it, and the understandings within the public sphere of the perceived social outcomes of jurisdictional REDD+.

This privilege of voice and representation perpetuates particular impact narratives and homogenizes certain groups while concealing others. This point should be taken into account when considering the contours of the debate over jurisdictional REDD+, or any other debate over proposed environmental governance strategies, especially those that are evolving in design. We argue that being aware of such privilege in the shaping of environmental narratives is important for understanding the dynamics of participation and representation, as well as for the types of narratives that are brought to the forefront of any debate. Similarly, it is necessary for understanding the reasons that certain voices take precedence in such debates.

Conclusion

If implemented, jurisdictional REDD+ will have profound implications for the social equity, rights, and welfare of people in California, Chiapas, and Acre, with the greatest impacts on vulnerable or marginalized groups, including both those who live near California point source emitters and the poorest of the poor in the developing world, whose culture and livelihoods depend most on the local forests in question and their resources (Fearnside 2003).

We analyzed this public debate, using environmental narrative analysis, to understand which social concerns are prevalent; which actors are involved; the extent to which actors who will likely be impacted by jurisdictional REDD+ have had voice or are represented within the debate; and the legitimacy that such representation holds.

We found that an expanding array of actors have become involved in this particular jurisdictional REDD+ debate, communicating an evolving set of social equity considerations into the climate change mitigation negotiating arena, including promoting new narratives in the REDD+ debate, such as the impacts on the use of offsets for people in the developed world in addition to people in the developing world. We also found that the power of certain actors to shape narratives over perceived social impacts affected (and will continue to affect) the strategies' ultimate design; public opinion and therefore the political climate of the state in relation to jurisdictional REDD+; and the willingness of the state to move forward with such a strategy. All of this will continue to have consequences for vulnerable actors who will likely be impacted by jurisdictional REDD+, but who do not have a direct voice in the debate. We argue that this is due to the politicized nature of the public sphere and the unequal power politics inherent in a representative democracy. Thus, actors with the privilege of voice shape narratives and public discourses over the impact of jurisdictional REDD+ on social equity, bringing certain social equity concerns to the forefront of public consciousness while obfuscating others. This has and will continue to result in particular climate mitigation strategies and associated outcomes for specific groups, including those who are most vulnerable (and often voiceless).

There is an increasing trend toward the use of flexibility mechanisms such as REDD+ in climate change mitigation strategies at different governance scales. The state of California offers an unusual subnational context within which to understand debates over REDD+ for climate change mitigation policies. Given that California is the only jurisdiction in the world that is currently considering REDD+ offsets as part of its climate change mitigation strategy, insights from this study will be useful to understand social equity implications of proposed REDD+ at other scales and contexts, and inform other contexts in which REDD+ is being proposed and used, potentially making them more effective and equitable. They may also help to shed light on particular impacted groups in those contexts, whose voices may or may not be heard or accurately represented within the environmental narratives being deployed.

In particular, the analysis highlights the highly unequal landscape within which public policy debates take place, and the potential for democratic processes to lack effective mechanisms for the representation of the interests of those who have the most at stake in relation to climate policy. Acknowledging this inequity is a first step toward implementing more effective measures for a fairer representation of a diversity of voices in debating and implementing more inclusive environmental policies, especially when these might have consequences for "distant others" and unborn generations.

Acknowledgements

The authors would like to thank Dr. Shirley Fiske and Dr. Stephanie Paladino for their insightful comments that have improved this chapter. We also acknowledge the helpful comments of anonymous reviewers. The usual disclaimers apply.

Notes

1. In the United States, this includes (among others) economy-wide GHG reduction legislation in Hawaii, Minnesota, Rhode Island, and Massachusetts. It also includes task forces and collaborations such as the Governors' Climate and Forests Task Force and the Western Climate Initiative.
2. California Environmental Protection Agency Air Resources Board. News Release. "California and Quebec Sign Agreement to Integrate, Harmonize their Cap-and-Trade Programs." http://www.arb.ca.gov/newsrel/newsrelease.php?id=508 (accessed July 10, 2014).
3. Governors' Climate & Forests Task Force. http://www.gcftaskforce.org/about/gcf_overview/ (accessed July 10, 2014).
4. http://www.coderedd.org/letter-of-support/#.U794DKggbEV (accessed July 10, 2014).
5. http://www.nature.org/ourinitiatives/regions/northamerica/unitedstates/california/newsroom/renowned-indigenous-leaders-environmental-experts-unite-to-address-global-ef.xml (accessed June 29, 2014).
6. http://www.coderedd.org/letter-of-support/#.U794DKggbEV (accessed July 10, 2014).
7. http://www.forbes.com/sites/skollworldforum/2013/07/22/california-again-leads-the-way-this-time-with-forest-carbon-offsets/ (accessed June 29, 2014).
8. http://forests4climate.org/forests-as-climate-solution/ (accessed January 10, 2015).
9. http://www.oilwatch.org/en/documents/declaraciones-en/579-california-don-t-let-shell-roast-the-planet (accessed January 10, 2015).
10. http://libcloud.s3.amazonaws.com/93/7f/a/834/Factsheet_Risks_of_REDD_in_Californias_cap_and_trade.pdf (accessed January 12, 2015).
11. http://www.theafricareport.com/Columns/big-red-card-for-california-redd.html (accessed January 10, 2015); https://intercontinentalcry.org/ten-of-the-worst-redd-type-projects-that-affect-indigenous-peoples-and-local-communities/ (accessed January 10, 2015); http://www.globalinfo.nl/Nieuws/california-and-chiapas-advancing-land-grab-policies.html (accessed January 10, 2015); http://www.bailiffafrica.org/no-redd-in-africa-network-writes-letter-to-california-officials-opposing-the-states-redd-inclusion/ (accessed January 10, 2015).
12. http://www.ienearth.org/international-outcry-against-californias-forest-offset-scam/ (accessed January 10, 2015).
13. http://articles.latimes.com/2013/nov/14/opinion/la-oe-ramo-cap-and-trade-california-20131114 (accessed June 25, 2014).
14. http://libcloud.s3.amazonaws.com/93/ca/b/2271/Letter_to_Governor_and_ARB_re_CA_REDD_final.pdf (accessed June 25, 2014).
15. http://libcloud.s3.amazonaws.com/93/d9/7/637/Issue_Brief_California_Air_Resources_Board_REDD.pdf (accessed June 29, 2014).
16. http://www.oilwatch.org/en/documents/declaraciones-en/579-california-don-t-let-shell-roast-the-planet (accessed January 13, 2015).
17. http://www.yesmagazine.org/issues/what-would-nature-do/should-chiapas-farmers-pay-the-price-of-californias-carbon (accessed January 3, 2015).
18. http://news.mongabay.com/2013/0507-redd-california.html (accessed January 3, 2015).

19. http://www.nature.org/ourinitiatives/urgentissues/global-warming-climate-change/how-we-work/creating-incentives-to-stop-deforestation.xml (accessed January 3, 2015); http://www.forbes.com/sites/skollworldforum/2013/07/22/california-again-leads-the-way-this-time-with-forest-carbon-offsets/ (accessed January 3, 2015).
20. http://www.edf.org/climate/redd (accessed January 3, 2015).
21. http://www.leginfo.ca.gov/pub/13-14/bill/sen/sb_0601-0650/sb_605_cfa_20130813_085646_asm_comm.html (accessed January 3, 2015).
22. http://www.cacurrent.com/storyDisplay.php?sid=7028 (accessed January 3, 2015).
23. http://www.leginfo.ca.gov/pub/13-14/bill/sen/sb_0601-0650/sb_605_cfa_20130813_085646_asm_comm.html (accessed January 3, 2015).
24. Interview with Senator Ricardo Lara's office, October 21, 2014.
25. http://leginfo.legislature.ca.gov/faces/billStatusClient.xhtml (accessed January 10, 2015).
26. http://leginfo.legislature.ca.gov/faces/billNavClient.xhtml?bill_id=201120120SB535 (accessed January 10, 2015).

References

Agarwal, Anil and Sunita Narain. 1991. *Global warming in an unequal world.* New Delhi: Centre for Science and Environment.

Air Resources Board (ARB). 2014. *First update to the climate change scoping plan, 136.* Sacramento: State of California.

Beymer-Farris, Betsy A. and Thomas J. Bassett. 2012. The REDD menace: Resurgent protectionism in Tanzania's mangrove forests. *Global Environmental Change* 22, no. 2: 332–341.

Boyd, Emily. 2009. Governing the clean development mechanism: Global rhetoric versus local realities in carbon sequestration projects. *Environment and Planning A* 41: 2380–2395.

Büscher, Bram, Sian Sullivan, Katja Neves, Jim Igoe and Dan Brockington. 2012. Towards a synthesized critique of neoliberal biodiversity conservation. *Capitalism Nature Socialism* 23, no. 2: 4–30.

Center for Evidence-Based Conservation (CEBC). 2009. *Guidelines for systematic review in conservation and environmental management, version 3.1.* Bangor: Bangor University and Centre for Evidence-Based Conservation.

Corbera, Esteve, Katrina Brown and W. Neil Adger. 2007. The equity and legitimacy of markets for ecosystem services. *Development and Change* 38, no. 4: 587–613.

Fearnside, Philip M. 2003. Conservation policy in brazilian amazonia: Understanding the dilemmas. *World Development* 31, no. 5: 757–779.

Forsyth, Tim. 2003. *Critical political ecology.* London: Routledge.

Forsyth, Tim and Andrew Walker. 2008. *Forest guardians, forest destroyers: The politics of environmental knowledge in northern Thailand.* Seattle: University of Washington Press.

Hiraldo, Rocío and Thomas Tanner. 2011. Forest voices: Competing narratives over REDD+. *IDS Bulletin* 42, no. 3: 42–48.

Keohane, Nathaniel. 2013. *Environmental Defense Fund comments on draft recommendations by the REDD Offset Working Group (ROW).* Washington, DC: Environmental Defense Fund.

Macdonald, Kenneth Iain. 2010. The devil is in the (bio)diversity: Private sector "engagement" and the restructuring of biodiversity conservation. *Antipode* 42, no. 3: 513–550.

Mcdermott, Constance L., Lauren Coad, Ariella Helfgott and Heike Schroeder. 2012. Operationalizing social safeguards in REDD+: Actors, interests and ideas. *Environmental Science & Policy* 21: 63–72.

Mcdermott, Melanie H. and Kate Schreckenberg. 2009. Equity in community forestry: Insights from north and south. *International Forestry Review* 11, no. 2: 157–170.

Nepstad, Daniel, Derik Broekhoff, Greg P. Asner, Ludovino Lopes, Michelle Passero, Peter Riggs, Rosa Maria Vidal, Steve Schwartzmann, Toby Janson-Smith, Tony Brunello and William Boyd. 2013. *California, Acre and Chiapas: Partnering to reduce emissions from tropical deforestation: Recommendations to conserve tropical rainforests, protect local communities and reduce state-wide greenhouse gas emissions.* The REDD Offset Working Group. http://www.arb.ca.gov/cc/capandtrade/sectorbasedoffsets/row-final-recommendations.pdf.

O'Brien, K., S. Eriksen, L. P. Nygaard and A. Schjolden. 2007. Why different interpretations of vulnerability matter in climate change discourses. *Climate Policy* 7: 73–88.

Osborne, Tracey (2013). Beyond Safeguards: A Critique of Carbon Markets for REDD+. Blog post on the University of Arizona Public Political Ecology Lab website. http://ppel.arizona.edu/?p=264 (accessed April 1, 2014).

Peskett, Leo, D. Huberman, E. Bowen-Jones, G. Edwards and J. Brown. 2008. Making REDD work for the poor. A Poverty Enviornment Partnership Report, http://www.odi.org/sites/odi.org.uk/files/odi-assets/publications-opinion-files/3451.pdf (accessed June 28, 2014).

Picq, Manuela. 2013. Will California fall into the REDD trap? *Al Jazeera English*, May.

Pirard, Romain. 2012. Market-based instruments for biodiversity and ecosystem services: A lexicon. *Environmental Science & Policy* 19–20: 59–68.

Roe, Emery M. 1991. Development narratives, or making the best of blueprint development. *World Development* 19, no. 4: 287–300.

Scoones, Ian. 2009. Livelihoods perspectives and rural development. *Journal of Peasant Studies* 36, no. 1: 171–196.

Simon, Cecilia. 2013. *ROW draft recommendations comments*. Toluca, Mexico.

Strassburg, Bernardo B. N., Bhaskar Vira, Sango Mahanty, Stephanie Mansourian and Adrian Martin. 2012. Social and economic considerations relevant to REDD+. In *Understanding Relationships between Biodiversity, Carbon, Forests and People: The Key to Achieving REDD+ Objectives: A Global Assessment Report*, eds Parrotta, J. A., Wildburger, C. and Mansourian, S., 83–114. Vienna: IUFRO World Series.

11

DOING REDD+ WORK IN VIETNAM

Will the New Carbon Focus Bring Equity to Forest Management?

Pamela McElwee

Attention to land-based carbon management has become an urgent global issue in the past 10 years, particularly in the development of "avoided deforestation" policies, referred to as Reduced Emissions from Deforestation and forest Degradation, or REDD+. Pilot programs to prepare countries for REDD+ readiness are now emerging in many different nations, funded by bilateral and multilateral donors, and involving new institutions like the UN-REDD program and the Forest Carbon Partnership Facility (FCPF) of the World Bank (Cerbu et al. 2011). Yet key questions have been raised about how REDD+ will actually work, given that nations themselves will determine much of the on-the-ground activity toward meeting international benchmarks (Corbera and Schroeder 2011). Further, many of the REDD+ readiness projects being implemented focus on different interests reflecting the wide variety of donors supporting such actions.

Given these heterogeneous approaches to REDD+ and the high diversity of countries that plan to participate, it is unclear if REDD+ will actually reduce carbon emissions from deforestation in a cost-effective way, which was the original goal of the policy. Additionally, is it realistic to hope that REDD+ can fundamentally change unsound forest management regimes that have dominated in tropical countries for much of the past 100 years? A final question surrounds the social impacts of REDD+ approaches: can REDD+ do more than just conserve carbon? Many organizations have asserted that REDD+ activities need to be combined with co-benefits, such as biodiversity conservation or sustainable development, and using REDD+ to tackle poverty among forest dwellers has been a commonly proposed approach (Luttrell et al. 2013; Tacconi et al. 2013). In other words, can REDD+ motivate more participatory, livelihood-positive benefits for marginalized forest peoples?

These are ambitious hopes, and I explore in this chapter how realistic they may be by looking at the development of REDD+ in one developing country that has long struggled to reconcile sustainable forest management with the needs of a growing and relatively poor population. By following how REDD+ readiness activities have unfolded in Vietnam over the past 5 years, I ask questions regarding the relative prioritization of non-carbon goals in Vietnam's REDD+ process. In this chapter, I assess three key topics that will need to be addressed with regard to how REDD+ can focus on the needs of forest-using communities. First, I look at whether participatory mechanisms for local involvement in forest management have been included in REDD+ priorities. Second, I examine how the question of livelihoods have been addressed by local policymakers, and how benefits might be used to improve local conditions, especially for the poorest. Finally, I examine how safeguards are being developed to potentially guard against abuses of rights for those participating in or affected by REDD+.

My initial conclusions from this assessment of Vietnam's situation is that existing mechanisms to address participation and livelihoods, as well as the requirement that there be safeguards in place, are currently insufficient to spark much-needed reforms in an intransigent state forest management system. While much global attention has focused on the potential of market mechanisms like REDD+ to endanger local livelihoods through exclusion from resources (e.g., Corbera 2012), such concern may be focused on the wrong elements of REDD+. Indeed, the so-called market aspects of REDD+ are in some ways a red herring, as it is likely that much REDD+ funding will continue to arrive in the form of bilateral and multilateral development aid for the foreseeable future, as is the case currently for Vietnam. Yet even in this type of non-market funding situation, there is thus far insufficient attention to key concerns surrounding participation and livelihoods.

Methods

In this chapter I use fieldwork I have been conducting in Vietnam since 2008 on the emergence of REDD+, especially my participation in a number of stakeholder workshops and meetings on policy among both national and local actors, along with surveys of local households in one province where REDD+ readiness work has been piloted since 2010 (Lam Dong province in the south-central area of the country in an upland tropical forest). I also look at how participation, livelihoods and safeguards have been incorporated in the development of the first two Provincial REDD+ Action Plans (PRAPs), whereby local provinces have taken on the work of determining how they are likely to implement REDD+ (Lam Dong province and Dien Bien province in the northwest of the country; see Figure 11.1). In the following sections of this chapter, I look at how questions surrounding participation, livelihoods and safeguards have been discussed in global REDD+ negotiations, and how these are being addressed in REDD+ readiness projects on the ground in Vietnam.

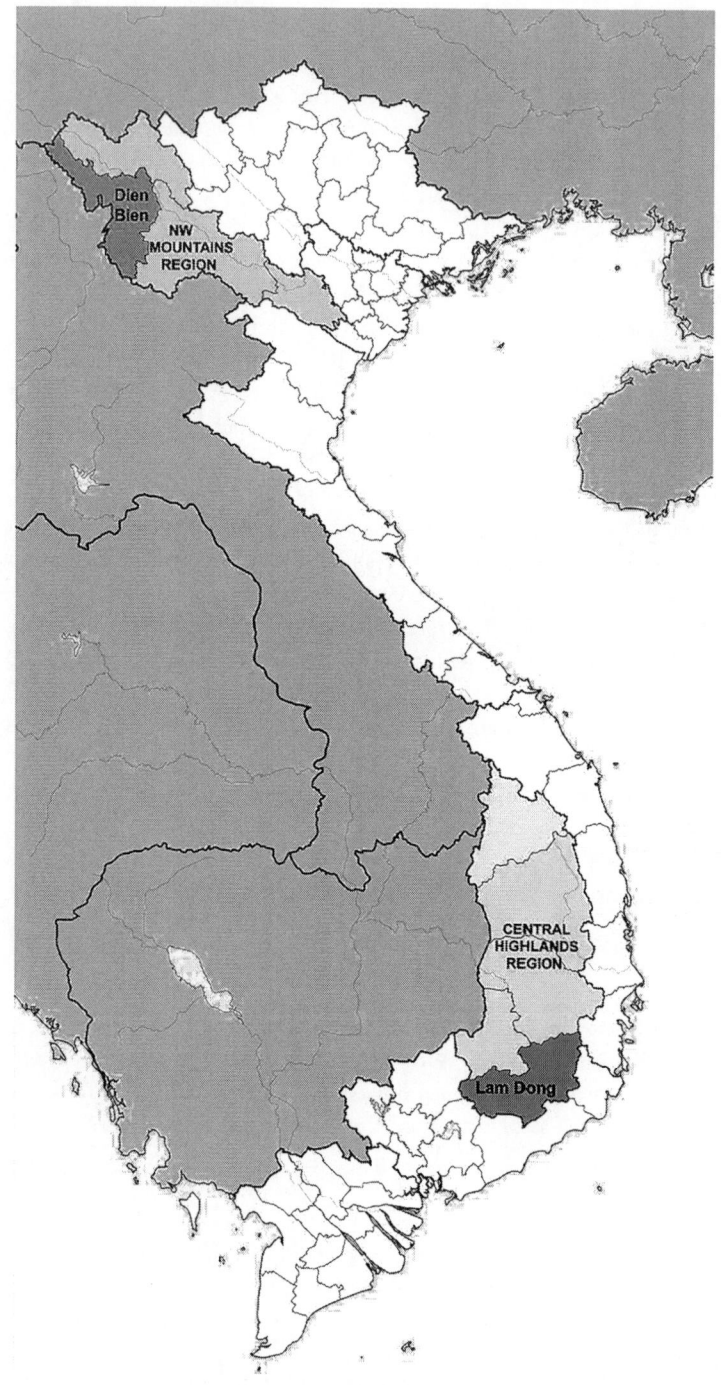

FIGURE 11.1 Map of Vietnam with provinces discussed in chapter highlighted.

Participation and Livelihoods in Global REDD Policies

Formal negotiations over REDD+ have been underway since the 2007 Bali meeting of the Conference of the Parties (COP), where the concept was endorsed for the first time by the signatories of the Framework Convention on Climate Change (UNFCCC) (for a comprehensive review, see Corbera et al. 2010; Agrawal et al. 2011; Gupta et al. 2013). Following the Cancun COP in 2010, the working group on Long Term Cooperative Action agreed to support the development of REDD+ and encouraged countries to begin to contribute to future implementation by taking a number of steps (see Box 11.1). As the Cancun statement indicates, formulating REDD+ and other forest policies in a participatory way, or how REDD+ might facilitate positive impacts on the lives of the poorest forest dwellers, did not receive formal attention. Rather, the idea of safeguards was chosen as a stand-in for these larger questions of equity and benefits from REDD+.

BOX 11.1 THE CANCUN STATEMENT ON REDD

At the Cancun Meetings of the UNFCCC, countries were encouraged to begin to develop:

(a) A national strategy or action plan;
(b) A national forest reference emission level and/or forest reference level or, if appropriate, as an interim measure, subnational forest reference emission levels and/or forest reference levels, in accordance with national circumstances, and with provisions contained in decision 4/CP.15, and with any further elaboration of those provisions adopted by the Conference of the Parties;
(c) A robust and transparent national forest monitoring system for the monitoring and reporting of the activities referred to in paragraph 70 above, with, if appropriate, subnational monitoring and reporting as an interim measure, in accordance with national circumstances, and with the provisions contained in decision 4/CP.15, and with any further elaboration of those provisions agreed by the Conference of the Parties;
(d) A system for providing information on how the safeguards referred to in annex I to this decision are being addressed and respected throughout the implementation of the activities referred to in paragraph 70, while respecting sovereignty

(Paragraph 71, Cancun Agreements, http://cancun.unfccc.int)

In order to begin setting up local action plans, monitoring systems and safeguards, many bilateral and multilateral donors have been funding REDD+ readiness pilot projects since 2009. These include the World Bank's FCPF and the

UN's UN-REDD programs, as well as the Norwegian Development Agency, which has been a large supporter of bilateral REDD+ readiness actions, including pledges of $1 billion to Indonesia, $250 million to Guyana, and $30 million to Vietnam, among other countries. In addition, some voluntary carbon accounting projects involving the private sector have also begun to operate in anticipation of REDD+ financing in the future.

However, questions of good governance, particularly in the form of formal arrangements for participation in the development of REDD+ policies, have not been well addressed in most country readiness plans for REDD+, according to early analysis. Despite the fact that many donors, such as UN-REDD, have called for clear systems of information access and local participation (UN-REDD 2013), reports to date have indicated that participation has generally been weak in pilot activities, with many communities only consulted, rather than being involved in a systematic manner in all aspects of REDD+ planning (Hall 2012; Brown 2013). Procedural equity, in which affected communities are instrumental in the development of natural resources programs, has long been an elusive goal for many governments in the global South (McDermott et al. 2013). To date, there is no clear UNFCCC guidance on how local participation or equity should be fostered or promoted through REDD+, leaving this question to individual projects and county programs to tackle (Krause et al. 2013; White 2013; Sunderlin et al. 2014). As a result, many national-level REDD+ readiness projects have primarily proceeded in a top-down fashion, and have focused mostly on technical issues, such as carbon monitoring, paying little attention to structural changes that may be needed in forest sectors. A 2011 report noted that in a review of forest governance in Indonesia, Ghana, Mozambique, Tanzania and Vietnam, REDD proposals have been

> over-hasty, formulaic and barely credible plans that could do more harm than good . . . [in the form of] fast-developing national REDD strategies that focus on how to count and monitor carbon rather than how to bring about the major policy and capacity changes needed to be "ready" for REDD. All are based on the idea that with enough money over two to four years, a top-down, government-led process will improve governance and give forest-based practitioners what they need to guarantee emissions reductions and qualify for REDD payments.
>
> *(IIED 2011, p. 5)*

Livelihoods have similarly not been a major topic of discussion at UNFCCC meetings to hammer out REDD+ standards. Livelihoods have largely been equated with discussion of benefit sharing; that is, how to get money to people who undertake forest conserving activities (Lawlor et al. 2010; Luttrell et al. 2013). Many REDD+ proponents have rather simplistically assumed that, all other things being equal, a land use that provides the most money will be the one that the farmer chooses; thus REDD+ discussions have often focused on quite basic models of forest area, carbon prices and opportunity costs (e.g., see Strassburg et al. 2009). But livelihoods are about more than income; they are about how individuals and households manage a portfolio of actions to support household welfare and achievements, including, though not entirely limited to, income stream management. To date, the experience of forest carbon projects on livelihood indicators is mixed; some carbon projects have increased smallholder incomes, diversified livelihoods and built capacity and skills, while other projects have had minimal or negative impacts (Boyd et al. 2007; Caplow et al. 2011; Reynolds 2012; Lawlor et al. 2013). Unfortunately, many REDD+ readiness projects have downplayed these challenges in favor of mostly technical discussions of setting carbon prices and covering opportunity costs of participation, assuming that livelihood gains will follow (Milne 2012).

Indeed, evidence to date indicates that rather than explicit attention to participation and livelihoods, most discussion on the social aspects of REDD+ at the global level have focused on how REDD+ projects will use safeguards to ensure participants' rights are protected and there are no adverse impacts on involved communities and households (Chhatre et al. 2012; Visseren-Hamakers et al. 2012). These safeguards include use of such actions as free, prior and informed consent (FPIC) in advance of REDD+ planning. The COP at Cancun in 2010 agreed to the principle of safeguards, although details were lacking; many COP participants found the final decision too weak, as it only requires from participating nations "a system for providing information" on how governments are addressing the problem of safeguards in REDD+. The Subsidiary Body on Scientific and Technical Advice (SBSTA) has been working through possible approaches for reporting on safeguards in the future, but guidance is still somewhat unclear. Consequently, different REDD+ projects have developed their own approaches to safeguards, including the UN-REDD's Principles & Criteria (P&C); the World Bank's FCPF Strategic Environmental and Social Assessment (SESA); and the Climate, Community and Biodiversity Alliance (CCBA) REDD+ Social & Environmental Standards (SES).[1]

Yet while these safeguard standards all refer to the idea that local communities must be involved in REDD+ development, experience on the ground suggests that many of the private carbon projects certified by CCBA, for example, failed to meet stated goals for participation and information access (Suiseeya and Caplow 2013). Further, the limited focus of many safeguards policies at the project level has primarily been on preventing abuses (a "do no harm" approach) rather than bottom-up suggestions on how to enhance local forest-based livelihoods as part of a multifaceted sustainable forest management strategy (a "do more good" approach) (McDermott et al. 2012; De La Fuente and Hajjar 2013). Given this lack of attention to participation and livelihoods as integral components of safeguards, it is not surprising that many countries like Vietnam, which are attempting to initiate REDD+ activities, have had problems prioritizing these issues, as I explore later.

Forest Management Challenges in Vietnam

Before reviewing how REDD+ projects are addressing participation and livelihoods in Vietnam, I provide a brief overview of how previous eras of forest management and policy have addressed these questions. Importantly, the state has long been the dominant actor in Vietnam's forest sector. Shortly after the Democratic Republic of Vietnam (DRV) was founded in 1954, forest policy was developed for the complete nationalization of the forest estate and the establishment of state-owned logging companies to manage these lands. The nationalization of forests was extended to the South after 1975, during reunification of Vietnam at the conclusion of the Vietnam War (McElwee 2016). In 1986, the ruling Communist Party began to liberalize the economy and move to more market-oriented planning. Land and forest laws were revised in the early 1990s, and at that time it was believed that issuing long-term lease rights for households to use forest land, rather than continued state management, would result in better protection of forests and expansion of tree planting. However, this decentralization was top-down, and local participation (with a few exceptions) was mostly limited to receiving poor-quality lands for reforestation. The state retained control over much of the best forest land, such as those in national parks and reserves, and the land allocation policies only ended up providing individual household rights to less than one-third of the total forest estate; various state organs (including the Ministry of Agriculture, state-owned logging companies, local provinces, the army and so forth) continue to control the rest (Nguyen 2006).

Ambitious forest plantation programs have expanded forest cover in recent years, although deforestation remains problematic in many protected reserves. Further, communities as legal entities control forest rights in only a small number of provinces, amounting to less than 1 percent of the total forest land area (Nguyen 2006). Finally, the benefits of forestry have been unevenly received, with many households receiving very little of their income from forest sources despite living in heavily forested areas (Thuan et al. 2006; McElwee 2010). Conflicts between forest-using communities and state forest managers continue on an almost daily basis in many areas of Vietnam (To et al. 2013).

Development of REDD Policy in Vietnam

A national REDD+ steering committee was established in Vietnam in early 2011, facilitated by the Ministry of Agriculture and Rural Development; a special REDD+ office, which will coordinate with the UNFCCC, was established within the Vietnam Administration of Forestry. A National REDD+ Network was set up in 2009 for NGOs and donors to offer advice to the REDD+ readiness process. This REDD+ Network has several subcommittees that have been tackling issues such as governance; monitoring, reporting and verification (MRV); financing and benefit distribution systems (BDS); and local implementation

(Pham et al. 2012). The government of Vietnam approved a National REDD+ Action Plan for 2011–2020 in summer of 2012, which encourages the development of provincial action plans, pilot projects and legal frameworks, stating that the National REDD+ Program

> will contribute to reducing emissions from deforestation and forest degradation (REDD), to promoting forest conservation, sustainable forest management and the enhancement of carbon stocks, jointly comprising REDD+, and at the same time improving the livelihoods of the rural population in Vietnam.
>
> *(Hang et al. 2011, p. 12)*

At least 17 different donor-funded pilots are underway in assorted provinces to publicize REDD+, conduct carbon baseline measurement, and perform other activities. By far the largest donor to the readiness process has been Norway, which has pledged more than $30 million total in two phases to the Vietnam UN-REDD program; development agencies of the United States, Japan, Germany and Finland are the other major bilateral donors.

Can these new REDD+ projects tackle some of the past barriers to successful forest conservation in Vietnam? To answer this question, I examine how REDD+ projects are being developed by different actors, and how these projects have dealt with key issues of participation, livelihoods and safeguards.

Participation: Can REDD+ Increase Local Involvement in Forest Decision-Making?

Forest management in Vietnam has long been dominated by the state. Government forest offices retain great control over both a significant land area that is directly state managed, as well as the right to be involved in private forest land management, such as through taxing and regulating the sale of forest produce. Much of the local forest sector in Vietnam still operates on what is known as a *xin-cho* model, which means "ask-give," and which is a legacy of the long socialist era. Local people and subnational governments ask for resources, which the central state grants (or not), but local initiative is highly stymied by these norms of waiting for central government approval. Increasing the initiative and participation of local actors has been a goal of donor projects in the forestry sector for nearly 20 years, with mixed results (Wode and Huy 2009). Despite a strong push in the 1990s for attention to community forestry, for example, it still remains the case that less than 1 percent of the forest area is managed by communities with firm land use rights. (This also means that communities are also unlikely to be able to receive carbon payments collectively, at least for the foreseeable future, due to these legal constraints [UN-REDD Vietnam and MARD 2010].) This is in contrast to relatively strong communal and indigenous tenure rights in much of Latin America, for example.

REDD+ projects in Vietnam thus operate in a climate of uncertain responsibilities and roles for local communities. Outside of direct project stakeholders, there is little awareness of REDD+ among local peoples and the general public (Pham et al. 2012), and civil society actors are relatively weak and have little voice in REDD+ debates, where the state plays the leading role (Di Gregorio et al. 2013). The development of local policies and PRAPs in two provinces so far have demonstrated that most activities surrounding REDD+ are considered to be the responsibility of provincial forest departments, who only rarely engage with demands of local households or communities. In the development of the Dien Bien province PRAP, for example, authorities asserted that they had held some meetings with local communities during the development of the policy, but it was not clear how this participation influenced the outcomes of the process, or if these meetings were simply a way to pass information downward. Furthermore, the final decision on which communes (the lowest level of state administration) would receive REDD+ support and funding will be made by the province on the basis of areas with high forest extent, high deforestation rates and the potential for afforestation, not on local willingness or enthusiasm to undertake REDD+ activities (Dien Bien Forest Department, personal communication, 2014).

The primary mode by which local participation is being integrated into REDD+ activities in Vietnam is through implementation of FPIC agreements, which have been introduced by the UN-REDD project. UN-REDD has praised Vietnam as the first country to successfully implement FPIC for REDD+ (UN-REDD 2014). Yet questions remain about how transparent and fair such FPIC consultations have really been, and if they truly count as full participation. The concept is of FPIC is very hard to understand in Vietnamese and primarily has been translated as a type of community consultation, rather than bottom-up participation. For example, in pilot trials of FPIC in Lam Dong province, village level meetings held by the UN-REDD project to get consent for REDD+ activities have been held but were very short (only two hours), and only 45 minutes were allocated for questions and answers after awareness-raising activities (mostly centered on what climate change was, and how forests affect climate, with frequent use of the metaphor that forests are like the lungs of the earth—see Figure 11.2) before the villagers had to make the decision whether to consent to REDD+ (Nguyen et al. 2010). Communities were not presented with any information on the possible risks and costs of participation (such as changes in agricultural practices that they might have to made in response to REDD+) that might have allowed them a fuller range of consent options. Rather, villagers in group meetings were asked general questions like "Do you want your forests to be conserved?" (Lam Dong Forest Department, personal communication, 2014). Not surprisingly, this was supported by most people, since the question did not refer to any costs that might be incurred in forest conservation or how it might be carried out.

There was also little variation between communities in the options presented for participation in an FPIC process: votes were held collectively (usually with a

FIGURE 11.2 Poster explaining "Trees Are the Lungs of the Earth" outside a forest ranger station in Lam Dong Province, Vietnam.

non-anonymous show of hands) to say yes or no to REDD+. But there were few other possibilities for communities to propose local inputs into the REDD+ activity development process (UN-REDD consultant, personal communication, 2013).

Livelihoods: Can REDD+ Increase Household Incomes and Diversify Livelihoods from Forests?

Donors in Vietnam have emphasized the linkages between poverty alleviation and forestry that might be addressed by REDD+. Yet current REDD+ readiness activities have paid insufficient attention to how REDD+ could be used to fund specific poverty activities. For example, the national REDD+ Action Plan proposes a general attention to "forest-based livelihoods" but offers no practical ideas or suggestions of what types of livelihoods might be encouraged and how financing would be used (Hang et al. 2011).

Two major structural issues stand out as particular livelihood challenges for REDD+. The first is the channeling of REDD+ money and attention through provincial forest departments. These have never had a history of success in attending to local livelihoods, as they are dominated by professional foresters with little training or interest in economic, sociological or anthropological approaches to natural resources. As provinces develop their local action plans for

REDD+ (PRAPs), the forest departments that are managing this process have devoted little attention to questions of poverty and livelihoods. One of the least elaborated aspects of the two existing PRAPs are the sections discussing how lost livelihoods might be compensated for if REDD+ implementation requires changes in forest use. Little livelihood data is presented in either document, in contrast to fairly detailed data on forest carbon content.

The second challenge is that REDD+ activities on the ground are so far primarily targeting ethnic minority households, presumed to practice shifting (or swidden) cultivation and to be agents of deforestation, to the exclusion of other drivers of deforestation, such as state logging and state-driven coffee and rubber expansion. The two existing PRAPs primarily discuss the impact of local ethnic minority households on forest resources. Yet significant non-local, non-ethnic minority drivers of deforestation, namely the conversion of forests for rubber plantations in Dien Bien and for greenhouse export-oriented agriculture in Lam Dong, are not addressed in either province's PRAP. This attention to the poorest households as responsible for forest loss, while wealthier and more connected individuals and companies are not examined, is potentially worrisome.

In terms of national policy in Vietnam, livelihoods in the context of REDD+ have mostly been addressed through discussion of a benefit distribution system (BDS) (UN-REDD Vietnam and MARD 2010; Sikor et al. 2012). So far, indications are that a national REDD+ fund will be set up, which would disburse finances downward to provincial funds, which would then decide how to distribute to local beneficiaries. How to ensure conditionality (that is, that payees only get the money if the forest protection is delivered) remains problematic in national discussions. Trial consultations in local provinces on benefit distributions systems also reveal wide disparities between groups as to how benefits might be shared, such as if they should be in cash or in kind (Sikor et al. 2012). To date, no payments have yet occurred in any major REDD+ readiness pilot in Vietnam. This has been a source of disappointment in local areas, especially in Lam Dong province, where such pilots are now nearly five years old. As one provincial official said in an interview,

> Phase one of REDD has had a lot of talk, but not much action . . . REDD we've discussed for five years and there is no money. We've promised people we will give them money in the future but it's not clear that we will be able to do so. And that is very dangerous.
> *(Lam Dong official, personal communication, 2014)*

Will REDD Safeguards Be Sufficient?

Like other countries, Vietnam has been discussing setting up a legal safeguards system (known in Vietnamese as *đảm bảo an toàn*) for REDD+, and it is clear that the minimal decisions on safeguards that were passed at the Cancun COP are now affecting country implementation. In a recent draft report, a proposed national safeguards roadmap for Vietnam is quite legalistic and does not move far from the basic Cancun safeguards framework. Local provinces are likely to follow the lead of national authorities on this: officials in Dien Bien province told me that safeguards should be established nationally, and thus these were not included in consideration of their local PRAP.

Currently, the draft national safeguards roadmap refers primarily to ensuring a consistent legal environment and assuring information access on REDD+, but proposes little else to improve participation, equity or development as a required goal or outcome of REDD+ projects (SNV and VNFOREST 2013). For example, the roadmap suggests matching REDD+ safeguards to other requirements under commitments such as the Convention on Biological Diversity, but the reporting obligations for these other conventions are quite weak and have few formal requirements. In interviews with policy-oriented non-governmental organizations (NGOs) in Vietnam, activists expressed concern that existing international reporting requirements for conventions are minimal, and that the reports issued are usually short and written by consultants for the sake of meeting requirements, rather than to effect policy or implementation on the ground. These NGOs are concerned that REDD+ safeguards reporting to the UNFCCC will similarly be formulaic. They fear that reporting will include neither serious consideration of how to protect vulnerable peoples from abuse in REDD+ projects, nor holistic consideration of how to use REDD+ to promote more equitable livelihoods in development (Vietnam NGO, personal communication, 2014).

How might a more active level of engagement with reporting and safeguards be achieved? The safeguards roadmap refers to the need to provide access to information for REDD+, but according to one NGO activist interviewed, this is a weak safeguard unless people know how to make use of public information. For example, some countries use public advocates who help people navigate access to public information.

Further, it is not clear what mechanisms will be in place to enforce statutory rights in REDD+, for instance through grievance mechanisms. Vietnam does not have a tradition of using citizen lawsuits to enforce existing environmental laws, as is the case in other countries. One NGO worker suggested that REDD+ would be much strengthened if it included the ability to sue government forestry departments if safeguard regulations are not followed. Such a mechanism has not traditionally been available in Vietnam, and should it be implemented for REDD+, the activist stated it could have a positive effect on other areas, such as pollution law enforcement.

Conclusions: Making REDD+ Participatory and Pro-poor

To what degree will REDD+ be able to make good on the many high hopes that have been placed on it? The potential signs that REDD+ is proceeding with insufficient attention to participation and livelihoods, and with weak and unenforceable safeguards, should be worrying news. In places where REDD+ and carbon valuation are driven by top-down processes, whether by donors or states, the outcomes are likely to be less satisfactory than in places with bottom-up and genuine participation (Schroeder and McDermott 2014). Angelsen (2013) expresses pessimism that donor-led funding can lead to substantial policy reform, without lower-level buy-in, and REDD+ is likely to be no exception. Disappointments on both global and local scales may be common outcomes. As one local official in Lam Dong said to me, getting involved in REDD+ is like "grabbing for the sky," in that it seems a far off and impossible task to meet both international obligations and local expectations (Lam Dong Forest Department, personal communication, 2014).

For Vietnam, it is not yet clear how REDD+ can be a positive driver for change in the forest sector. Overall, in the discussions about REDD+, and in the development of local policies and pilots to date, very little attention has been paid to the poverty and social aspects of REDD+ implementation. Despite lip service from donors in documents for REDD+ readiness, the actual development of provincial REDD+ action plans has largely ignored issues of participation, livelihoods or safeguards. Nor has attention been given to the risks that might be incurred if poor people are induced to make land use changes in response to carbon markets that restrict their production of food, or if new forms of exchange and marketization are introduced to areas unfamiliar with them. There has been far more focus on technical issues, such as establishing baseline levels of carbon emissions, than to long-term social monitoring of the household-level effects of REDD+ payments and land use changes.

The lack of strong support from the UNFCCC from the earliest stages for a unified safeguards approach is potentially burdensome for national and subnational levels that are seeking guidance, and the minimal devotion to key social concerns in the Cancun and Warsaw frameworks has meant that these issues are treated fairly simplistically at country levels. For example, in Vietnam, the idea of participation, which encompasses a potentially large range of possible actions, is primarily reduced to the idea of doing a FPIC consultation. Sustainable livelihoods approaches, which might reveal a range of possible activities, are primarily reduced to the idea of having a formal BDS. Are FPIC and BDS likely to truly empower citizens to play more engaged roles in forest management? So far, at least in Vietnam, the indication is that FPIC is inadequate and participation is too rote, while livelihood indicators for BDS systems are weak and monitoring of livelihoods is challenging.

Even the establishment of legal safeguards may not be sufficient to overcome these challenges. Additional initiatives are also potentially needed, such as the creation of baselines to understand livelihood changes, and of reporting mechanisms to track changes in health or education among REDD+ participating communities. Overall, the jury is still out on how REDD+ will be able to be a tool for the betterment of local forest-using communities, but continued attention to the issues of participation, livelihoods and safeguards is surely needed.

Acknowledgements

The research for this chapter was made possible by a grant from the National Science Foundation Geography and Regional Science Division for the project "Downscaling REDD Policies in Developing Countries: Assessing the Impact of Carbon Payments on Household Decision-Making and Vulnerability to Climate Change in Vietnam" (grant #11028793). My Vietnamese collaborators on this project were additionally supported by the Economy and Environment Program for Southeast Asia (EEPSEA) for fieldwork in 2011, and since 2012, they have also supported by a US Agency for International Development (USAID) Partnership for Enhanced Engagement in Research grant: "Research and Capacity Building on REDD+, Livelihoods, and Vulnerability in Vietnam: Developing Tools for Social Analysis of Development Planning." Our deepest thanks to all these funders for their generous support. Thank you also to the Vietnamese partners who contributed to the collection of data outlined in this chapter: Nghiem Phuong Tuyen, Le Thi Van Hue, Tran Huu Nghi, Vu Thi Dieu Huong, Dao Minh Truong, Le Trong Toan, Ha Thi Thu Hue, Ha Thi Tu Anh, Nguyen Viet Dung, Nguyen Xuan Lam, and Nguyen Hai Van.

Note

1. More details on these guidelines can be found at UNREDD- http://www.un-redd.org/Multiple_Benefits_SEPC/tabid/54130/Default.aspx; Forest Carbon Partnership Fund- http://www.forestcarbonpartnership.org/sites/forestcarbonpartnership.org/files/Documents/PDF/Nov2011/FCPF%20Readiness%20Fund%20Common%20Approach%20_Final_%2010-Aug-2011_Revised.pdf; andCCBA—http://www.climate-standards.org/ccb-standards/.

References

Agrawal, Arun, Daniel Nepstad and Ashwini Chhatre. 2011. Reducing emissions from deforestation and forest degradation. *Annual Review of Environment and Resources* 36, 1: 373–396.
Angelsen, Arild. 2013. *REDD+ as performance-based aid: General lessons and bilateral agreements of Norway*. WIDER Working Paper No. 2013/135. Tokyo: World Institute for Development Economics Research, United Nations University.
Boyd, Emily, Peter May, Manyu Chang and Fernando Veiga. 2007. Exploring socioeconomic impacts of forest based mitigation projects: Lessons from Brazil and Bolivia. *Environmental Science and Policy* 10, 5: 419–433.
Brown, Michael. 2013. *Redeeming REDD: Policies, incentives and social feasibility for avoided deforestation*. London: Earthscan.

Caplow, Susan, Pamela Jagger, Kathleen Lawlor and Erin Sills. 2011. Evaluating land use and livelihood impacts of early forest carbon projects: Lessons for learning about REDD. *Environmental Science and Policy* 14, 2: 152–167.

Cerbu, Gillian, Brent Swallow and Dara Thompson. 2011. Locating REDD: A global survey and analysis of REDD readiness and demonstration activities. *Environmental Science and Policy* 14, 2: 168–180.

Chhatre, Ashwini, Shikha Lakhanpal, Anne Larson, Fred Nelson, Hemant Ojha and Jagdeesh Rao. 2012. Social safeguards and co-benefits in REDD+: A review of the adjacent possible. *Current Opinion in Environmental Sustainability* 4, 6: 654–660.

Corbera, Esteve. 2012. Problematizing REDD+ as an experiment in payments for ecosystem services. *Current Opinion in Environmental Sustainability* 4: 612–619.

Corbera, Esteve, Manuel Estrada and Katrina Brown. 2010. Reducing greenhouse gas emissions from deforestation and forest degradation in developing countries: Revisiting the assumptions. *Climatic Change* 100, 3: 355–388.

Corbera, Esteve and Heike Schroeder. 2011. Governing and implementing REDD. *Environmental Science and Policy* 14, 2: 89–99.

De La Fuente, Teresa and Reem Hajjar. 2013. Do current forest carbon standards include adequate requirements to ensure indigenous peoples' rights in REDD projects? *International Forestry Review* 15, 4: 427–441.

Di Gregorio, Monica, Maria Brockhaus, Tim Cronin, Efrian Muharrom, Levania Santoso, Sofi Mardiah and Mirjam Büdenbender. 2013. Equity and REDD+ in the media: A comparative analysis of policy discourses. *Ecology and Society* 18, 2: 39. doi:10.5751/ES-05694-180239.

Gupta, Joyeeta, Nicolien van der Grijp and Onno Kuik. 2013. *Climate change, forests and REDD: Lessons for institutional design*. London: Routledge.

Hall, Anthony. 2012. *Forests and climate change: The social dimensions of REDD in Latin America*. Cheltenham: Edward Elgar.

Hang, Nguyen, Wulf Killmann, Xuan Phuong Pham and Eveline Trines. 2011. *Viet Nam national REDD+ program: Background document*. Hanoi: UN-REDD Programme.

IIED. 2011. *Just forest governance—for REDD, for sanity*. London: International Institute for Environment and Development.

Krause, Torsten, Wain Collen and Kimberley Nicholas. 2013. Evaluating safeguards in a conservation incentive program: Participation, consent, and benefit sharing in indigenous communities of the Ecuadorian Amazon. *Ecology and Society* 18, 4, 1. doi:10.5751/ES-05733-180401.

Lawlor, Kathleen, Erin Madeira, Jill Blockhus and David Ganz. 2013. Community participation and benefits in REDD+: A review of initial outcomes and lessons. *Forests* 4, 2: 296–318.

Lawlor, Kathleen, Erika Weinthal and Lydia Olander. 2010. Institutions and policies to protect rural livelihoods in REDD regimes. *Global Environmental Politics* 10, 4: 1–11.

Luttrell, Cecilia, Lasse Loft, Maria Fernanda Gebara, Demetrius Kweka, Maria Brockhaus, Arild Angelsen and William Sunderlin. 2013. Who should benefit from REDD+? Rationales and realities. *Ecology and Society* 18, 4: 52. doi:10.5751/ES-05834-180452.

McDermott, Constance, Lauren Coad, Arielle Helfgott and Heike Schroeder. 2012. Operationalizing social safeguards in REDD+: Actors, interests and ideas. *Environmental Science and Policy* 21: 63–72.

McDermott, Melanie, Sango Mahanty and Kate Schreckenberg. 2013. Examining equity: A multidimensional framework for assessing equity in payments for ecosystem services. *Environmental Science and Policy* 33: 416–427.

McElwee, Pamela. 2010. Resource use among rural agricultural households near protected areas in Vietnam: The social costs of conservation and implications for enforcement. *Environmental Management* 45, 1: 113–131.
McElwee, Pamela. 2016. *Forests are gold: Trees, people and environmental rule in Vietnam.* Seattle: University of Washington Press.
Milne, Sarah. 2012. Grounding forest carbon: Property relations and avoided deforestation in Cambodia. *Human Ecology* 40, 5: 693–706.
Nguyen, Quang Tan. 2006. Forest devolution in Vietnam: Differentiation in benefits from forest among local households. *Forest Policy and Economics* 8, 4: 409–420.
Nguyen, Quang Tan, Luong Thi Truong, Nguyen Thi Hai Van and K'Tip. 2010. *Evaluation and verification of the free, prior and informed consent process under the UN-REDD Programme in Lam Dong Province, Vietnam.* Bangkok: RECOFTC—The Center for People and Forests.
Pham, Thu Thuy, Moira Moeliono, Nguyen Thi Hien, Nguyen Huu Tho and Vu Thi Hien. 2012. *The context of REDD+ in Vietnam: Drivers, agents and institutions.* Bogor: Center for International Forestry Research.
Reynolds, Travis. 2012. Institutional determinants of success among forestry-based carbon sequestration projects in Sub-Saharan Africa. *World Development* 40, 3: 542–554.
Schroeder, Heike and Constance McDermott. 2014. Beyond carbon: Enabling justice and equity in REDD+ across levels of governance. *Ecology and Society* 19, 1, 31. doi:10.5751/ES-06537-190131.
Sikor, Thomas, Adrian Enright, Nguyen Trung Thong, Nguyen Vinh Quang and Vu Van Me. 2012. *Piloting local decision making in the development of a REDD+ compliant benefit distribution system for Viet Nam.* Hanoi: UN-REDD Programme.
SNV and VNFOREST. 2013. *Roadmap for environmental and social safeguards for Vietnam's National REDD+ Action Programme: Gap analysis of existing policies, laws and regulations.* Hanoi: SNV Netherlands Development Organization and Vietnam Administration of Forestry (VNFOREST).
Strassburg, Bernardo, Kerry Turner, Brendan Fisher, Roberto Schaeffer and Andrew Lovett. 2009. Reducing emissions from deforestation—the "combined incentives" mechanism and empirical simulations. *Global Environmental Change* 19, 2: 265–278.
Suiseeya, Kimberly and Susan Caplow. 2013. In pursuit of procedural justice: Lessons from an analysis of 56 forest carbon project designs. *Global Environmental Change* 23, 5: 968–979.
Sunderlin, William, Anne Larson, Amy Duchelle, Ida Resosudarmo, Thu Ba Huynh, Abdon Awono and Therese Dokken. 2014. How are REDD+ proponents addressing tenure problems? Evidence from Brazil, Cameroon, Tanzania, Indonesia, and Vietnam. *World Development* 55(March): 37–52.
Tacconi, Luca, Sango Mahanty and Helen Suich. 2013. The livelihood impacts of payments for environmental services and implications for REDD. *Society and Natural Resources*, 26, 6: 733–744.
Thuan, Dinh Duc. 2006. *Forestry, poverty reduction and rural livelihoods in Vietnam.* Hanoi: Labour and Social Affairs Publishing House.
To, Xuan Phuc, Phan Dinh Nha, Pham Quang Tu and Do Duy Khoi. 2013. *Mau thuan dat dai giua cong ty lam nghiep va nguoi dan dia phuong (Conflicts over land between state forest companies and local people).* Hanoi: Forest Trends.
UN-REDD. 2014. *Ensuring inclusive, transparent and accountable national REDD+ systems: the role of freedom of information.* Geneva: UN-REDD.
UN-REDD Vietnam and MARD. 2010. *Design of a REDD compliant benefit distribution system for Viet Nam.* Hanoi: UN-REDD and Ministry of Agriculture and Rural Development.

Visseren-Hamakers, Ingrid, Constance McDermott, Marjanneke Vijge and Benjamin Cashore. 2012. Trade-offs, co-benefits and safeguards: Current debates on the breadth of REDD+. *Current Opinion in Environmental Sustainability* 4, 6: 646–653.

White, D. 2013. A perfect storm? Indigenous rights within a national REDD+ readiness process in Peru. *Mitigation and Adaptation Strategies for Global Change* 19: 657–676.

Wode, Bjoern and Bao Huy. 2009. *Vietnam: Study on state of the art of community forestry in Vietnam*. Hanoi: Deutsche Gesellschaft für Internationale Zusammenarbeit (GTZ).

SECTION IV
REDD, Rights, and Equity

12

RENEGOTIATING REDD

Beyond Social Safeguards to Social Contracts

Michael Brown

When Rousseau ([1754] 2005, 33) warned, "you are undone if you once forget that the fruits of the earth belong to us all, and the earth itself to nobody," he could have been presaging the chorus of critiques leveled by the opponents of the program mechanism known as REDD.[1] With its particular focus on the commoditization of carbon, REDD, and the latest iteration REDD+, raises an intriguing question: should and can carbon as it occurs in tropical forests with all its complexity become a tradable market asset?

REDD too frequently perpetuates a top-down planning process, rather than a collective exercise, as Ostrom (1990) advises is necessary for resource management in places where property is communally owned and competing claims may exist. In it, frontline forest communities remain marginal planners and decision-makers. While it is expected by those actually holding decision-making power that the people on the ground (who assume the actual risk in avoiding deforestation) will participate wholeheartedly, there is little to suggest that the palette of economic incentives offered to them is sufficient on a project-by-project basis. I suggest that this is due to the absence of any credible social contract that convincingly binds stakeholders together in REDD.

In this chapter I argue that more focus on social contracts and social feasibility is needed if REDD is to work. Social contracts will need to clarify roles and responsibilities, along with the nature and timing of incentives and oversight of the process. This is a complex endeavor, as neither communities nor their principal interlocutors in any REDD social contracting are far from homogenous. I come to this conclusion from 30 years of working primarily across sub-Saharan Africa, with forays in the Caribbean, South Pacific, South Asia, and Latin America. My work has been in natural resource management, agriculture, anti-corruption and good governance,[2] landscape-level forest and biodiversity conservation, and integrated conservation and development. I have led the design and implementation of programs and projects, oftentimes under US Agency for International Development (USAID) funding, while working for nongovernmental organizations (NGOs) and for-profit firms. I have focused on methods for effective multi-stakeholder participation that emphasize the poor.

Recent Grounds for Optimism and Pessimism in REDD

As it has been designed, REDD, the most recent international effort to avoid deforestation and mitigate global climate change, has little chance of succeeding. Much of the critique of REDD has focused on the credibility of measurements and reporting that rely upon baselines of historical deforestation rates (Karsenty, Vogel, and Castel 2014) and the plausibility of avoiding deforestation where offsets (at best) are produced, rather than actual reductions in greenhouse gas emissions (Friends of the Earth 2008). My concerns lie in the weakest part of REDD overall: the infeasibility of REDD policies, programs, and projects to effectively engage with forest communities, particularly at scale. This has been particularly the case during the preliminary readiness phase. These efforts include international and national planning efforts backed by the United Nations Framework Convention on Climate Change (UNFCCC) and its principal partners,[3] particularly the UN REDD program (UN-REDD) and the World Bank, as well as parallel initiatives in which carbon is valued in voluntary markets.[4]

I broadly label issues pertaining to REDD feasibility involving frontline forest communities as "social feasibility."[5] Since frontline communities will of necessity be the principal stewards of any additional carbon sequestered through avoided deforestation, REDD needs to work for them in both the short and long term. In the short term, this includes simply getting activities off the ground with some funding and aspiration. *In the long term, it involves questions of sustainability, inclusive of institutional capability, policies, and incentives provided.* Currently, the imperative to succeed generates rose-tinted outlooks that are outpacing any verifiable indication that, when sustainability is considered in detail, REDD is on the right track on the ground (Brown 2013).

Despite the backing of intergovernmental organizations, national governments and their development agencies, numerous conservation-related big international nongovernmental conservation organizations (conservation BINGOs), and a number of for-profit intermediary firms and investment banks, there is little empirically to indicate that REDD programming is any further along on the pathway to sustainability than biodiversity conservation or development programming. This is noteworthy. While there has not been any explicit academic or applied research on this particular topic, it appears to me that much of the intellectual apparatus, along with best practices and standards, are drawn from what BINGOs and associated standard-setters have established. While supported further by an array of private sector intermediaries providing additional institutional weight, such as certification, validation, and verification organizations, much in fact remains hypothetical due to the weak evidence base underpinning biodiversity and forest conservation programming.

The perspective of one prominent donor on the state of knowledge in biodiversity conservation is noteworthy:

> USAID's programs are based on a number of prominent hypotheses in the field of conservation and development that aim to protect biodiversity and natural resources while improving human well-being. Although these hypotheses are used widely by conservation organizations and development agencies, *little information exists about their efficacy* [emphasis added]. Despite long-standing programs in priority biodiversity regions, there is still a lack of long-term datasets and baselines to analyze the effectiveness of conservation efforts.
> *(USAID 2012, 160)*

REDD draws on best practices from conservation and development (Brown 2013), such as those formulated for sustainable forestry by the Forest Stewardship Council (FSC 2011). If the credibility of FSC's standards remain at issue on the sustainable forest management side (see Griscom, Ellis, and Putz 2014) and on the social side (see Venne 2007; Marx et al. 2012), it is reasonable to hypothesize that the firmness of the foundational underpinnings of REDD itself are suspect as well. And while Tollefson, Gale, and Haley (2009) fairly describe the high degree of complexity in FSC standard setting (and thus, any related REDD standard setting based on FSC best practices), social issues arguably remain among the most difficult for FSC to address, particularly in remote forest contexts where the transparency and accountability of logging companies working under non-transparent government agreements is often low (Brown 2013).[6]

Just as biodiversity conservation is a complex endeavor involving social, political, institutional, economic, and scientific issues, REDD operates in an equally complex environment. While optimism and a sense of control over how REDD is being rolled out is present, scratching a bit beneath the surface raises legitimate reasons for concern.

FPIC, Social Safeguards, and Implausible Assumptions

Current applications of free, prior, and informed consent (FPIC) guidance, along with other social safeguards, are insufficient to enable REDD projects to become socially feasible (Brown 2013), though they may superficially satisfy aspects of the "Four S's of Social Soundness"—issues of scale, systems models, stakeholder engagement, and sustainability—as well as application of the Nature, Wealth and Power paradigm extended through USAID (see USAID, LEAF, and FCMC 2012). Social feasibility, as I conceive it, is not explicitly employed in REDD, development, or biodiversity conservation programming, which arguably do not drill down to the specific demands that a feasibility analysis would require. Social feasibility involves issues at multiple scales. It goes considerably beyond FPIC and other social safeguards in considering what is *specifically needed* for projects to work and become sustainable from the perspective of local communities. In the absence of social feasibility, achieving sustainability is problematic, even if the Four S's of Social Soundness covering the big picture are addressed.

FPIC in itself does not explicitly consider feasibility issues either, while social safeguards are primarily designed to minimize potential harm to people from forced evictions, inequitable benefit sharing schemes, and inadequate consultation. In an overview of the topic, the Swedish Society for Nature Conservation (SSNC) (2013, 24, 26) observes in regard to one of the most recognized standards for validation and verification of REDD projects, the Climate, Community, Biodiversity (CCB) Alliance standard:

> The CCB has a problem with the consistent interpretation and enforcement of a standard . . . [due to] lack of proper and workable definitions and insufficient guidance on how principles, concepts and terms are to be interpreted and assessed . . . [and that] the CCB standard does not provide sufficient guidance in terms of determining what is an acceptable FPIC process.[7]

With no clear criteria for when free, prior, and informed consent is attained, FPIC becomes subjective in the eyes of the project developer and certifier. Seemingly, so long as checklist verification of gender-balanced consultation and a series of exercises occurs (e.g., programming content is presented to concerned people; a participatory rural appraisal exercise is conducted, based on the project developer's theory of change; and concerned people are requested to provide feedback on activities within a set period of time), then FPIC certification has often been achieved—formally, at least (see Brown 2013, 176–178).

While rights-based approaches and civil society critiques of REDD have correctly focused on inadequate safeguard processes, they too do not answer the question of what REDD needs to do to actually work on the ground. Nor do the plethora of civil society critiques available in the invaluable REDD-Monitor information portal,[8] for example, go beyond reiterating discontent with the consultation process instituted by UN-REDD in many countries and national-level activities supported by World Bank funding.

The categorical rejection of carbon markets by the most persistent critics of REDD (see Friends of the Earth 2009) leaves little space for those communities who would be willing to explore using carbon as a commodity and markets as a mechanism for avoiding deforestation through REDD. And that is the rub: public domain evidence is lacking that the principal participants driving REDD programming have come up with a framework to approach this convincingly, where local people buy in to the process, sustain that participation, and do not pay undue opportunity costs in doing so.

As these communities may bear principal responsibility for forest resource stewardship under REDD, their full-fledged participation is crucial to REDD's success. Since best practices and standards are promoted by major development and conservation thought leaders and their financial backers, few challenge the direction and logic under their imprimatur. While social safeguards provide frontline communities with some nominal security against unintended harm that could emerge, there is little beyond assumptions about rising future carbon asset values that may benefit communities participating in REDD. On the surface, impoverished communities facing changes to land use patterns are being asked to bear inordinate risk to their livelihoods, with the probability of positive, sustainable returns on their investment highly uncertain.

REDD Pilots

Strategies and tools currently employed in the piloting phase of REDD project designs remain unconvincing if sustainability is the principal evaluation criterion (Brown 2013, 147, 151–153, 171–185, 213). Presently, those stakeholders driving REDD standards and programming do not appear to be as attentive to project sustainability as they are to the launching of pilot activities that meet funding criteria. To achieve sustainability, REDD projects *must* be feasible from the beginning. On the other hand, to be funded, REDD projects must be deemed certifiable. Arguably, the bar set for sustainability, were it squarely addressed, is more demanding than for obtaining funding. This has led to an array of "good enough" or "optimal ignorance" planning decisions, where intellectually appealing, yet fundamentally vague, theory of change approaches are used as principal funding justifications. SSNC (2013, 15–23) presents case studies illustrating how key social information gathering was deferred by auditors, even while CCB certification was granted in REDD projects in Cambodia, Indonesia, and Kenya.

Examples of REDD pilot projects that are supported through both voluntary and national initiatives are scattered across the literature. Naughton-Treves and Day, for example, present a review of early lessons learned from REDD projects supported by USAID. They note:

> *It is important to remember that all of the REDD projects reported herein on are essentially pilots, and still in very early stages.* Nonetheless, it is becoming increasingly evident that implementation of REDD is likely to be more complex than the optimistic early estimates suggested. It has become a standard operating assumption that clear land tenure is a pre-requisite for participation in REDD projects, to reduce risks to permanence of the carbon storage. *But for better or worse, most tropical forests are considered property*

> of the state rather than of the individuals or communities living in them, and there are often competing/overlapping claims on forest resources, such as logging and mining concessions.
>
> (Naughton-Treves and Day 2012, iv; emphasis added)

Not only is clarifying land tenure a cost to be incurred in REDD project development, it is embedded (or not) in a broader social contract between the state and citizenry. Throughout much of the world, Africa particularly, land tenure is often vested de jure in the state but de facto by citizens as they go about their business—that is, until the state cuts a deal with some private entities in which the prevailing modus operandi on the ground may be disrupted through state prerogatives. In these cases, it is fair to say that the social contract binding the state to its citizenry may be weaker than its perceived prerogatives to negotiate with third parties based on the formalized rights with which it has vested itself. In situations like this, social feasibility will be difficult to achieve.

Benefit sharing is one explicit element in REDD, though definitions or criteria for operationalization often remain less than fully transparent. The Tanzania Natural Resource Forum (2011) notes that REDD can only work if those reducing deforestation and forest degradation, and those bearing the costs, receive substantial and sustainable direct benefits, including carbon payments. Yet the specifics for how carbon finance will generate direct benefits remains vague at various programming levels. The nature of "substantial" and "sustainable" payments remains ill-defined, with much to be learned from the pitfalls of integrated conservation and development projects, and payment for ecosystem services experience (see Blom, Sunderland, and Murdiyarso 2010).

On the voluntary side, where innovation potential may be greatest, Zwick (2013, 1) notes that landscapes, another bell and whistle in REDD, are inspiring excitement:

> The excitement in Warsaw over landscapes took two forms: some were relieved to finally see it front-and-center, while others were pleasantly surprised and even shocked to see that these approaches existed at all. . . . Those dual reactions highlight a disconnect that has always existed between the UNFCCC and the voluntary carbon markets. As early as the Bali climate talks in 2007, and probably long before that, REDD project developers warned that the negotiators charged with creating a REDD governing regime were ignoring the way REDD was taking shape on the ground. . . . Five years on, the track record in voluntary carbon bears that out: with few exceptions, voluntary REDD projects work by helping local people develop sustainable livelihoods.

In counterpoint to this, in July 2013, prior to the Warsaw UNFCCC meeting, UN-REDD posted an article entitled "Pilot Projects versus National Policy in the REDD Arena" (Gari 2013, 1), which established the following relationship between voluntary pilots and national programming:

> Pilot projects for REDD, no matter if well designed or even if earning carbon credits, will prove an insufficient effort if they do not influence national development policies and institutions . . . Furthermore, pilot projects often broadcast a disparity of methods in designing and implementing REDD activities, for instance when defining reference levels, carbon rights or benefit-sharing arrangements. Such disparity . . . will later cause controversy: for instance, project beneficiaries may enter complaints, or even grievances, if they feel their project is more difficult or yields fewer benefits than another. Such disparity will also undermine the ability of governments—and their prerogative—to establish national standards and policy for REDD.

These positions raise more questions than answers. On the one hand, Zwick's statement implies that there are empirical datasets that broadly support the point that voluntary projects are helping local people to develop sustainable livelihoods. This is not the case beyond anecdotal accounts. Nor as the statement implies have improved livelihoods been broadly observed. The most that can be said for many projects is that they have *aspirational rhetoric* related to improving livelihoods, sustainably.

Nor is it clear that poor forest dwellers are *more responsible* for driving deforestation than policies that create incentives for private sector interests to clear large tracts of land (e.g., for oil palm plantations) or enable government officials to act with impunity in misusing REDD resources. Clearly, keeping a focus on smallholders as *the* problem justifies funding to modify dysfunctional farmer-level behaviors; that is, it is more politically palatable than combating administrative corruption in avoided deforestation.

On the other hand, Gari's (2013) defensive posture regarding voluntary REDD innovations misses the major point in turn—REDD participatory planning at international and national levels has disappointed and evoked widespread discontent, and is part of the reason why voluntary REDD efforts have flourished. Neither social contracts nor social feasibility is evinced at national levels either, where programming exhibits more disconnect than coordination.

Social Contracts Must Become Central if REDD Is to Work

Social contracts have several centuries of conceptual underpinning from political philosophers such as Hobbes, Locke, Rousseau, Rawls, and Nozick.[9] In REDD, I believe that social contracts are essential not only from a rights and social justice perspective, but also for reasons of utility. In allowing ourselves to avoid articulating a well-defined social contract as a framework for action, REDD stakeholders perpetuate uncertainties and ambiguities about how the program will work in practice, objectively considering risks and opportunities.

By assuming that REDD can be successfully implemented as any other top-down development intervention—through incentives, participation, and nominal buy-in, and satisfying some basic needs—we misconstrue the extent to which risk and opportunity costs to local stakeholders may be impacted by unsuccessful REDD activities. The implications of a failed REDD activity go far beyond those of a poorly maintained water delivery system or a shoddily constructed clinic. Where customary rights holders are being asked to invest in generating common pool and individual benefits, demanding behavior changes impacting wealth and security, as well as future development prospects in the event of failure, quality information, transparency, and analytical capability are needed to achieve FPIC. For this type of risk assumption, a negotiated social contract is paramount. Contracts should stipulate the nature of information, transparency, and analysis needed, along with how decisions are to be made impacting communities stewarding REDD resources. No such negotiation framework currently exists.

In corporate social responsibility contexts, White (2007, 3) notes that

> the rules of the business-government-civil society engagement—which is integral to the social contract—are being rewritten not through some formal, centralized mechanism, but as a result of the pressures of shifting societal expectations about business' role in society. It is a process fraught with contradictions and uncertainties.

In the REDD world, where values are placed on abstract commodities such as carbon and customary resource use among peoples with claims to the land, trees, and soils in which this carbon is harbored, contradictions and uncertainties abound between de jure rights and de facto use patterns. Yet without an enabling social contract that conceptually and procedurally frames objectives, rights, and obligations in complex, contested situations, REDD stands little chance of succeeding. Even if successfully launched, as information eventually becomes available to enable more objective assessment of the opportunity costs to local stakeholders REDD may engender, discontent with the terms of REDD projects may well emerge to undermine sustainability. Barring militaristic interventions, perhaps echoing what has been reported among the Kenyan Sengwer (No REDD in Africa Network 2014), nation-states and REDD planners will have difficulty implementing REDD if local people do not buy into its premises and plans.

Issues requiring negotiation include: incentives in tenure rights; guaranteed payments; community capacity building; insurance in the event the terms of agreements are not fulfilled and means of objectively assessing this; how future carbon payments are to be calculated, including all deductions for transaction costs on the side of project developers (known sometimes as the pro rata international development costs); and what the government is prepared to sign in terms of stewarding or guaranteeing agreements, enabling tenure security so that the definition of property rights is clear enough to enable parties to negotiate in good faith, and recourse in the event of any non-compliance.

Assuming that REDD simply needs to offer up a market mechanism based on efficiency and premised on payments to poor farmers and resource users using opportunity cost calculations that discount their real opportunity costs to development is shortsighted (see Karsenty and Ezzine de Blas 2014). Simplistically applying payment for ecosystem and environmental services models to resolve natural resource management questions that require cooperation, negotiation, and incentives is a surefire recipe for failure, regardless of which investors or thought leaders support the process.

Precedents for Social Contracts

Most discussion in international development approaching the concept of social contracts has been related to *social protection*. Ambiguity remains as to what social protection in the contexts of international development might specifically refer to (Hickey 2011). More often than not, there is little sense as to how the terms of the social protection are arrived at or whether local stakeholders themselves perceive the terms, or the ends sought, as adequate or acceptable. In the natural resources management sector, useful conceptual frameworks, such as USAID's Nature, Wealth and Power (Anderson et al. 2013), do not, for example, include explicit guidance for how to work out the balance between these three key pillars in a potential social contract. Nor is it clear how power relations impact the design and implementation of social protection modalities.

Social safeguards, in turn, are generally applied by a range of intervening stakeholders in development and conservation contexts to reduce the risk of negative impacts from specific activities driven by external interests. While social safeguards are usually conceived as protecting people engaged in REDD activities, the line between such safeguards and broader social protection is not always explicit in developing country contexts, particularly for land and resource tenure. For example, while national constitutions may recognize the rights of indigenous or other local peoples to land and forest resources, there may not be any accompanying operational guidance—legislation or regulations—to enact these rights. This can lead to situations, such as in the Democratic Republic of the Congo and Indonesia, where disputes between communities and the government over local forest concessions and resource extraction abound. In these contexts, respective citizenries may feel that people's rights are not protected or that specific project-level activities do not involve appropriate safeguards, even if projects nominally incorporate safeguards and national policies champion them as well.

Increasingly, the use of *performance-based contracting* principles have also been considered as a vehicle to help promote wins for the environment, the general good, and individual household welfare. Payments for environmental services

(PES) could be considered as one form of performance-based contract, with REDD a policy program that embraces various strategies to transfer cash to resource managers based on verifiable emissions reductions in forests or other carbon-sequestering natural resources. While in PES arrangements, external stakeholders may arbitrarily fix a price for a service rendered, the premise most often in REDD is that payments will reflect either voluntary or (ultimately) compliance market pricing. Forest stewards at the community level thus participate in a capitalist venture, where the assumption is that benefits of stewarding forest conservation activities will exceed all incurred costs. In situations where compliance markets do not function and voluntary market activities remain circumscribed to a very small number of participants,[10] however, banking on markets is a risky endeavor, especially for economically and politically marginalized peoples.

While each of these concepts—social protection, social safeguards, and performance-based contracting—covers an important aspect of what could be addressed in programs concerned with social feasibility, they are piecemeal approaches that do not provide a holistic framework for the wide range of issues needing to be addressed in REDD. This is why more all-embracing social contracts are key.

Envisioning Social Contracts in REDD

There is no precise formula for what social contracts would constitute in REDD, or how one would go about the process of initiating and implementing them. A closer look at recent thinking on social contracts in the international development arena, such as Hickey's (2011) work, however, could provide guidance for considering how social contracts in REDD might be approached. Social contracts can focus on promoting a specific contractual approach to social protection interventions themselves, including conditional cash transfers for performance of certain activities. Social contracts could also be part of a broader effort to locate social protection within a more binding set of relationships and agreements as a means of ensuring their political sustainability. In REDD, there are multiple issues of legitimate concern to frontline communities, for example, governance arrangements, benefits sharing schemes, performance monitoring, adaptive management triggers, and of course, social safeguards.

To move to formalized social contracts, sufficient political will is needed. Absent alternative funding, this will prove challenging. Will it be feasible for international development agencies to support this type of initiative and what would the role of various stakeholders to the social contracting process be? These questions will need answering on a case-by-case basis.

The vision of the social contract that I am referring to in this chapter straddles two aspects pertaining to individuals. Following Berlin (1969), *negative rights* refers to the rights of people to be protected from overt sources of harm (e.g., security or failed development initiatives setting people back), whereas *positive rights* refer to those broader sets of goods and capacities that enable people to flourish. In providing social safeguards against harm, while ostensibly offering opportunities for livelihood enhancement, REDD attempts to do both. Rational choice contractarianism holds that in the context of a social contract, rational agents are presumed to maximize their advantage or self-interest, which is also an apparent underpinning premise of REDD. Under a rights-based approach, "people are presumed to be motivated by a concern for treating people fairly" (Black 2001, 116–117).

Given that social contracts would be determined by bargaining processes between governments, social groups, and citizens within specific contexts, it seems implausible that donor agencies can be the main facilitators in establishing a new paradigm for implementing this in REDD. This raises a number of challenges for donors, particularly concerning their engagement with issues of sovereignty, ownership, and working in more politically attuned ways with country systems, political discourses, existing policy channels, and competing rights claims over resources. Projects such as USAID's BIOREDD+ appear to offer some hope for community engagement. The just completed project is one of USAID's flagship climate change mitigation initiatives. The project works with Afro-Colombian and indigenous communities on the Pacific Coast of Colombia to prepare eight REDD+ projects involving 19 communities that are the proponents and direct owners of the projects (Greg Minnick, personal communication, 2015). BIOREDD+ may offer insights as to the degree to which community ownership coupled to outside facilitation through a USAID-implementing partner with no vested stake in the carbon credits (Chemonics International, in this case) offers a model for enhanced feasibility and FPIC through donor funded projects.

However approached, both rational choice contractarianism and broader social protection concerns will come to play in any REDD social contract negotiations. Figuring out how to do this—for example, piloting social contract negotiations in REDD—should become a high priority for all concerned with the sustainability of REDD and avoided deforestation outcomes.

The basis of the social contract I am suggesting, from either a national planning process level or a specific project level in local jurisdictions, is the following:

1. It must be multi-stakeholder negotiated, with local communities appropriately represented.
2. The participants in the contract need to be identified specifically in each case. Government and communities are the two key pillars, yet, without an entity able and willing to finance a negotiated outcome between the two key stakeholders, nothing will be achieved. This third pillar is to be identified among potential impact investors, an untapped group, who believe in the mission of avoided deforestation, have risk capital available to bring to

the table, see that current strategies are not working, and acknowledge that multi-stakeholder negotiation is needed.
3. Technical agencies like UN-REDD and the World Bank are best placed as subsidiary agents to support national processes driven by governments and frontline forest communities. Their role is both technical and financial. At present they maintain overwhelming leverage over all funding and implementation aspects, and in the case of the World Bank through its Forest Carbon Partnership facility is "creating new property rights to carbon that impinge upon existing statutory and customary rights of Indigenous Peoples and local communities, without clear safeguards or measures to prevent conflicts and negative impacts on community rights" (Rights and Resources Initiative 2014).
4. Representation of the necessary stakeholder groups to enable a negotiated process to work must be determined locally, taking ethnic and intra-community variation into account, and can be the object of external facilitation.
5. The social contract focuses on principles—who has a right to participate, how benefits sharing will be decided upon, how oversight will occur, how authorities and management responsibilities will be attributed, what types of recourse stakeholders will have in case of dispute or conflict, and how steering committees and rules will be established. More specific details will however be needed. This will require leadership and vision.

At present, none of these five points is applicable in REDD currently, nor is there any indication that planners have thought about them.

Conclusions

In *Redeeming REDD* (Brown 2013), I argue that contrary to the technocratic project that REDD has largely become, its efficacy will be determined by the quality of social relationships that various groups of people weave while building viable REDD initiatives. If REDD is to become sustainable, the groups that will be involved—local and indigenous peoples, local and national government agencies, international donor agencies, national and international investors, local and international NGOs, standards setters and auditors of carbon—must collectively and cooperatively bring more strategic thinking to the table than present indicators suggest. Superficial FPIC pronouncements and validation from NGO or private sector verification bodies with economic stakes in advancing REDD will not produce sustainability on the ground.

Just as conservation is a social and political process (Brechin et al. 2002), REDD will require social collaboration and political coherence to succeed (See Locatelli et al. 2008). The first step in launching such a process is establishing a credible social contract binding government, communities, and financiers at national and lower jurisdictional levels. This gap will perpetually paralyze REDD unless corrected.

The issue will become more acute if community analytical capacities are strengthened, so that in the course of conducting participatory rural appraisals and FPIC exercises, communities are given the chance to assess the costs, benefits, and risks of signing on to REDD.[11] With enhanced information availability, if not transparency, communities are understanding more about how the development world operates. Given a chance to reach an informed decision, if people assess that REDD offers a favorable option to the status quo or to what they perceive as potentially available in time, then a basis for social feasibility can be established. On the other hand, REDD will increasingly be rejected, to the detriment of planners and the global environment if feasibility is not perceived.

Yet, why would the approach to social contracts and feasibility change at this point?

The obvious answer is that while civil society and NGO opposition to REDD remains vibrant, forces for the status quo are strong. These forces, represented by UN-REDD, large conservation organizations, voluntary standard setters, and bilateral development agencies, have the means to propel REDD programming forward on the basis of current perceptions of adequacy. There is no compelling reason to believe that the situation will change, barring financial support from countervailing sources, which presently remains speculative.

Because this chapter will largely be read by social scientists, my thoughts are focused on what social scientists can do to enhance the prospects of REDD working for impoverished forest communities, who are customarily targeted in REDD programming.

1. REDD premises can be challenged, and more thought can be given to the opportunity costs faced by local stakeholders participating in REDD, versus the principal focus on social safeguards, which while worthy has not considered opportunity costs.
2. More attention can be paid to thresholds for FPIC—that is, we need to develop a firm understanding whether the standard used to verify FPIC by certifying bodies, using information in the public domain, is credible when considering what informed consent should constitute, given community complexities.
3. The scope for impacting what is required for REDD to become socially sustainable (and socially sound) can be made by social scientists. Consideration of both social contracts and social feasibility as a means to achieve social sustainability is merited. Conversely, applied research could consider whether current strategies relying on consultation, participatory rural appraisals, social safeguards, and Climate, Community and Biodiversity Alliance (2014) certification to provide disinterested, third-party verification of REDD programming are sustainable and socially sound (see Russell 2012).

Social scientists have an opportunity to go beyond what to date have been their principal preoccupations in REDD: safeguards, benefit sharing, and FPIC—all worthy pursuits. By focusing more on social contracts and social feasibility, social scientists can catalyze transformational change through REDD that will contribute to poverty reduction and sustainable natural resource management. By focusing more on establishing a framework for approaching social contracts, social scientists can contribute to positive, transformational change in poverty reduction and natural resource management.

Whether it will be academic anthropologists or professional development anthropologists who engage, *someone* has to start asking more uncomfortable questions about the assumptions underpinning REDD. More reflection needs to be placed on whether current best practices such as FPIC and social safeguards are really good enough. And if they are deemed good enough by decision-makers, what are the criteria used to judge it so, beyond reflecting consensus among vested stakeholders with overriding self interest in perpetuating the status quo?

Most simple of all, social scientists could consider whether anything resembling a negotiated social contracting process in REDD currently exists or if a more formalized social contracting is worthy. To leave it to vested stakeholders to enhance social feasibility for local peoples involved in REDD is overly optimistic. Here, I argue, both social scientists and impact investors, who have the ability to leverage significant resources to serve as game changers as to how complex development challenges are approached, have potentially important roles to play in advancing social contracts as a key precondition for achieving socially feasible REDD programming.

Notes

1. I will employ the term REDD (Reduced Emissions from Deforestation and Forest Degradation) to refer to all iterations of the concept of carbon mitigation through the preservation of standing forest carbon stocks, also known as "avoided deforestation." REDD+ has become "the mitigation option with the largest and most immediate carbon stock impact in the short term per hectare and per year globally as the release of carbon as emissions into the atmosphere is prevented" (see UNFCC 2014). Many multilateral initiatives such as the UN-REDD Programme, the Forest Carbon Partnership Facility (FCPF), and Forest Investment Program (FIP), hosted by the World Bank, have been supporting it. So too have national development agencies like USAID and the Norwegian Agency for Development Cooperation. A description of US support for the REDD architecture and finance is found at http://www.usaid.gov/sites/default/files/documents/1865/2010-USG-SL-REDD-Strategy-Brochure.pdf. REDD. Projects are also being implemented by conservation and development NGOs on a decentralized and voluntary basis. Across all REDD projects is a commitment to the sustainable management of forest carbon stocks applying safeguards, which "are policies or mechanisms, sometimes expressed as principles, intended to 'do no harm,' and sometimes also to 'do good,'" (USAID, LEAF, and FCMC 2012).
2. Recently the terms "governance" and "good governance" are being increasingly used in development literature. See http://www.unescap.org/resources/what-good-governance, for example.
3. For the full list of partners, see http://unfccc.int/adaptation/knowledge_resources/databases/partners_action_pledges/items/5005.php?nwp=org.

4. See Ecosystem Marketplace's "Maneuvering the Mosaic: State of the Voluntary Carbon Markets 2013" for the most complete overview of where REDD fits in voluntary market activities: http://www.forest-trends.org/vcm2013.php.
5. Social feasibility goes beyond what USAID parlance refers to as "social soundness." Social feasibility as used here requires that evidence be forthcoming and that actions and policies work as intended, not simply that they do no harm. The burden is placed on upstream design of feasible actions, rather than assuming that safeguards and adaptive management based primarily on monitoring and evaluation will generate feasibility or soundness.
6. The impact of FSC standards on REDD can be ascertained from the following quote from Ben Vickers, UN-REDD Technical Advisor:

> FSC has more potential clout than it takes credit for, especially in general forest policy debates. It is the only international organization with social criteria for forest management, and can use this as an entry point—as a gold standard—especially now with the massive influx of finance linked to the climate change and REDD debates. This is a once in a generation opportunity and mustn't be missed. (FSC 2011)

7. For specific examples of where the standard appears to fall short, See the Mai Ndombe project implementation report and separate monitoring plan submitted for meeting the FPIC standard under the CCB Alliance requirements, for example: https://s3.amazonaws.com/CCBA/Projects/Mai_Ndombe_REDD_Project/Mai+Ndombe+Final+CCB+PIR.pdf and https://s3.amazonaws.com/CCBA/Projects/Mai_Ndombe_REDD_Project/Mai+Ndombe+CCB+Final+Monitoring+Plan.pdf. SSNC (2013, 15–24) also provides examples of where the standard falls short.
8. See http://www.redd-monitor.org/.
9. See Replogle (1989) for a discursive analysis of social contracts from a rights-based perspective.
10. Peters-Stanley and Gonzalez (2014, vii) note "the limited scale of purely voluntary action alone was increasingly evident last year, when three of every four offsets transacted were sold to preexisting clients," meaning few participants are entering any carbon offsets markets.
11. The Community Options Analysis and Investment Toolkit (COAIT) offers guidance on how community analysis was approached in one USAID-funded program in central Africa, the Central African Regional Program for the Environment from 1998–2006 (see Bonis Charancle et al. 2009).

References

Anderson, J., M. Colby, M. McGuahey, and S. Mehta. 2013. *Nature, wealth and power 2.0: Leveraging natural and social capital for resilient development*. Washington, DC: US Agency for International Development.

Berlin, I. 1969. *Four essays on liberty*. Oxford: Oxford University Press.

Black, S. 2001. The rational and the fair (bilateral exchange, multilateral contexts, social contract). *Pacific Philosophical Quarterly* 82 (2): 115–144.

Blom, B., T. Sunderland, and D. Murdiyarso. 2010. Getting REDD to work locally: Lessons learned from integrated conservation and development projects. *Environmental Science & Policy* 13 (2): 164–172.

Bonis Charancle, J.M., A.M. Tiani, M. Brown, G.A. Akwah, Z. Mogba, G. Lescuyer, R. Warne, and B. Greenberg. 2009. Strengthening local analytical capability: Community options analysis and investment. In *Search of common ground: Adaptive collaborative* management in Cameroon, eds. M.C. Diaw, T. Aseh, and R. Prabhu, 275–300. Bogor, Indonesia: Center for International Forestry Research (CIFOR).

Brechin, S., P. Wilshusen, C. Fortwangler, and P. West. 2002. Beyond the square wheel: Toward a more comprehensive understanding of biodiversity conservation as social and political process. *Society and Natural Resources* 15: 41–64.

Brown, M. 2013. *Redeeming REDD: Policies, incentives and social feasibility for avoided deforestation.* New York: Routledge.

Climate, Community and Biodiversity Alliance. 2014. *CCB standards.* http://www.climate-standards.org/ccb-standards/.

Forest Stewardship Council (FSC). 2011. FSC certification—a gold standard promoting social rights of Asian smallholders. *Interview with Ben Vickers.* June 28. https://mbasic.facebook.com/notes/forest-stewardship-council-fsc/fsc-certification-a-gold-standard-promoting-social-rights-of-asian-smallholders-/10150242829739872/.

Friends of the Earth. 2008. *REDD myths: A critical review of proposed mechanisms to reduce emissions from deforestation and degradation in developing countries.* December. http://www.elaw.org/system/files/15+foei+forest+full+eng+lr.pdf.

Friends of the Earth. 2009. Sub-prime carbon: Rethinking the world's largest new derivatives market. http://libcloud.s3.amazonaws.com/93/77/4/452/SubprimeCarbonReport.pdf.

Gari, J. 2013. Pilot projects versus national policy in REDD arena. *UN-REDD Programme Newsletter* 39 (June/July). http://www.un-redd.org/Newsletter39/PilotProjectsversusNationalPolicy/tabid/129673/Default.aspx.

Griscom, B., P. Ellis, and F. Putz. 2014. Carbon emission performance of commercial logging in East Kalimantan, Indonesia. *Global Change Biology* 20 (3): 923–937.

Hickey, S. 2011. *The politics of social protection: What do we get from a 'social contract' approach?* Chronic Poverty Research Center Working Paper 216 (July). Manchester: Institute for Development Policy and Management, University of Manchester.

Karsenty, A. and D. Ezzine de Blas. 2014. Du mésusage des métaphores: Les paiements pour services environnementaux sont-ils des instruments de marchandisation de la nature? In *L'instrumentation de l'action publique,* eds. C. Halpern, P. Lascoumes, and P. Le Galès, 161–89. Paris: Presses de Sciences Po.

Karsenty, A., A. Vogel, and F. Castel. 2014. Carbon rights, REDD and payments for environmental services. *Environmental Science and Policy* 35 (January): 20–29.

Locatelli, B., Kanninen, M., Brockhaus, M., Colfer, C.J.P., Murdiyarso, D., and Santoso, H. (2008). Facing an uncertain future: How forests and people can adapt to climate change. Forest Perspectives no. 5. CIFOR, Bogor, Indonesia.

Marx, A., M. Maertens, J. Swinnen, and J. Wouters. 2012. *Private standards and global governance: Economic, legal and political perspectives.* Northampton, MA: Edward Elgar.

Naughton-Treves, L. and C. Day, eds. 2012. *Lessons about land tenure, forest governance and REDD: Case studies from Africa, Asia and Latin America.* http://www.nelson.wisc.edu/ltc/docs/Lessons-about-Land-Tenure-Forest-Governance-and-REDD.pdf.

No REDD in Africa Network. 2014. *Kenya preparing for REDD in the Embobut Forest and forcing Sengwer People "into extinction".* January 31. http://www.no-redd-africa.org/images/pdf/sengwer-evictions-redd.pdf.

Ostrom, E. 1990. *Governing the commons: The evolution of institutions for collective action.* Cambridge: Cambridge University Press.

Peters-Stanley, M. and G. Gonzalez. 2014. *Sharing the stage: State of the voluntary carbon markets 2014.* Washington, DC: Forest Trends' Ecosystem Marketplace.

Replogle, R. 1989. *Recovering the social contract.* Totowa, NK: Rowman & Littlefield.

Rights and Resources Initiative. 2014. *Status of forest carbon rights and implications for communities, the carbon trade, and REDD investments.* http://www.rightsandresources.org/documents/files/doc_6594.pdf.

Rousseau, J. J. [1754] 2005. *A discourse on the origin of inequality and a discourse on political economy.* Stilwell, KS: Digireads.

Russell, D. 2012. Principles of social and environmental soundness form natural resource management (NRM) for REDD. Presented at the Forest Carbon, Markets and Communities training workshop on social and environmental soundness of REDD, Bangkok, November.

Swedish Society for Nature Conservation (SSNC). 2013. *REDD plus or REDD "light"? Biodiversity, communities and forest carbon certification.* http://www.naturskyddsforeningen.se/sites/default/files/dokument-media/REDD%20Plus%20or%20REDD%20Light.pdf.

Tanzania Natural Resource Forum. 2011. *REDD realities: Learning from REDD pilot projects to make REDD work.* http://www.tnrf.org/files/e-REDD%20Realities.pdf.

Tollefson, C., F. Gale, and D. Haley. 2009. *Setting the standard: Certification, governance, and the Forest Stewardship Council.* Vancouver: University of British Columbia Press.

United Nations Framework Convention for Climate Change (UNFCC). 2014. *Methodological guidance for activities relating to reducing emissions from deforestation and forest degradation and the role of conservation, sustainable management of forests and enhancement of forest carbon stocks in developing countries (SBSTA).* http://unfccc.int/methods/redd/methodological_guidance/items/4123.php.

US Agency for International Development (USAID). 2012. *Request for proposal SOL-OAA-12–0000–50, measuring impact.* April 17. Washington, DC: USAID.

USAID, Lowering Emissions in Asia's Forests (LEAF), and Forests, Carbon, Markets, and Community (FCMC). 2012. Southeast Asia regional training workshop: Social and environmental soundness in REDD programs and projects. Bangkok. November 5–9.

Venne, M. 2007. *An analysis of social aspects of Forest Stewardship Council forest certification in three Ontario case studies.* MA thesis, Wilfred Laurier University, Waterloo, ON.

White, A. 2007. *Is it time to rewrite the social contract?* April. http://www.tellus.org/publications/files/BSR_AW_Social-Contract.pdf.

Zwick, S. 2013. Unpacking Warsaw, part two: Recognizing the landscape reality. *Ecosystem Marketplace.* December 5. http://www.ecosystemmarketplace.com/pages/dynamic/article.page.php?page_id=10089.

13

A WIN-WIN SCENARIO?

The Prospects for Indigenous Peoples in Carbon Sequestration: REDD Projects in Brazil

Janet Chernela and Laura Zanotti

Proponents of international programs to commoditize carbon sequestered through "avoided deforestation" portray it as a strategy that is favorable for conservation, development, and equity outcomes. By reducing greenhouse gas (GHG) emissions from deforestation, the programs promise to contribute to climate change mitigation and biodiversity conservation. By financially rewarding carbon storage, they promise to provide needed income for rural development. By transferring financial resources from northern temperate nations to southern tropical ones, proponents hope to address inequities in global wealth distribution. Advocates of the approach hold it to be a win-win situation in which three objectives—environmental protection, rural economic development, and global equity—are achievable simultaneously.

In this chapter we consider the prospects that the international carbon sequestration mechanism known as REDD (Reduced Emissions from Deforestation and forest Degradation) holds for indigenous peoples of the Brazilian Amazon. Our view is nuanced, as we weigh potential advantages and disadvantages to the approach. We conclude that evaluations cannot be made without taking into consideration the financial, social, and policy contexts in which the projects are to take place. We argue that the outcomes of REDD projects are largely dependent upon factors that are *external* to it. The specifics of any REDD proposal, therefore, can only be evaluated in the context of the conditions in which it will occur—conditions that include policies that would favor or impede project outcomes and may not be known at the time of planning. These include policies relating to consent procedures as well as government policies regarding international REDD funding. Moreover, our discussion recognizes the problem, often obscured in discussions of REDD, that evaluations cannot be assumed to be equivalent for proponents and stakeholders alike. For this reason, any totalizing evaluation, including win-win, must be regarded as premature. We suggest, instead, that a more useful analytic approach is to consider the specific financial and policy conditions that contribute to the success of a REDD project. By considering local evaluation and outside influence we should be able to more accurately assess, predict, and explain project outcomes and their potential costs, benefits, and impacts.

To explore these issues we consider three REDD projects involving indigenous peoples and lands in the Brazilian Amazon. The first is the Suruí Carbon Project, proposed by the Paiter-Suruí in the western Amazon. The second is the Cinta Larga Avoided Deforestation Project, proposed by the private investment company, Viridor, involving the Cinta Larga in the western Amazon. Lastly, the third is the Xingu SocioEnvironmental Carbon Project (XSEC), proposed by a consortium of nongovernmental organizations (NGOs) for almost 20 diverse indigenous peoples in the southeastern Amazon. The project proposals, in different stages of development, differ in significant ways, including size of carbon stocks, modes of project participation, market strategies, and the percentage of indigenous territories covered by the REDD program. We will use these cases to identify relevant criteria, evaluating each according to (1) conservation of nature, (2) economic payoff, and (3) social equity.

Comparing the three projects, we explore the way different procedures, policies, and practices affect outcome, increasing or decreasing the likelihood that any single objective will be met. We conclude that whether objectives will be met depends upon the stipulations of the proposal as well as the conditions in which these should emerge.

REDD: A Brief History

In 1997 the Kyoto Protocol called the attention of the international community to the high contributions of greenhouses gases from forest conversion, but provided no measures to reduce them.[1] At the 2005 meeting of the Conference of Parties in Montreal, the governments of Costa Rica and Papua New Guinea proposed a system of carbon credits in which financial resources from wealthier northern nations would be transferred to developing southern states to create incentives to reduce deforestation. The trading scheme would benefit tropical forest countries with high carbon stocks while also permitting industrial producers to maintain high emissions rates. The proposal was advanced in 2007 when the Conference of Parties agreed on a mechanism to accomplish these goals, launching the UN collaborative initiative on Reducing Emissions from Deforestation and Forest Degradation in Developing Countries, known by its acronym, REDD. At its 2009 and 2010 meetings in Copenhagen and Cancun, respectively, the mechanism was augmented to include the conservation and sustainable management of forests and enhancement of forest carbon stocks; this expanded version would be referred to as REDD+.

A number of funding initiatives have been established to receive and distribute REDD-related funds. The major multilateral funds are the UN-REDD Programme, the World Bank's Forest Carbon Partnership Facility, and the World Bank's Forest Investment Programme. In addition, several dedicated climate funds have been established through bilateral arrangements from source governments, like Norway, to tropical nations. Principal among these is the Amazon Fund, which accounts for a substantial portion of the financial flow toward REDD projects.

With 60 percent of the Amazon basin, the largest tropical rainforest in the world, Brazil is considered by many analysts to be the world's most important supply-side player in negotiations toward a climate framework that includes reducing deforestation emissions. The nation is not only a major producer of GHG; it is also a principal holder of carbon stocks in its vast standing forests. With high rates of deforestation as well as large stocks of carbon, Brazil is well positioned to receive carbon funds. In fact, the majority of international REDD+ funding for 2012 took the form of pledges to Brazil's Amazon Fund, totaling twice those of the World Bank's Forest Carbon Partnership Facility and five times those of the UN-REDD Programme (Overseas Development Institute 2014).

Brazil has been slow to adopt a plan for marketing credit for REDD+. In 2008 President Lula da Silva proposed Brazil's National Climate Plan, which committed the country to a 70 percent reduction in national deforestation over 10 years. Although draft laws have been presented in the national congress, none has been accepted. At the same time, the Environment Ministry is developing a proposal for a national REDD+ policy, and has held preliminary consultations with Amazon states and NGOs (Schwartzman et al. 2013).

Despite Brazil's reluctance to officially recognize an international market for carbon credits, it has made substantial advances in programs that compensate producers for ecosystem services such as carbon sequestration. In the northern Amazonian states, a number of REDD-like projects were instituted in the early 2000s, including the Bolsa Floresta program, which rewards traditional communities for their commitment to stop deforestation. Proambiente, one of Brazil's earliest programs of payment for ecosystem services, has been cited as a model for REDD schemes. These projects laid the legislative and political bases for further projects in reducing deforestation. In this regard, Brazil has been a pioneer.

The Role of Indigenous Lands in Carbon Sequestration

Today, about 91,000 indigenous people live in 654 federally recognized indigenous territories (IT) in Brazil. ITs constitute 12 percent of the national territory and 22 percent of the Amazon region, or about 1 million km^2 (ISA/Forest Trends 2010). In Amazonia the amount of land in Indigenous Territories exceeds all other types of protected areas, amounting to 955,710 km^2 (22 percent), compared to only 351,105 km^2 (8 percent) in Strict Protected Areas and 534,226 (12 percent) in Sustainable Use Protected Areas.

Moreover, indigenous lands in Brazil are protected by government mandate as well as the active monitoring presence of the indigenous inhabitants. Brazil's 1988 constitution confers exclusive usufruct rights to indigenous peoples on the basis of non-transferable, "original" rights. The land is owned by the nation-state, which the law stipulates is responsible for its protection. A federal Indian affairs agency, the National Indian Foundation (FUNAI), is charged with the role of representing the rights and interests of indigenous peoples and may call in federal enforcements if federally protected rights are violated. These legal protections, when enforced, provide a level of "excludability" not found in other forms of protected areas.

Indigenous territories have been more effective than other protected area types in maintaining standing forests and sequestering carbon. According to a 2007 study (Saatchi et al. 2007), biomass on indigenous lands accounts for 30 percent of total carbon stocked in the Brazilian Amazon (47 PgC [petagrams of carbon]). For this reason, among others, government agencies, scientific researchers, financial investors, conservation organizations, and indigenous peoples are carefully considering the implications of REDD projects in indigenous territories.

Modes of Participation

Projects of reduced deforestation entail the contributions of many actors and substantial resources. The initiator of a project is customarily designated the "project proponent" whereas those providing the carbon stocks are designated as "stakeholders." The Amazonian Institute of Sustainable Development (IDESAM) and The Nature Conservancy (TNC) define the proponent as the lead implementing institution that "holds the rights over the generated REDD credits and will be the main entity responsible for implementing project activities" (Cenamo et al. 2010, 25). The project proponents of 26 subnational forest carbon projects listed by May and Millikan (2010) for the Brazilian Amazon were distributed this way: 28 NGOs; seven government agencies; two research institutions; one for-profit carbon trading firm; and one indigenous community. Stakeholders are those affected by the project and constrained by its guidelines. In projects involving their own lands, indigenous peoples are designated stakeholders; in some instances, as we will see, they are also proponents.

Three Cases

The Suruí Carbon Fund

The case of the Paiter-Suruí in the western Brazilian state of Rondônia provides an important example in discussions of REDD and indigenous peoples. An estimated 1,231 Paiter-Suruí occupy 247,845 ha of submontane open and dense ombrophilous forests in Rondônia and Mato Grosso states. The objective of their

REDD proposal, the first in Brazil by an indigenous organization, is to prevent 13,575 ha of tropical rainforest from being cleared in the territory by 2038, and thereby avoid the emission of 7,258,352 tCO_2e^2 (IDESAM and Metareilá 2011).

Since the 1800s, the Suruí peoples have been impacted by rubber-gatherers, ranchers, and miners (Mindlin 1985). Migration into the region intensified in the 1960s, when the federal government created incentives for small-scale farmers from the developed south to resettle in the western Amazon territories. Ongoing irregularities and violence among the Suruí in the 1960s contributed to a removal of the federal agency charged with their protection.

The Suruí territory was formally demarcated in 1976 and legally registered in 1983 as the Sete de Setembro Indigenous Area. Despite its legal status, Suruí land continued to be invaded by loggers and ranchers. In the 1980s a new development project along highway BR-364 brought in over 200,000 migrants per year, which led to armed clashes and eventually resulted in the Suruí losing half of their territory (Hemming 1978; Mindlin 1985, 2011). Suruí communities then reorganized to situate villages strategically along the territorial edges to prevent incursion and protect remaining timber stocks (Vitel et al. 2013). However, invasions into the territories continue into the present, and the Suruí remain skeptical that government agencies will fulfill promises or remove illegal invaders. Following a series of government-financed development projects that resulted in severe deforestation and population decline, the Suruí began to engage in market activities such as coffee production and cattle ranching. They also began to organize on behalf of their well-being.

In 1988 the Suruí created the Metareilá Organization of the Paiter Indigenous People (Associação Metareilá). The organization continues to take public stands in favor of conservation of natural resources and in opposition to the illegal sale of lumber. It declares logging an environmental offense and evicts loggers from the territory. In addition to reforestation and conservation, the organization invests in training and education, sending young Suruí to study outside the area (Von Mittelstaedt 2010). The first university graduate was Almir Narayamoga Suruí, chief of the Gamep clan, who is a principal author of the Suruí Forest Carbon Project. Almir has since garnered important experience, working with the World Bank in 1996 to improve indigenous participation in a socio-ecological zoning plan, and with the Amazon Conservation Team (ACT-Brasil) in 2006 to institute a Suruí 50-year plan for sustainability. ACT is a Virginia-based organization that specializes in training indigenous peoples in the use of geographic information system (GIS) technology to map their lands for purposes of demarcation, monitoring, and registering cultural and biological resources. As part of the project, Suruí mappers were trained in the use of cameras, GPS (Global Positioning System) equipment, GIS systems, laptops, and Internet tools such as Google Earth (Gaspar and Machado 2007). The Suruí have already begun to use Google Earth to detect and expel wood poachers and gold prospectors (Von Mittelstaedt 2010).

A simulation of potential consequences showed that the Suruí could lose 30 percent of their forest within 50 years, and all of it by the end of the twenty-first century, if they took no protective action (Von Mittelstaedt 2010). Chief Almir approached the Washington-based firm Forest Trends to seek assistance in managing and reforesting Suruí lands. The result, based on the 50-year land management plan, is the Suruí Forest Carbon Project (Chernela 2011). The objective of the project is to protect 13,575 ha from being cleared by 2038, thereby avoiding the emission of more than 7 million tCO_2e (IDESAM and Metareilá 2011), and to sell the carbon credits on the voluntary carbon market. The project is expected to contribute to the sustainable development of the Suruí lands by building local capacity to manage forestry operations and establish an information technology center within Suruí land (ISA and Forest Trends 2010). At the 2009 COP 15 in Copenhagen, representatives announced a partnership between the US Agency for International Development (USAID), the Suruí, and Forest Trends, a US-based NGO, to organize a fund to sell carbon credits generated in Suruí lands. USAID is one of several donors to the Suruí REDD preparation process (USAID 2014).

Associação Metareilá is the named proponent of the carbon sequestration project. The role of the Associação Metareilá is to administer the project, overseeing the allocation of all resources to the Paiter-Suruí. Four additional NGO partners contribute expertise: (1) Kanindé, the Associação de Defesa Etnoambiental (Association for Ethno-environmental Defense), a Brazilian-based, not-for-profit, civil society organization working to defend indigenous rights and promote sustainable business practices; (2) Amazon Conservation Team (ACT-Brasil, Equipe de Conservação da Amazônia), discussed earlier; (3) IDESAM, the Instituto de Desenvolvimento Sustantável do Amazonas (Amazonian Institute of Sustainable Development), a Brazilian NGO providing technical capacity, responsible for quantifying carbon emissions and modeling carbon emissions reduction scenarios; (4) Forest Trends, a non-profit organization that promotes the marketing and investment of forest product initiatives. The Suruí project continues to evolve, but is hampered by continuing uncertainties surrounding the rights of indigenous peoples to develop carbon projects, sell carbon credits, and receive REDD benefits directly.

In April 2012, the Suruí became the first indigenous group in the world to win international certification for a forest carbon conservation project to reduce GHG emissions. The Suruí Forest Carbon Project was validated under both the Verified Carbon Standard (VCS) and the Climate, Community and Biodiversity (CCB) standard, the dominant standards for accrediting REDD programs. Another important step was made in September 2013, when the Suruí announced a partnership with Natura Cosméticos, Latin America's largest cosmetics maker, to sell 120,000 tons of carbon offsets from the Suruí Forest Carbon Project (Business Wire and Forest Trends 2013). This arrangement provides the Suruí with a client for their carbon offsets. The arrangement also allows Natura to meet the self-imposed emission reduction goals of its Natura Carbon Neutral Initiative (Business Wire and Forest Trends 2013).

Although this arrangement is a major boost to the Suruí project, concerns remain over the falling price of carbon credits and the long-term stability of the market. This deal represents only a fraction of the intended sequestration of the Suruí project, which aims to sequester over 5 million tons of carbon dioxide over the next 30 years, but it is a significant step toward the Surui goal.

The Cinta Larga Avoided Deforestation Project

The Cinta Larga live in 34 communities spread out over four indigenous areas in Mato Grosso and Rondônia states in western Brazil. The territories total 27,000 km^2 (2.7 million ha). The present population of 1,700 is about one-third its size of 5,000 in 1968.

Like the Suruí, the Cinta Larga suffered ongoing intrusions into their territories. Throughout the nineteenth and twentieth centuries, waves of rubber tappers, loggers, miners, ranchers, and land speculators poured into their lands. Invasions reached a peak with the inauguration of Brazil Highway 364 into the western Amazon and resulting atrocities conducted by rubber and diamond companies in the 1950s and 1960s (Mindlin 2011). The Figueiredo Report (1967), commissioned by the Ministry of the Interior, revealed a history of torture, slavery, sexual abuse, land theft, and mass murder. The report found the federal agency then charged with protection of indigenous peoples—the Serviço de Proteção aos Índios (SPI)—responsible for breaches of human rights and for the extermination of entire tribes. The exposure led to the closure of the agency and its replacement by FUNAI in 1969. FUNAI attraction posts in the mid-1970s drew the Cinta Larga out of the forest and brought disease. Within three decades the Cinta Larga were reduced to a third of their former population (Leroy 2003).

Between 1989 and 1991, four contiguous Cinta Larga territories were demarcated. Despite the federal demarcation of the lands, conflicts continued to arise over the Cinta Larga territories. In 1999, news of diamonds drew thousands of illegal prospectors to the area. The Brazilian Intelligence Agency and the federal police estimate that over US$20 million in diamonds are illegally extracted from the Roosevelt Reserve each year (Toni et al. 2011). Today 1,600 Cinta Larga are divided into four contiguous demarcated indigenous territories: Aripuanã Park (1,603,250 ha), Roosevelt (230,826 ha), Serra Morena (147,836 ha), and Aripuanã (750,649 ha). Together, the Cinta Larga territories total about 2.7 million ha.

In late 2009, the Florida-based firm, Viridor Carbon Services, initiated negotiations with the Cinta Larga for an avoided deforestation project (Chernela 2011). In February 2010, Viridor representatives met with 62 Cinta Larga. On the basis of that meeting, Viridor writes, "Indigenous leadership is actively participating in all phases of development to ensure protection of their concerns and traditions" (Viridor 2013). The proposed project allows Viridor rights to commercialize the full extent of Cinta Larga territories —100 percent of the 2.7 million ha belonging to the Cinta Larga (Viridor 2013). Viridor presented the Cinta Larga with a 50-page legal contract valid for 50 years (Toni et al. 2011), and has offered to pay US$375,000 to the Cinta Larga in advance (Toni et al. 2011). The promise of these funds prompted the Cinta Larga to expedite authorization of the project by FUNAI. The Cinta Larga appealed to the Federal Public Attorney (MPF) who, in turn, called for caution (Rondônia Ao Vivo 2010).

In 2012, a contract laying out the terms for an avoided deforestation project with Viridor Carbon Services was signed by a Cinta Larga man called Chief Marcelo. In return, Chief Marcelo was given an advance payment of two trucks (Salomão 2012). It is not known whether Chief Marcelo has any authority to enter into such a binding arrangement. FUNAI described the signing of the contract this way: "[It] immobilizes the whole area and was negotiated by only a few individuals of the community, without the consent of all community members" (Nery et al. 2013, 12).

Although describing themselves as an independent carbon trading company based in Ponte Vedra Beach, Florida, Viridor Carbon Services appears to be a subsidiary or shell company belonging to Viridor Waste Management, a waste collection and landfill disposal firm incorporated in Britain and owned by the Pennon Group (Viridor 2013). The company, one of the largest waste disposal firms in the industry, is facing complaints from residents, governments, and environmental agencies over importing waste, negative impacts on recycling, toxic leakage, and air pollution.[3] In a policy context of carbon credits, Viridor can extend credits from its Amazon holdings to pay for the costs of pollution incurred in its GHG-emitting European facilities.

So far, FUNAI has declined to endorse the contract, questioning the legality of the arrangement and expressing concerns that its design may prevent the Cinta Larga from carrying out basic subsistence activities, such as clearing forests for planting, without seeking permission from the company. Since FUNAI's earlier refusal to endorse the contract, Viridor made some alterations, and now describes the project as being "developed as a partnership between the Cinta Larga people and Viridor Carbon Services" (Viridor 2013). Nowhere are the Cinta Larga shown to be legal partners.

In late 2012 Viridor Carbon Services announced that it was "actively seeking partners" to finance the Cinta Larga project, which it describes as the largest development project of avoided deforestation in indigenous communities (Salomão 2012). There is no evidence of it having been approved by either standardizing agency to date. This project is particularly concerning to observers because of the history of exploitation and discrimination that the Cinta Larga have experienced, and the lack of transparency from Viridor. However, the Cinta Larga have several advocacy organizations, including a Cinta Larga Council (CPCL) and the Coordination of Indigenous Organizations of the Cinta Larga Peoples (Patjamaaj) working on their behalf. They also maintain relations with COIAB, the Coordination of the Indigenous Organizations of the Brazilian Amazon, and Kanindé, the indigenous rights and sustainability organization working with the Suruí. The project remains contentious.

The Xingu SocioEnvironmental Carbon Project (XSEC)

The Xingu SocioEnvironmental Carbon Project (also called the Xingu Indigenous Lands Ecosystem Services Project), which we refer to as XSEC, is a large project located in the southeastern Amazon in the states of Pará and Mato Grosso. It aims to conserve the landscapes of over 25 indigenous groups living in the vicinity of the Xingu River basin.

The project combines three indigenous protected area complexes—one consisting of four territories of Gê-speaking Kayapó (105,833 km^2, population 7,500); the Xingu National Park (26,420 km^2, population 3,170); and the Panará reservation (4,880 km^2, population 170). Together, these federally recognized units comprise a continuous north-south corridor of some 14 million ha, containing two endangered biomes with some of the richest biodiversity in the world. When joined with the recently created 5-million ha protected region to the north, the Terra do Meio, the combined areas total 22 million ha (Schwartzman and Zimmerman 2005).

The region's indigenous peoples, whose lands are legally guaranteed to them, play a critical role in preserving the region's biodiversity in a context of relentless pressure from mining, logging, ranching, and export agriculture. Most recently, large soy plantations have pushed ranchers and colonists ever closer to the reserve (Turner 1995; Zanotti 2009).

The XSEC involves several types of participation. The project builds on three long-standing initiatives: one involving the Kayapó, Conservation International (CI) and the International Conservation Fund of Canada (ICFC); one between the peoples of the Xingu Park and the Brazilian NGO Instituto Socioambiental (ISA); and one between the US-based Environmental Defense Fund (EDF) and the Panará peoples. The five proponent NGOs bring important conservation, technical, and legal expertise to the endeavor. Among them are four international organizations: CI, the EDF, the Wild Foundation, and the ICFC. CI, one of the largest environmental organizations in the world, has been working among the Kayapó since 1992 but reduced its activities to the Kayapó fund. EDF has worked with the Panará for three decades. ICFC and the Wild Foundation maintain partnerships with three Kayapó NGOs: the Instituto Raoni, the Instituto Kabu, and the Protected Forest Association.

The two Brazilian NGOs, the Environmental Research Institute of Amazonia (Instituto de Pesquisa Ambiental da Amazonia, IPAM) and the Social and Environmental Institute (Instituto Socioambiental, ISA) have long-term ties to the groups of the Xingu and to governments, NGOs, and international financial institutions. IPAM is a research institute that engages in scientific activities associated with the Amazonian forest. ISA has worked with indigenous peoples of the basin for many years. For example, ISA and the Xingu Lands Indigenous Association (ATIX), an indigenous NGO that it helped to establish, collaborate on projects involving territorial monitoring and control, bilingual education, and sustainable economic alternatives.

Proponents hope to secure compensation for reductions in deforestation through the voluntary carbon market, including short-term donations, and, if made available, through internationally recognized REDD markets. This project, unlike the two other projects explored in this chapter, approaches deforestation in accordance with the Brazilian government's National Plan on Climate Change. While this strategy may cause delays and sacrifice a certain amount of flexibility, it aims to bring stability to the project by complying with established national guidelines.

In the highly complex and multiple-partner XSEC project, each community will design its own management plan, delineating the zones to be included and excluded from the REDD project. Between 2008 and 2009, the proponent NGOs held 26 regional and village level meetings on climate change, forests, and REDD+ in the indigenous territories (Stephan Schwartzman, personal communication, April 24, 2012). As a consequence, some villages are beginning to develop management plans. Lands to be included in the integral plan will be subject to regulation and monitoring; benefits will be calculated accordingly. In lands outside the REDD designation, ordinary activities, including cutting and burning forests for gardens, may continue.

Discussion

The three project proposals reviewed here exemplify three different approaches to REDD in indigenous lands. Although all are at different phases of the development process, the projects provide an opportunity to consider some of the factors that affect project outcome. It will become apparent that the outcome of a project is largely a function of its characteristics vis-à-vis the context in which it will operate, including national priorities, financial markets, and international environmental policies. It is also apparent that outcomes depend upon the capacity of local groups to assert their agency in different aspects of proposal development, including the ongoing legal structures that may ensure ownership and distribution of benefits and rights. Because these contexts differ, totalizing judgments of REDD projects, whether all good or all bad, are immaterial. In this section, we will briefly compare projects for internal features of size and proportion dedicated to carbon sequestration, decision-making structure, and financial prospects. We then evaluate these features for possible conservation, economic return, and equity outcomes.

Comparison of Proposed Projects

The wide variation in REDD projects is one of the obstacles to across-the-board generalizations. The three considered here, for example, present a number of alternatives. The Suruí case is unusual in that it is designed by the Suruí people.

This contrasts with the Cinta Larga case, where the design is controlled entirely by the private company Viridor, and the Xingu case where the design is a dialogic project between NGOs and indigenous peoples. In the Suruí case, the total amount of land is the smallest, as is the percentage of land dedicated to carbon sequestration. It is surpassed in land area by the Cinta Larga proposed project, and even more so by the Xingu proposed project. In the latter cases the percentage of land dedicated to carbon sequestration has not been determined.

If we consider the projects according to criteria outlined at the outset—conservation, economics, and equity—we find a number of procedures that point to likely outcomes. As we will see, however, the outcomes are largely dependent upon external political and financial conditions that are often outside the control of proponents or stakeholders. We now consider the prospects for benefits in each of the three areas claimed by protagonists of REDD: conservation, economic return, and equity.

Conservation Benefit

The potential for positive conservation outcomes varies across the proposed projects. If the conservation benefit is a simple function of the size of the unit set aside for carbon sequestration, and if we hold landscape features constant, the XSEC should offer the greatest conservation benefit, followed by the Cinta Larga, and then the Suruí. The XSEC integrated REDD proposal provides an extraordinary opportunity for conservation. The huge complex of legally ratified, indigenous territories, with active surveillance, should provide effective protection against predatory encroachment. Moreover, proponents of this project purport to bring together important resources and skill in conservation science and practice.

It must be kept in mind, however, that only a fraction of the indigenous territory will be set aside for carbon storage. The size and location of the parcel subject to the constraints of avoided deforestation must be made by the forest dwellers and must take into consideration long-term resource needs. In the Suruí case, the only example where the project design reflects the producers' land use plans over the long term, the land designated within the carbon project is a 12,217 ha parcel (IDESAM 2011). This fraction of land, amounting to no more than 5 percent of Suruí territory, contrasts with the 100 percent identified in the Cinta Larga/Viridor project. If we assume the same 5 percent apportionment in the cases where the sizes of the set-asides have not yet been determined, the portion of land designated for avoided deforestation and carbon sequestration would be dramatically reduced: 270,000 ha in the Cinta Larga case and 595,428 ha in the XSEC case. Projected carbon amounts would be adjusted accordingly.

Although internal pressure may be a threat, a far greater threat is external pressure from mining, logging, ranching, and export agriculture in the region. The sources of greatest pressure lie in the vicinity of the XSEC project, where two major highways run alongside the indigenous areas and where the Xingu River is threatened with a series of dams. While the XSEC holds the greatest promise for conservation, it is also the project with the largest number of impediments and threats to its existence.

Economic Benefit

A market for emissions is created through a system of credit allowances in which GHG emissions are limited (capped), and monetary value is awarded to avoided emissions. Where carbon is commoditized, businesses that exceed their emission quotas can buy the extra allowances and those who reduce emissions can sell their unused allowances as carbon credits. For sellers of carbon credits, economic gain will be directly proportional to the amount of carbon sequestered and its worth on the carbon market. Net economic gain will have to take into account costs, risks, and terms of distribution.

The projects considered here illustrate three different marketing strategies: in the Cinta Larga/Viridor case the proponent is a private company; in the Suruí case the indigenous peoples are both proponents and stakeholders who sell to buyers from the voluntary market; in the XSEC case the NGO proponents appeal to indigenous stakeholders and government officials alike to create a nationally recognized indigenous REDD area and conservation corridor.

Owning carbon rights in the Brazilian Amazon is a cost-effective strategy for a producer like Viridor, who may exceed its emission allowance and find it advantageous to purchase credits from those who "reduce emissions." The Cinta Larga/Viridor proposal is the only one with ready access to an existing but unstable market. From the standpoint of economic potential, the Cinta Larga project promises investors the most attractive return. However, the political climate for the Cinta Larga/Viridor is the most unreliable, as it may not receive approval from the required Brazilian national agencies charged with the protection of indigenous peoples.

The Suruí Forest Carbon Project is an indigenous-led project with powerful financial and political allies. With an estimated projection for carbon sequestration of no more than 26 million tons (Saatchi et al. 2007), its economic potential is the lowest of the three projects. Yet even this modest amount of income would greatly benefit the Suruí, who have historically had difficulties in sustainable entry into markets (Vitel et al. 2013). In the Suruí case, moreover, the costs of the preparation phase have been paid by international aid. Since there is little or no accrued indebtedness, risks are minimized, at least in this early phase. The risk quotient was reduced substantially when the Suruí made a deal to sell some of their carbon. The financial success of the Suruí project depends upon the future of international and national climate policies that are required for robust carbon trading.

The potential for generating carbon credits is greatest in the XSEC where the total amount of estimated carbon exceeds 600 million tons. Its financial reward, therefore, should be highest. However, its economic return is tied to decisions regarding Brazil's national REDD policy, which continues in abeyance, now three years after the proposal's formulation. Transaction costs are high in all cases, but are highest for the XSEC, where the technical costs are proportional to the vast amount of land and numbers of consultations. A sizable investment must be made in technical expertise, information sharing, and training—costs that subtract from profits. Transaction costs must be borne before the economic returns are in, creating additional economic risk and complicating investment. From this standpoint, while it affords the greatest conservation promise, of the three projects the XSEC investment is perhaps the most costly and involves the most risk.

Equity

The projects are further complicated by the existing uncertainties about the rights of indigenous peoples to develop carbon projects, sell carbon credits, and receive REDD benefits directly. The organization of participant roles in REDD development distinguishes project proponents from stakeholders. According to the definition provided by Cenamo et al. (2010), the project proponent is the entity that holds the rights over REDD credits generated by the project and the principal entity responsible for implementing project activities. A potential dilemma may arise, then, when a project proponent holds the rights over the carbon credits but the owners of the traded assets (stakeholders) are excluded from the decision-making process.

The rights of indigenous peoples to their own resources, and the commodification of those resources, is the subject of Section 231 of the Brazilian Constitution of 1988. The text, however, retains important ambiguities and is subject to differing interpretations. The matter of carbon rights brings these ambiguities into relief. For example, a major concern is whether indigenous peoples are recognized under Brazilian law as the owners of the carbon sequestered in their forests. Different interpretations of the constitution have led analysts to arrive at different conclusions (Barbieri 2008; Ramos 2011). Research into the question carried out by the law firm Baker & McKenzie, on behalf of the Suruí case, concluded that indigenous peoples could indeed enter into contractual agreements concerning carbon storage and carbon emissions from their forests (ISA and Forest Trends 2010). No position on this can be conclusive, however, until it is tested in the courts.

Assuming for the moment that environmental and market factors are constant, the amount of land available for carbon sequestration is the single best determinant of economic return. Since the total amount of land in any single case is given, an inherent trade-off exists between land set aside for carbon trading and land to be used for other purposes. It is here that equity and economic return may diverge. The interests of those that would maximize monetary profit without concerns for other land use are at odds with those who depend upon the land for their present and future livelihoods. For this reason, it is of utmost importance that indigenous peoples have full decision-making abilities and power to protect their rights. While it remains unclear what role indigenous groups will have on the international level at climate change conventions, at least at the subnational and local levels there are opportunities for exercising agency in the process. Standards for assessing adequate levels of information sharing are vague. Discussions of the process point out the potential pitfalls when "consultation" is taken to mean approval by the few rather than full participation by the population to be affected.

The Cinta Larga proposed project appears to hold the greatest risk for the indigenous participants because all of the carbon rights to their lands are invested in the project, which is under the authority of the outside entity, Viridor. If land use decisions in the case of the Cinta Larga are to be made by outsiders whose single criterion is profits from carbon credits, the Cinta Larga may find themselves subject to constraints that they did not anticipate. This is a special concern in the absence of safeguards protecting signers who may not be adequately informed.

The Suruí project, in which the indigenous peoples are simultaneously proponents and stakeholders, is the only case in Brazil that includes indigenous peoples as decision-makers at every level of project development. Whether we refer to equity as an ethical or procedural matter—that which is fair and just—or a financial matter, referring to ownership in an enterprise, the Suruí project is the only one of the three where the owners of the land are also the owners of the carbon credits and active designers of the project.

The degree of equity for the indigenous participants in the XSEC project will be largely determined by the local information sharing and planning processes. Because the project involves several sets of actors, including many that are located great distances from one another, the possibility remains that only a few select community members will be actively involved in shaping the project. It remains a concern among the proponents that the process in which REDD projects are enacted not marginalize the indigenous peoples who are key players in the process. It is incumbent upon the proponents, who mediate information to the indigenous communities, to ensure that safeguards are met as this promising REDD+ project moves forward.

REDD projects pose new challenges to indigenous communities who will have to learn new technical and intergovernmental argots in order to engage in REDD projects (Chernela 2014). What carbon is and how it is connected to standing forests and local revenue sources may still be confusing for communities and individuals that are more distant from decision-making processes about the carbon projects and policies. It is therefore the responsibility of proponents to ensure that indigenous stakeholders are thoroughly familiar with the terms of all contractual arrangements, and their alternatives, before agreements are sought. It is also important that policies are in place to ensure just enactment of projects with ongoing transparency, clearly stated expectations, and agreed-upon goals.

Conclusions

Claims that carbon sequestration is a win-win proposition are premature. This is due, in part, to the fact that the success of any REDD strategy depends upon specific policy frameworks and other external conditions in which it will occur. In comparing three projects that provide contrastive strategies, we have explored how the performance of each is highly variable, depending upon political and economic scenarios that are outside the control of project participants. While these external factors are difficult to predict, they play a significant role in determining outcomes.

The outcome of any REDD program, then, will depend upon the policy environment of the future. Contrary to the simplistic stances that carbon sequestration in indigenous lands benefits conservation and people alike, we have shown that the extent to which any outcome will reach fruition, or any specific aspect of the arrangement succeed, depends upon external conditions. Ongoing uncertainties include carbon rights and ownership; payment for environmental services; natural resource management governance, rules, and regulations; representation in international climate negotiations; and protocols for consultation, or "free, prior, and informed consent" (Alangui et al. 2010; Schroeder 2010; Chernela 2011; Mahanty and McDermott 2013). Propositions that treat REDD projects as static and independent of context perform a disservice by overlooking factors that, while external to projects, are critical to outcomes.

Another complicating factor, often overlooked by analysts, is perspective. A project's success may be measured differently by proponents and stakeholders, further inhibiting facile assessments of project outcomes. In REDD projects, economic and conservation outcomes are inseparable, since the selling of carbon credits accomplishes both. For those whose overriding stake is profit, as is the sole or leading concern of the proponents, the most efficient means of obtaining economic return is to maximize the amount of marketable carbon credits. Since the economic return is a function of the amount of carbon stored, the project proponent with no stake in the land benefits from complete preservation. Any subtraction for other uses or services constitutes a trade-off. For those whose livelihoods depend upon the same forests, the options are different. In REDD projects in indigenous lands, forested land must be divided into a portion set aside for carbon credit generation and a portion dedicated to food production, living space, and other uses.

Allocations must be viable in the long term. If the amount of land designated to subsistence production proves inadequate for growing populations, indigenous peoples may find themselves facing severe limitations, as land formerly available may become off-limits. These trade-offs do not exist in the same way for proponents and outside investors for whom carbon credits are the sole consideration and who experience none of the local constraints brought about by the project.

The objectives of conservation, like those of economic return, prioritize standing forests and carbon stocks. In order to create optimal conditions for forest preservation, the interests and needs of indigenous residents, who vigilantly protect the lands in which they live, must be a concern for proponents. Payment for ecosystems services would compensate the owners of lands and provide them with incentives to maintain standing forests. Only by these means may the three objectives of REDD—conservation, revenue, and equity—be met. The matter points out the extent to which all three outcomes—equity, conservation, and economic payoff—are inextricably related to one another and to the policy decisions and practices of the project. Contrary to the simplistic stances that carbon sequestration benefits local peoples, we have raised questions about whether the goals and priorities of proponents and stakeholders are assumed to be equivalent. We have argued that neither the goals of a REDD project nor the stakes in a project are likely to be the same for all participants.

Acknowledgements

An earlier version of this paper was presented by Janet Chernela at the 2011 conference, ICARUS II—Climate Vulnerability and Adaptation: Marginal Peoples and Environments at the University of Michigan, Ann Arbor. The authors wish to thank Beto Borges, Steve Schwartzman, Barbara Zimmerman, and Rob Riker for their generosity in providing documentation, conversation, and insights.

Notes

1. Since 2007, estimates of amount of greenhouse gas emissions from deforestation have fluctuated between 15% and 20% (Alencar et al. 2010; CBO 2012).
2. The abbreviation tC02e is shorthand for tonnes of carbon dioxide equivalent, a measure that allows a comparison of greenhouse gases in terms of their global warming potential.
3. Viridor Carbon was established in 2009 as a:

> full-service project development firm that partners with . . . entities to document, register, and commercialize carbon sequestration and other environmental benefits accruing from long-term property conservation commitments or sustainable forestry management practices. (Viridor 2013)

References

Alangui, Wilfredo, Grace Sibido, and Ruth Tinda-an. 2010. *Indigenous Peoples, Forests, and REDD Plus: State of Forests, Policy Environment and Way Forward*. Philippines: Tebtebba Foundation.

Alencar, Ane, Osvaldo Stella Martins, Andre Nahur, Daniel Nepstad, Anrea Cattaneo, Tracy Johns. 2010. Brazil's Emerging Sectoral Framework for Reducing Emissions from Deforestation and Degradation and the Potential to Deliver Greenhouse Gas Emissions Reductions from Avoided Deforestation in the Amazon's Xingu River Basin. *EPRI*. http://www.epri.com/abstracts/Pages/ProductAbstract.aspx?productId=00000000000102160 (accessed September 9, 2014).

Barbieri, Samia Roges Jordy. 2008. *Os Direitos Constitucionais dos Índios e o Direito á Diferença, Face ao Princípio da Dignidade da Pessoa Humana*. Coimbra: Edições Almedina, SA.

Business Wire and Forest Trends. 2013. Amazonian People and Cosmetics Giant Work to Save Endangered Rainforest and Slow Climate Change. *Business Wire*, September 10, 2013. http://www.marketwatch.com/story/amazonian-people-and-cosmetics-giant-work-tosaveendangered-rainforest-and-slow-climate-change-2013-09-10 (accessed September 9, 2014).

CBO. 2012. Deforestation and Greenhouse Gases. *US Congressional Budget Office*. http://www.cbo.gov/publication/42687 (accessed September 1, 2014).

Cenamo, Marioano Colinia, Mariana Nogueira Pavan, Marin Thereza Campos, Ana Cristina Barros, and Ferdanda Carvahlho. 2010. *Casebook of REDD Projects in Latin America*. Brasília: Instituto de Desenvolvimento Sustentável do Amazonas and The Nature Conservancy.

Chernela, Janet. 2011. Structures of Participation: Indigenous Peoples in Two Projects of Reduced Deforestation (REDD) in the Brazilian Amazon. *Climate Vulnerability and Adaptation: Marginal Peoples and Environments. ICARUS: Initiative on Climate Adaptation Research and Understanding through the Social Sciences*. http://www.icarus.info/wp-content/uploads/2011/06/Chernela.pdf (accessed May 21, 2004).

Chernela, Janet. 2014. Fire and Ice: Talking about Carbon in the Brazilian Amazon. *Practicing Anthropology* 36, 3: 17–21.

Gaspar, Renata, and Altino Machado. 2007. ACT Intermediates Partnership between the Suruí Ethnic Group and Google. June 13. http://www.amazonia.org.br (accessed June 11, 2011).

Hemming, John. 1978. *Red Gold: The Conquest of the Brazilian Indians*. London: Macmillan.

IDESAM (Instituto de Conservação e Desenvolvimento Sustentável do Amazonas), and Metareilá (Associação Metareilá do Povo Indígena Suruí). (2011). Projeto de Carbono Florestal Suruí. IDESAM, Manaus. http://www.idesam.org.br/wp-content/uploads/2013/04/PCFS_PDD_portugues_V1.pdf (accessed October 2, 2016).

ISA/Forest Trends. 2010. *Avoided Deforestation (REDD) and Indigenous Peoples: Experiences, Challenges, and Opportunities in the Amazon Context*. São Paulo: Instituto Socio-Ambiental and Forest Trends.

Leroy, Jean-Pierre. 2003. Violência, o Meio Ambiente e os Índios. *CIMI*. http://www.midiaindependente.org/pt/blue/2003/12/269505.shtml (accessed March 3, 2011).

Mahanty, S., and C.L. McDermott. 2013. How Does 'Free, Prior and Informed Consent' (FPIC) Impact Social Equity? Lessons from Mining and Forestry and Their Implications for REDD. *Land Use Policy* 35: 406–416.

May, Peter H., and Brent Millikan. 2010. *The Context of REDD+ in Brazil: Drivers, Agents, and Institutions*. CIFOR Occasional Paper 55. Indonesia: Center for International Forestry Research.

Mindlin, Betty. 1985. *Nós Paiter: os Suruí de Rondônia*. Petrópolis: Editora Vozes.

Mindlin, Betty. 2011. Suruí Paiter. *Institutio Socio Ambiental*. http://pib.socioambiental.org/en/povo/Suruí-paiter (accessed June 9, 2011).

Nery, Demain, Mariana Chistovam, Isabel Mesquita, Juliana Splendore, Osvaldo Stell, and Paulo Moutinho. 2013. *Indigenous Peoples and the Reducing Emissions from Deforestation and Forest Degradation (REDD+) Mechanism in the Brazilian Amazon: Subsidies to the Discussion of Benefits Sharing*. Brasília, DF: Amazon Environmental Research Institute.

Overseas Development Institute. 2014. The REDD Desk. http://theredddesk.org/countries/actors/overseas-development-institute (accessed April 21, 2014).

Pennon Group. 2013. Business Review. *Pennon Group*. http://www.pennongroup.co.uk/pennon/en/corebusinesses/viridor/businessreview (accessed April 4, 2013).

Ramos, Alcida Rita. 2011. Os Direitos Humanos dos Povos Indígenas no Brasil. In *Desafios aos Direitos Humanos no Brasil Contemporâneo*, eds. Bjorn Maybury-Lewis and Sonia Ranincheski, 65–81. Brasil: Verbena Editora.

Rondônia Ao Vivo. 2010. MPF recomenda à Funai que assessore índios Cinta Larga em contrato de créditos de carbono. *Jornal Electronico* 4 de Junho.

Saatchi, S. S., R. A. Houghton, R. C. Dos Santos Alvala, J. V. Soares, and Y. Yu 2007. Distribution of Aboveground Live Biomass in the Amazon Basin. *Global Change Biology* 13: 816–837.

Salomão, Marta. 2012. Cacique troca compromisso de venda de crédito de carbono por caminhonetes. *Estado de Sao Paulo*. http://estadao.com.br (accessed April 9, 2012).

Schroeder, H. 2010. Agency in International Climate Negotiations: The Case of Indigenous Peoples and Avoided Deforestation. *International Environmental Agreements-Politics Law and Economics* 10, 4: 317–332.

Schwartzman, Stephan, André Villas Boas, Katia Yukari Ono, Marisa Gesteira Fonseca, Juan Doblas, Barbara Zimmerman, Paulo Junqueira, Adriano Jerozolimski, Marcelo Salazar, Rodrigo Prates Junqueira, and Maurício Torres. 2013. The Natural and Social History of the Indigenous Lands and Protected Areas Corridor of the Xingu River Basin. *Philosophical Transactions of the Royal Society* 368, 1619: 1–12, doi:10.1098/rstb.2013.0308.

Schwartzman, Stephan, and Barbara Zimmerman. 2005. Conservation Alliances with Indigenous Peoples of the Amazon. *Conservation Biology* 19, 3: 721–727.

Toni, Fabiano, Isadora A. R. Ferreira, and Igor N. R. Ferreira. 2011. Adapting to emerging institutions: REDD+ projects in the territories of the Suruí and Cinta-Larga Indigenous Peoples. ICARUS Conference, May 2011, Ann Arbor, MI. http://www.icarus.info/wp-content/uploads/2011/05/PanelTheme2aPresentations.pdf (accessed May 21, 2014).

Turner, Terence. 1995. An Indigenous People's Struggle for Socially Equitable and Ecologically Sustainable Production: The Kayapó Revolt Against Extractivism. *Journal of Latin American Anthropology* 1: 98–121.

USAID. 2014. About Brazil. http://www.usaid.gov/brazil (accessed July 30, 2014).

Viridor. 2013. Brazil. *Viridor Carbon Services*. http://viridor.net/project-brazil.php (accessed September 19, 2013).

Vitel, Claudia Suzanne Marie Nathalie, Gabriel Cardoso Carrero, Mariano Colini Cenamo, Maya Leroy, Paulo Mauricio Lima A. Graça, and Philip Martin Fearnside. 2013. Land-Use Change Modeling in a Brazilian Indigenous Reserve: Construction of a Reference Scenario for the Suruí REDD Project. *Human Ecology*. http://goo.gl/7tM5va (accessed March 7, 2014).

Von Mittelstaedt, Juliane. 2010. Using the Internet to Save the Rainforest: How an Amazonian Tribe Is Mastering the Modern World. June 8. Spiegel Online; Druckervision. http://www.spiegel.de/international/world/0,1518,druck-698511 (accessed June 12, 2011).

Zanotti, Laura. 2009. Economic Diversification and Sustainable Development: The Role Non- Timber Forest Products Play in the Monetization of Kayapó Livelihoods. *Journal of Ecological Anthropology* 13: 26–41.

14

EQUITY CONCERNS DURING REDD+ PLANNING AND EARLY IMPLEMENTATION

A Case from Malawi

Heather M. Yocum

As REDD+ and "REDD-ready" voluntary carbon offset projects are implemented in Malawi, they encounter cultural and geographic particularities that create new equity concerns and exacerbate existing tensions surrounding forest access and land use in and around the project area. These concerns span the entire process of carbon offset production, including the social components of REDD+ projects and the technical activities necessary for project validation. The case studies that follow demonstrate the importance of considering the impacts of the social programs planned as part of a project as well as the technical activities that precede carbon sales.

This chapter presents two specific examples of social justice and equity issues arising from voluntary carbon offset projects in protected areas in the Northern Region of Malawi. First, the implementation of REDD+ will mean changes in land use and access to resources not only on forest land within the project boundaries, but also on community land outside of the project zone where tree planting and other livelihood projects are planned. Communities and state agencies have different interpretations of how these tree planting activities will impact land use, crop production, and human-animal conflict. Second, technical processes such as biomass surveys and ongoing monitoring, reporting, and verification (MRV) activities may violate cultural norms or traditional belief systems when these activities occur in culturally important forest spaces—such as graveyards—or involve trees with spiritual and ritual significance. This case study examines voluntary carbon projects that are in the planning and early implementation stages, in which planners have completed the initial carbon measurements and started to roll out the various livelihood projects planned as part of the overall carbon conservation strategy.

In general, REDD+ and other voluntary carbon projects can pose multiple social and environmental justice concerns for communities. The process of obtaining informed consent from target communities is subject to different power dynamics between actors, which can result in coercion of some people by project planners or community leaders (Milne and Adams 2012). In some cases, the consent process fails to involve members of the target communities in the project in any substantive way (McElwee 2012; McElwee this volume). Even after the initial consent process concludes, people may not fully understand the full range of costs and benefits of carbon projects (Chernela 2014; Yocum 2013). In some cases, violence has erupted between affected communities and those policing project areas (Checker 2009). Some communities have lost access to forest resources that store carbon, while others have been forcibly removed from within project areas in order to maximize carbon sequestration potential (Beymer-Farris and Bassett 2012).

The latest guidelines for REDD+ and some voluntary standards have begun to address more comprehensive concerns regarding consent, benefits distribution, resource access, and respect for cultural values (Laughlin et al. 2013; Peskett and Todd 2012; REDD+ Social and Environmental Standards Initiative 2012). Voluntary projects, however, including for-profit, private sector projects in countries that do not have or are still developing national REDD+ strategies and guidelines—like those discussed in this chapter—are required to meet the specifications of their chosen validator, which may or may not be as stringent as REDD+ guidelines. These guidelines are presented as best practices and recommendations that provide the flexibility needed to adapt consent procedures for individual project and in-country scenarios (Laughlin et al. 2013; UN-REDD Programme 2010). Thus, there is a significant amount of interpretation and adaptation on the part of carbon developers, project planners, and community representatives as they attempt to introduce the project to target communities and obtain their consent for the project.

In addition to these concerns, the social and ecological context poses additional problems for achieving project outcomes while equitably distributing benefits and risks associated with carbon projects. The standardization of consent and technical measurement procedures can interact in unexpected ways during the early phases of project implementation (Milne 2012), reinforcing the importance of attending to the specific historical, social, and ecological context for each project area. In Malawi, members of target communities are concerned about receiving compensation for carbon storage services and maintaining their access to forest resources; however, they are also concerned about how the project will impact their ability to manage forest spaces outside of the reserve, as well as to observe traditional belief practices involving spaces or objects that might be targeted for project activities.

These observations are based on 12 months of ethnographic research in forest and game reserves in the Northern Region of Malawi between 2009 and 2012, including in-depth interviews and participant observation with stakeholders from target communities, nongovernmental organizations (NGOs), the Department of National Parks and Wildlife (DNPW), the Forestry Department, donor agencies, and a for-profit carbon development company.

Background and Research Sites

This case study considers voluntary carbon projects in two protected areas in Malawi's Northern Region: Vwaza Marsh Wildlife Reserve and Mkuwazi Forest Reserve. Vwaza Marsh Wildlife Reserve protects a perennial wetland, a lake, and the surrounding forested hills. This area is an important breeding area for migratory birds and waterfowl and is also home to elephant, buffalo, hippopotamus, crocodile, and numerous species of antelope (Department of National Parks and Wildlife 2004). The Mkuwazi Forest Reserve is located near the shores of Lake Malawi and protects a hill and the low-lying forests that surround it. One of the last stands of broadleaf evergreen forest in the area, the 17 km^2 reserve provides an important habitat for several species of culturally important trees as well as rare primates, several species of birds found nowhere else in Malawi, and at least one endemic species of butterfly (Department of Forestry 2009).

The Vwaza and Mkuwazi protected areas are also important resources for the approximately 125,000 people who live within 5 km of the boundaries of these areas. People living near these areas rely primarily on firewood or charcoal to meet their energy needs, and are dependent on rain-fed agriculture to grow maize, legumes, pumpkins, fruits, and vegetables as well as cash crops of tobacco, groundnuts, soy, or cotton. The protected areas are important resources for firewood, building materials, and wild foodstuffs such as meat, eggs, fruits, vegetables, and mushrooms. Many people living near protected areas also sell products from the forest, including charcoal, timber, fiber, mushrooms, meat, and cultivated honey. Some of these activities, such as the collection of non-timber forest products, are permitted under a co-management agreement between the communities and the state authorities that manage the protected areas, while others, such as hunting and logging, are unlawful.

Protected areas in Malawi have a complex historical and social context, which has ramifications for the carbon projects being planned there. Land in Malawi falls into three categories: government land, which includes all designated protected areas; customary land, which is under the purview of traditional authorities and where villages and agricultural fields are located; and private land. The protected areas were carved out of customary land, and their expansion has been at the expense of local people, who were forced to abandon their homes and fields. In the early 1900s, the colonial administration began to establish game and forest reserves to protect valuable game animals and commercial timber species (Morris 2001; Zulu 2008, 2012). In the 1920s, villages were relocated from inside the newly established Mkuwazi Forest Reserve to adjacent areas (Interview, May

2012), and restrictions were put in place in Vwaza and Mkuwazi, and other protected areas to control hunting and cutting timber by non-white persons (Adams and McShane 1992; McCracken 2006; Young 1953). After gaining independence in 1964, the Malawian government expanded existing forest and game reserves and established several national parks (Zulu 2009). In the 1970s, Vwaza Marsh was declared a game reserve, and approximately 2,000 people were forcibly removed from within the new reserve boundaries, their homes dismantled to discourage them from returning (Department of National Parks and Wildlife 2004). Many residents living near Vwaza and Mkuwazi feel that the resources in the protected areas are rightfully theirs, and as such regularly enter the reserves—with or without permission from the authorities—to hunt, cut timber, or gather other resources. These coercive removals and ongoing disputes over access to resources are continued sources of tension between local residents and the state agencies that manage these areas.

In the 1990s, global recognition of the failure of this "fortress conservation" model (Brockington 2002) prompted a shift toward community-based natural resource management (CBNRM), which attempted to engage the community in the conservation of protected areas in exchange for financial or other in-kind benefits.[1] In Malawi, co-management agreements between the communities and the managing state agencies make provisions for community groups to collect non-timber forest products such as thatching grass, mushrooms, wild fruits and vegetables, and firewood, as well as limited amounts of timber for construction (Department of National Parks and Wildlife 2004). The agreements also allow for the communities to receive a portion of the revenue generated through tourism in these areas. These revenues have been used to finance small development projects, such as putting metal roofs on schools and purchasing concrete for the construction of new teachers' houses (Interview, February 2012). Although unauthorized entry into the protected areas is illegal, enforcement is difficult. The boundaries are porous, enforcement agencies are understaffed, and people living in adjacent villages routinely enter the protected areas outside of the rules dictated by the co-management agreements.

Carbon project planners have drawn on these co-management agreements when structuring benefits distribution plans for both the Vwaza and Mkuwazi areas. The carbon project in Vwaza was envisioned as part of a multimillion-dollar, multiyear conservation and development project, funded by USAID and implemented by Total LandCare (TLC), a conservation and development NGO based in Malawi with strong ties to the United States. As part of a strategy to create sustainable revenue streams for the communities living near the protected areas, TLC partnered with Terra Global Capital, a US-based, for-profit carbon development firm, to create a voluntary carbon offset project. During my fieldwork in 2011–2012, field teams were working to complete the initial biomass surveys, and a community organizer trained by Terra Global conducted initial introductory meetings with the target communities. The project was validated in 2014 through the Verified Carbon Standard and the Climate, Community, and Biodiversity Alliance, with the goal of integrating this project into a national REDD+ framework in the future. Project planners hope that revenues from carbon sales could provide money to buy necessities, replacing a reliance on natural resources with a cash economy, thereby maximizing conservation and development outcomes (Terra Global Capital 2013a, 2013b; Terra Global Capital and Total LandCare 2011).

The project in Mkuwazi was similarly conceived as a conservation and development project that would generate revenue for the seven communities living adjacent to the reserve in exchange for protecting the forest and reducing illegal logging. The project was originally funded in 2008 by USAID as part of their Community Partnerships for Sustainable Resource Management (COMPASS II) and implemented by the Malawian government, a consortium of Malawian NGOs, and Plan Vivo (Malawi Environment Endowment Trust 2009). Planners held community meetings to introduce the carbon project, measured carbon stocks, and trained community technicians to record tree growth to support future MRV requirements (Interview, April 2012). The Mkuwazi project stalled when the planners were unable to complete final certification to begin selling carbon.[2] Pending successful implementation of the Kulera Project, Terra Global Capital plans to expand its activities to Mkuwazi and other protected areas across Malawi.

Tree Planting and Land Use

Part of the "+" in REDD+ means that additional social and ecological benefits are planned as part of the carbon project. This means that not all project activities necessarily take place within the defined project area monitored for carbon sequestration. In Vwaza, additional livelihood and conservation activities will take place outside of the protected areas on community-owned land within 5 km of the protected area border. One such livelihood activity includes expanding woodlots[3] and sponsoring reforestation efforts in village forest areas to meet subsistence needs for firewood and building materials, thereby reducing pressure on the timber resources inside of the project zone and increasing the carbon stores—and earning potential—of the project.

Unlike neighboring countries, Malawi does not have wildlife management areas to serve as a buffer between the protected area and adjacent villages. In those countries, co-management arrangements near protected areas with buffer zones usually allow for limited use of resources from within this buffer area; likewise, some carbon projects site tree planting initiatives in these zones. In Malawi, however, there are no buffer zones between the protected areas and the fields and villages of the communities living nearby. In some areas in Vwaza, crops are cultivated to within feet of the wire fence that defines the protected area, and some villages are located just across the dirt access road, less than 30 feet from the boundary. If the project planners intend to plant trees in order to reduce the dependence on resources within the protected areas—and therefore within the carbon project zone—it is necessary to plant trees on customary land administered by traditional authorities.

Although tree planting projects have been taking place in these areas for a number of years, the carbon project would increase the scale and scope of tree planting. The project planners propose to plant or regenerate 5,600 ha of woodlots near Vwaza in the first 3 years (Terra Global Capital 2013b). The tree planting project will target community land that is not currently cultivated, aiming to regenerate forest areas controlled by village leaders and to establish or reforest woodlots within the village.

Villagers and state officials have different interpretations of the possible impacts of these tree planting activities. Project planners and state officials believe that increased tree cover outside of the protected area will not only provide more accessible fuelwood and building materials for local residents, but also serve as a de facto buffer zone to decrease unauthorized entrance into the protected area and reduce human-animal conflict. Villagers, on the other hand, worry that increased tree cover near their villages will entice more animals to leave the protected area and escalate incidences of crop predation by wildlife.

The issue of human-animal conflict was a common concern for both officials and villagers, and was a constant source of tension between managers and the communities. Near Vwaza, most incidences involved illegal hunting inside of the reserve or animals eating crops and destroying agricultural fields in surrounding village areas. Many villagers considered hunting as a way to add meat to a protein-poor diet, as well as a method to control animal populations (Yocum 2013). Hunting inside of any protected area is illegal, and the DNPW is involved in educational outreach to deter hunting, while game scouts and forest guards routinely remove wire snares, confiscate homemade firearms, and arrest those caught hunting or possessing bush meat. On the other hand, the destruction of crops by animals that leave the reserve is a constant concern for people living near the wildlife reserve. For example, during one week, elephants destroyed crops on four separate nights in a single village. In addition to eating food crops, large animals such as elephants, buffalo, and hippos can be particularly destructive when moving through fields since they can trample storage sheds and other infrastructure. These large animals are also dangerous when confronted, particularly at night.

Villagers expressed concern that trees and vegetation outside of the reserve would provide additional food and cover that would entice animals to leave the reserve more frequently or in larger numbers. One woman directly associated increased tree cover with increased crop predation, stating that "when we take care of those trees outside of the reserve, elephants come to eat those trees. As elephants eat those trees, they also eat maize" (Interview, May 2012). Another woman reiterated this sentiment, adding that "[project planners] tell us that we should plant trees. But elephants come and eat those branches of trees. So then how will we be helped?" (Interview, May 2012). A third woman worried that project planners had not anticipated how the destruction of trees by elephants would impact the total amount of carbon that would be sequestered and so people would receive lower payments overall. These statements express concern not only for potential increases in crop predation, but also fears that the destructive power of elephants could deprive people of anticipated revenues from carbon sales as well as the fuelwood and building timbers that the tree planting project was meant to address in the first place.

There was also fear that dangerous animals, particularly buffalo and poisonous snakes, would be attracted to village woodlots and forest areas if grasses and other vegetation were allowed to proliferate in these areas. One man worried that planting additional trees would pose risks not only to their crops but also to the safety of his family:

> [Planting trees] may invite some wild animals which will come and eat [the trees]. For our safety, we have to clear around our surroundings so that snakes should be scared away. So how can we be helped so that snakes and wild animals should not be coming closer to our homes?
>
> *(Interview, May 2012)*

When animals leave the protected areas, the need to protect crops as well as the opportunity to eat meat often prompts villagers to kill the animals. Residents reported that since most people cannot afford to eat meat regularly, they are tempted to kill animals that enter their fields. However, killing animals that leave the protected areas is also illegal.

Many residents felt they bore the burden in either situation: if they were caught killing animals either inside or outside of the reserve, they could be fined or jailed; however, when animals left the reserve and destroyed crops, the farmers rarely, if ever, received compensation from the DNPW or the Malawian government. Furthermore, community members were legally unable to do anything other than try to frighten the animals away or contact DNPW staff to come and deal with them. Particularly with larger animals such as elephants, hippos, and antelope—many of which are endangered and all of which are protected whether they are inside or outside the boundaries of the protected areas—people could face heavy fines or jail time for harming the animals in any way.

While many of the DNPW personnel are not unsympathetic to the hardships caused by crop predation, their priority is the conservation of the animals and habitat within the protected areas. Many of them expressed a deep concern for protecting these areas, both to conserve the animals and plants, but also to ensure that people living near the reserves could continue to draw on those resources in future generations. They see woodlots and tree planning as key steps in the conservation and development effort. The DNPW employees' livelihoods depend on the continued state management of these areas, so they have jobs and careers that are contingent upon continued efforts to conserve these areas, efforts that are increasingly tied to the carbon project.

In Vwaza, the multiple potential futures of the tree planting activities and what they mean for different actors may become a key source of conflict as the project progresses. In this case, concerns over the distribution of and access to resources may take unexpected forms. During the early stages of the carbon project, residents near Vwaza voiced their concern that the carbon project would impact their access to resources within the reserve; however, many residents were equally as concerned that the livelihood projects planned as part of the carbon project would further inhibit their ability to control wildlife that leave the reserve and damage crops and village infrastructure. The issue of animal predation of food crops was as important to residents as maintaining access to the resources within the reserve and ensuring that they were included in the benefits distribution plans for the project. Therefore, as REDD+ activities extend beyond the project area and increasingly include livelihood projects that may impact resources both within and outside of the project area, critical analysis must also extend beyond the immediate project area to include the full range of project impacts and the spaces in which they are experienced.

Cultural Norms and Sacred Spaces

Before any of the social projects related to the carbon project can take place, the project must be certified to sell carbon, requiring a full assessment of how much carbon the forests can sequester over the project period. The technical processes used to measure the carbon stocks for the project can exacerbate existing tensions regarding access to and use of resources. It can be difficult to balance the need to meet the technical standards for measuring carbon stocks mandated by international standards, company policy, donor expectations, or third-party verifiers with the need to respect cultural and local norms that dictate appropriate behavior within the project areas. This becomes more complicated given the time constraints and limitations posed by the validation and verification processes necessary to produce marketable carbon credits.

The technical activities used to create the initial baseline studies and to perform the ongoing monitoring, reporting, and verification (MRV) activities require routine field surveys throughout the life of the carbon project. The initial biomass plot survey activities include measuring the size of the trees within the plot, taking photos and GPS coordinates, and collecting samples of soil and non-tree vegetation to be analyzed for carbon content in the lab. Field teams also bury a piece of metal rebar in the soil, remove a portion of the bark of a large indicator tree, and paint other trees to create markers for each permanent sampling plot to facilitate future field surveys.

These required technical procedures are considered innocuous by project planners; however, these activities have very different meanings for communities who are often unaware of when the procedures are to take place or what they entail. Sample biomass plots might be located in or adjacent to graveyards, important cultural sites, or include individual trees that are important in traditional belief systems. In Malawi, the spirits of the deceased (*mizimu*) are associated with woodlands, graveyards, and wild animals, while living people are associated with villages, agricultural fields, and domesticated animals (Morris 2000a; Schoffeleers 1997). Mizimu are not considered to be devoid of life, but represent another phase of being that continues to be important to surviving relatives and community members. Graveyards are an important site in the transformation of deceased persons into ancestral spirits; likewise, particular trees are an integral part in practices that connect the living and spiritual communities. Ancestral spirits play an important role in the lives and welfare of the living through their association with rain and fertility, and are therefore accorded a great deal of respect. Certain tree species, such as *msolo* (*Pseudolachnostylis maprouneifolia*) and *mpoza* (*Annona senegalensis*) are particularly associated with the spirits of ancestors and so are accorded a degree of respect (Morris 2000b). In some areas, offerings of flour are left at the foot of these trees to promote successful hunting, and these trees are prized for their fruit and medicinal qualities (Morris 2000b).

Across the African continent, graveyards are located in wooded areas and are protected through cultural practices, making graveyards particularly attractive sites for biodiversity conservation efforts (Sheridan and Nyamweru 2008). In Malawi, people do not hunt, gather firewood or thatch, collect medicine, or cut down trees in these areas. Other than during funerals and certain initiation rites, access to graveyards is restricted and avoided. Anyone entering a graveyard is met with mistrust and suspicion, and potentially viewed as a witch or sorcerer seeking body parts of the deceased as activating substances for harmful medicines (Colson and Gluckman 1959; Morris 2000a; van Breugel 2001). Even in areas that have been heavily deforested or degraded, stands of trees surrounding the graveyards remain intact. Since these areas are already being conserved by the community, these biodiverse micro-hotspots are often targets for conservation and carbon investment.

Biomass surveys and MRV activities, when conducted in graveyards or involving particular tree species with high cultural value, may violate culturally appropriate modes of behavior and resource use in these areas. For example, cutting and removing vegetation from graveyards during the biomass surveys violates cultural traditions that mandate that all vegetation remain inside of these spaces in order to show proper respect for the ancestral spirits and maintain the appropriate balance between the worlds of the spirits and the living. Likewise, digging in graveyards for any reason other than to prepare a grave is considered suspicious and alarming behavior.

Furthermore, debarking and marking trees is associated with traditional divination rituals that are used to identify sorcerers. The Mkuwazi Forest Reserve derives its name from particular trees used in divination rituals. Mkuwazi is a cognate of the word *mwabvi* or *uhavi*, which refers to the poison ordeal, a ritual in which accused sorcerers are given a poisonous drink prepared from the bark and roots of the tree and other medicinal substances (Interview, May 2012). If the accused person vomits, then it is taken as proof of their innocence; death is proof of their guilt. Mkuwazi Forest Reserve is one of the last places that the *mkuwazi* or *mwabvi* tree still grows. People sometimes travel hundreds of kilometers to collect the bark from this tree for use in these divining rituals. The exact location of these trees within the reserve is a carefully guarded secret known only to those who are specially trained to administer the ritual, and only these individuals are allowed to remove bark or roots from these trees. Traditional leaders living near Mkuwazi expressed concern that marking or debarking these trees during routine project survey activities could compromise the efficacy of these rituals (Interview, May 2012).

Additionally, MRV activities can contradict the rules set forth in the co-management agreements in both the Mkuwazi and Vwaza areas and can reignite villagers' fears that the carbon project would lead to further loss of access to forest resources. Tree bark and roots are used for fiber or medicine, but forest and wildlife managers discourage people from removing bark and roots because excessive harvest can damage trees. As such, this activity is limited or disallowed in the co-management agreements. People living near the protected areas found it difficult to understand why field teams conducting biomass surveys were allowed to enter the protected areas and remove tree bark, roots, and surrounding vegetation, often with the help of the same forest guards and game scouts who enforced the co-management agreement and punished locals for similar activities. Furthermore, misunderstandings about these biomass surveys played into community fears that the carbon project would further restrict their access to forest resources and project benefits. In Vwaza, the pieces of metal rebar buried in the ground were mistaken by residents for machines that would strip the area of carbon resources to avoid compensating locals. Other residents thought project planners were burying landmines meant to harm villagers who illegally entered the protected areas.

Project timelines can increase the potential for such misunderstandings. While some validation standards—such as the Gold Standard of the Climate, Community, and Biodiversity Alliance used in Vwaza (2013)—specifically require the documentation of areas of high cultural value during project design and initial phases of project implementation, members of target communities might not be immediately forthcoming about the locations of these places. The location of *mwabvi* trees is not common knowledge because of the powerful substances that can be derived from them. Due to this secrecy, and the pejorative way that traditional beliefs about sorcery are often viewed by outsiders, communities are unlikely to disclose the precise location of these trees, and so their location is unlikely to appear on lists or maps of culturally important areas made during initial visits by project planners, even if that information is explicitly solicited.

Other important areas, such as graveyards, might be easily identifiable by biomass survey teams, but observing cultural expectations for proper conduct in those areas requires consultation with the affected communities.

In order to reduce misunderstanding and potential future conflicts, field teams need to communicate the procedures involved in marking and measuring the permanent biomass plots to the community and establish procedures that allow for ongoing engagement with community members regarding these activities. In addition, provisions could be made to solicit community feedback on the procedures and the locations involved in MRV activities prior to implementation. These could be periodically revisited, and if necessary, amended, throughout the life of the project. Doing so would help to alleviate residents' worries about what the carbon project will mean for them and their ability to maintain proper relationships between the living community and the wider community of ancestral spirits.

Conclusion

These case studies illustrate the need for project planners, researchers, and affected communities to consider the full range of activities necessary to produce REDD+ projects and carbon offsets and to work together to identify and evaluate the potential areas of conflict or violations of social and environmental justice. The ability of target communities to attenuate exposure to various risks is impacted by all aspects of the carbon project, including the additional livelihood projects associated with REDD+ and the technical procedures that take place prior to project validation. Resource access is one of these concerns, but people living in target communities in the Northern Region of Malawi were also concerned about their ability to deal with wildlife, manage relationships with ancestral spirits, and mitigate risks from sorcery. Additionally, the particularities of the local context reorder where and when equity concerns arise and the forms they take. Not all resource concerns will be confined to designated project areas or be voiced as access to forest resources; instead these concerns may center on resources found outside of the forest area, or on the cultural importance of the resources in question. Project planners and researchers should scrutinize the earliest phases of the project—long before any benefits are distributed to communities—to identify new forms of resource distribution or decision-making that could jeopardize attempts to incorporate social benefits into these programs, or that arise from efforts to include social benefits in carbon projects in the first place.

These case studies provide two important lessons for project planners and for those engaging in critical research about the impacts of REDD+ and other voluntary carbon projects. First, the social benefits in REDD+ projects do not necessarily alleviate the social and environmental justice issues related to carbon offset projects; rather, the livelihood programs associated with REDD+ projects can themselves introduce additional equity concerns. With the coming of

REDD+ and an increased focus on expanding and enhancing the social benefits portfolio of climate change mitigation ventures, increasing numbers of additional livelihood projects will include components or activities that take place outside of the official carbon project area. As such, both the carbon sequestration component and the social component of the project—whether within the designated project area or outside of it—need to be considered as potential sites for violations of social justice and equity. These social projects require the same attention and critical engagement as the carbon offsetting process.

Second, engagement with target communities must begin early, during the planning and early implementation phases of the project, and include initial technical activities. Although engagement with target communities at this early stage may seem superfluous, the potential for negative social impacts on communities already exists at the earliest stages of the project. Before social or ecological benefits accrue, the very act of measuring and monitoring carbon stocks can infringe on cultural belief systems. These activities may reignite tensions over the control and access to land and forests in question, particularly if communities associate these actions with past efforts to control resources. Project planners should communicate all aspects of the project—including details about the initial biomass surveys and related technical activities—to residents of target communities.

The issues discussed in this case study are problems that stem from the confluence of REDD+ activities and the existing social, historical context of land appropriation, use, and asymmetrical power relationships between rural Malawians and their government, foreign NGOs, and business enterprises involved in the carbon project. Under the second version of the Climate, Community and Biodiversity Alliance (CCBA 2008) under which the Kulera project was validated, there are no explicit guidelines concerning the technical measurement processes or potential impacts of project activities that take place outside of the project area. While the third edition of the CCBA guidelines (CCBA 2013) have made great strides toward addressing this, projects already in the validation pipeline are grandfathered into the standards in place when they began the process. In spite of the issues detailed in this case study, the Kulera project was only the third such project worldwide to achieve the triple Gold Standard certification from the CCBA, meaning that the project is recognized for providing exceptional benefits for the community, biodiversity, and for carbon capture. This demonstrates the very real gaps between what is considered adequate participation and social safeguard considerations in the certification and verification world and what might be necessary on the ground.

Ethnographic examples and in-depth case studies demonstrate how risks and benefits can be distributed temporally, geospatially, and socioculturally across the particular social and environmental context in which these projects are implemented. Both the carbon project and associated social projects will encounter the specific histories of land use and natural resource management in those areas.

Depending on the extent to which project activities exacerbate or mitigate existing inequalities between community members or between target communities and land managers, REDD+ projects may bring up a host of concerns, some of which will not be directly related to carbon sequestration or benefits distribution, areas where REDD+ project planners, advocates, and critics are more conditioned to focus.

It is necessary to attend to the way that equity concerns are voiced by target communities and project planners, as well as the underlying issues informing these concerns. Waiting to examine impacts after the project is well underway may miss important opportunities to create REDD projects that enhance social well-being instead of merely doing no harm or worsening living conditions for target communities. Examining the implementation of specific REDD-ready projects illustrates the importance of local context to the potential successes or failures of REDD+. Anthropology is uniquely positioned to generate such case studies through long-term, qualitative research with various stakeholders in order to demonstrate the ways in which carbon offsets can create new challenges for meeting conservation, poverty alleviation, and climate mitigation targets.

Acknowledgements

This research was supported by the US Fulbright program as well as Michigan State University's Graduate School, Center for Gender in Global Context, Center for Advanced Study in International Development, African Studies Center, and the Department of Anthropology. Many thanks to Duncan, Cheri, Mike, and Elton for collaborating with me on this project. This research would not have been possible without the help and input from people living in target communities and those working for the Forestry Department and DNPW. *Zikomo kwambili! Tawonga chomene!*

Notes

1. A full discussion of the reasons behind this shift and the questionable success of CBNRM lies outside of the scope of this chapter, but for more information, see: Berkes (2004, 2007), Blaikie (2006), Brockington (2002, 2004), Ribot et al. (2006), Schafer and Bell (2002), and Zulu (2008).
2. The Plan Vivo project in Mkuwazi stalled due to several factors, including the inability to locate an initial buyer for the carbon credits (interview, NGO worker, November 2010; Plan Vivo, personal communication, November 2010). The government of Scotland was the original buyer, but withdrew financial support in 2010–2012, when many bilateral donors withheld non-essential aid to protest the late President Dr. Bingu wa Mutharika's increasing suppression of civil and political rights, as well as his government's refusal to follow the International Monetary Fund's recommendation to devalue the Malawian currency by over 50% and allow it to float on foreign currency markets. Donor relations were briefly normalized after his death in April 2012. The global economic recession and falling price of carbon credits also contributed to the difficulties in locating another buyer.

3. Woodlots are areas where trees are planted and cultivated to meet needs for timber and firewood. They are usually located on land in or near the village, or on private land. Village forest areas are forested areas with natural vegetation located on customary land. They are usually located outside of the village and are under the authority of traditional authorities.

References

Adams, Jonathan S., and Thomas O. McShane. 1992. *The myth of wild Africa: Conservation without illusion*. Berkeley: University of California Press.

Berkes, Fikret. 2004. Rethinking community-based conservation. *Conservation Biology* 18, 3:621–630.

Berkes, Fikret. 2007. Community-based conservation in a globalized world. *Proceedings of the National Academy of Sciences* 104, 39:15188–15193.

Beymer-Farris, Betsy A., and Thomas J. Bassett. 2012. The REDD menace: Resurgent protectionism in Tanzania's mangrove forests. *Global Environmental Change* 22, 2:332–341.

Blaikie, Piers. 2006. Is small really beautiful? Community-based natural resource management in Malawi and Botswana. *World Development* 34, 11:1942–1957.

Brockington, Dan. 2002. *Fortress conservation: The preservation of the Mkomazi game reserve, Tanzania*. Bloomington: Indiana University Press.

Brockington, Dan. 2004. Community conservation, inequality and injustice: Myths of power in protected area management. *Conservation and Society* 2, 2:411.

Checker, Melissa. 2009. Double jeopardy: Pursuing the path of carbon offsets and human rights abuses. In *Upsetting the offset: The political economy of carbon markets*, Steffen Böhm and Siddharta Dabhi, eds., 41–56. London: MayFly Books.

Chernela, Janet. 2014. Fire and ice: Talking about carbon in the Brazilian Amazon. *Practicing Anthropology* 36, 3:17–21.

Climate, Community, and Biodiversity Alliance (CCBA). 2008. *Climate, community & biodiversity project design standards*, Second Edition. Arlington, VA: Climate, Community, and Biodiversity Alliance. https://s3.amazonaws.com/CCBA/Upload/ccb_standards_second_edition_december_2008+(1).pdf (accessed July 25, 2016).

Climate, Community, and Biodiversity Alliance (CCBA). 2013. *Climate, community & biodiversity standards*, Third Edition. Arlington, VA: Climate, Community, and Biodiversity Alliance. http://www.v-c-s.org/wp-content/uploads/2016/05/CCB_Standards_Third_Edition_December_2013.pdf (accessed July 25, 2016).

Colson, Elizabeth, and Max Gluckman, eds. 1959. *Seven tribes of British Central Africa*. Manchester: Manchester University Press.

Department of Forestry. 2009. *Mkuwazi Forest Reserve co-management plan and agreement*. Mzuzu, Malawi: Department of Forestry, Government of Malawi.

Department of National Parks and Wildlife. 2004. *Vwaza Marsh Wildlife Reserve master plan*. Lilongwe, Malawi: Government of Malawi.

Laughlin, Jennifer, Charles McNeill, Gayathri Sriskanthan, and Nina Kantcheva. 2013. "UN-REDD Programme Guidelines on Free, Prior and Informed Consent (FPIC)." *UN-REDD Programme Secretariat*. www.un-redd.org/Launch_of_FPIC_Guidlines/tabid/105976/Default.aspx (accessed November 4, 2015).

Malawi Environment Endowment Trust. 2009. "Plan Vivo Project Design Document: Forest Conservation in Nyika National Park and Mkuwazi Forest Reserve, Malawi." *Project Design Document: Plan Vivo*. http://www.planvivo.org/wp-content/

uploads/2009.05.15_MEET_Malawi_ PDD_PlanVivo1.pdf (accessed February 2, 2015; site now discontinued).

McCracken, John. 2006. Imagining the Nyika Plateau: Laurens van Der Post, the Phoka and the making of a national park. *Journal of Southern African Studies* 32, 4:807–821.

McElwee, Pamela D. 2012. Payments for environmental services as neoliberal market-based forest conservation in Vietnam: Panacea or problem? *Geoforum* 43, 3:412–426.

Milne, Sarah. 2012. Grounding forest carbon: Property relations and avoided deforestation in Cambodia. *Human Ecology* 40, 5:693–706.

Milne, Sarah, and Bill Adams. 2012. Market masquerades: Uncovering the politics of community-level payments for environmental services in Cambodia. *Development and Change* 43, 1:133–158.

Morris, Brian. 2000a. *Animals and ancestors: An ethnography.* New York: Berg.

Morris, Brian. 2000b. *The power of animals: An ethnography.* New York: Berg.

Morris, Brian. 2001. Wildlife conservation in Malawi. *Environment and History* 7, 3:357–372.

Peskett, Leo, and Kimberly Todd. 2012. *Putting REDD+ safeguards and safeguard information systems into practice.* UN-REDD Programme Policy Brief. Geneva, Switzerland: UN-REDD Programme Secretariat. http://www.un-redd.org/Newsletter35/Policy BriefonREDDSafeguards/tabid/105808/Default.aspx (accessed November 4, 2015).

REDD+ Social and Environmental Standards Initiative. 2012. *REDD+ Social and Environmental Standards,* Version 2. http://www.redd-standards.org/guidelines/redd-ses-guidelines-version-2/119-redd-ses-draft-guidelines-v2-5-april-2012-english/file (accessed November 4, 2015).

Ribot, Jesse C., Arun Agrawal, and Anne M. Larson. 2006. Recentralizing while decentralizing: How national governments reappropriate forest resources. *World Development* 34, 11:1864–1886.

Schafer, Jessica, and Richard Bell. 2002. The state and community-based natural resource management: The case of the Moribane Forest Reserve, Mozambique. *Journal of Southern African Studies* 28, 2:401–420.

Schoffeleers, J. M. 1997. *Religion and the dramatisation of life: Spirit beliefs and rituals in southern and central Malawi.* Kachere Monograph, no. 5. Blantyre, Malawi: Christian Literature Association in Malawi.

Sheridan, Michael J., and Celia Nyamweru, eds. 2008. *African sacred groves: Ecological dynamics and social change.* Oxford: James Currey.

Terra Global Capital. 2013a. *Project implementation report: Kulera landscape REDD+ project for co-managed protected areas, Malawi climate (Community & biodiversity standard),* Version 1–0. San Francisco, CA: Climate, Community, and Biodiversity Alliance (CCBA). https://s3.amazonaws.com/CCBA/Projects/Kulera_Landscape_REDD%2B_Project_For_CoManaged_Protected_Areas_Malawi/Verification/Kulera+CCB+PIR+v1–0+(1).pdf (accessed November 4, 2015).

Terra Global Capital. 2013b. *Kulera landscape REDD+ project for co-managed protected areas, Malawi: Project design document to the climate, community & biodiversity standard,* Second Edition. San Francisco, CA: Terra Global Capital. https://s3.amazonaws.com/CCBA/Projects/Kulera_Landscape_REDD%2B_Project_For_CoManaged_Protected_Areas_Malawi/CCB+PDD+Kulera+-+v2–0+(1).pdf (accessed November 4, 2015).

Terra Global Capital and Total LandCare. 2011. "Kulera Biodiversity Project—Malawi: Project Design Document for Validation under the Climate, Community & Biodiversity Standard—DRAFT." http://www.terraglobalcapital.com/press/OMC%20CCB %20PD%20V4%20Sept%2020 12.pdf (accessed November 4, 2015).

UN-REDD Programme. 2010. *Perspectives on REDD+*. Geneva, Switzerland: UN-REDD Programme Secretariat.
Van Breugel, J.W.M. 2001. *Chewa traditional religion*. Kachere Monograph, no. 13. Blantyre, Malawi: Christian Literature Association in Malawi.
Yocum, Heather M. 2013. *The price of trees: Producing carbon commodities and conservation in Malawi's protected areas*, PhD dissertation, East Lansing: Michigan State University.
Young, W. P. 1953. Memories of the Nyika Plateau. *Nyasaland Journal* 6, 1:45–52.
Zulu, Leo Charles. 2008. Community forest management in Southern Malawi: Solution or part of the problem? *Society and Natural Resources* 21, 8:687–703.
Zulu, Leo Charles. 2009. Politics of scale and community-based forest management in Southern Malawi. *Geoforum* 40, 4:686–699.
Zulu, Leo Charles. 2012. Neoliberalization, decentralization and community-based natural resources management in Malawi: The first sixteen years and looking ahead. *Progress in Development Studies* 12, 2–3:193–212.

15
LESSONS FROM COMMUNITY FORESTRY FOR REDD+ SOCIAL SOUNDNESS

Janis B. Alcorn

Introduction

Why is community forestry relevant to REDD+ and the carbon fix? Community forestry (CF) and CF institutions are already managing the forests that REDD+ aims to keep standing. Community forestry offers a strong foundation for jumpstarting REDD+. Socially sound approaches based on CF lessons can improve REDD+ outcomes and build resilience for adapting to climate change irrespective of the fate of REDD+.[1] Globally, deforestation rates inside community forests with strong legal recognition are dramatically lower than in forests outside those areas (Stevens et al. 2014). Securing the tenure of community forests is increasingly recognized as the key strategy for REDD+ and other climate mitigation programs.[2]

This chapter challenges the unidimensional image of community forestry and REDD+ as "people planting trees," and assesses evidence that reveals the strategic ways that indigenous peoples and forest-dependent communities have innovated and responded to externally driven changes relevant to REDD+. The chapter takes the perspective that communities are agents in their own development. It is based on meta-analyses of relevant academic, donor and nongovernmental organization (NGO) literature, including case studies, reports and other meta-analyses. Lessons from CF are assessed for recommendations to improve REDD+ social soundness, design and implementation.

Key Dimensions and Variants of Community Forestry

What is community forestry? Community forestry (CF) exists where communities or groups of people have *partial* to *full rights* over specific forests. This may include the rights to benefit from forests, as well as the authority to establish and enforce rules governing access and use of forests on the basis of formal territorial recognition, title or legitimate historical claims. CF can be viewed as a subset of community-based natural resource management (CBNRM) (Murombedzi 2010) or community-based conservation (CBC) (Western and Wright 1994). While more cross-fertilization between CBNRM, CBC and CF discourses, practices and research could be helpful for climate change adaptation programs (Chishakwe et al. 2012) and for REDD+, this chapter focuses only on the discourse and analysis of CF.

Community forestry has different dimensions when valued through the lenses of rights, labor, climate change and/or economics. Likewise, REDD+ outcomes depend on how communities and forests are valued. The carbon market and REDD+ programs give new financial incentives to national governments to assert sovereign claims over forests. The existence of this new forest-based asset could drive state-linked elites to use REDD+ funding or private carbon sales to threaten the tenurial security of communities. "Social forestry" (communities providing labor in state-owned forests) is being positioned in global REDD+ discourse as the progressive way to provide social benefits from REDD+[3] without recognizing community tenure over forests. This normative view of social forestry offers easy entrance for large-scale transactions initiated by interests of the global North seeking to offset their own greenhouse gas emissions. The carbon market and the international climate change agreements also create incentives for national elite networks to sell rights to carbon over large forest areas with the stroke of a pen. Carbon rights and forest rights are central to REDD+ and community forestry, and are hotly contested (Almeida et al. 2014).

"Discovered" versus "Designed" Community Forestry: Rights and Implications for Implementation

The rights dimension of CF can be highlighted by understanding CF as either "designed" or "discovered" (Seymour 1994). Designed CF has been introduced by development and agroforestry projects, and typically involves the introduction of seedling nurseries and reforestation[4] plans to be implemented by communities serving as labor. This designed framing ignores rights and dominates CF practice and discourse. Discovered[5] CF, on the other hand, has been self-generated and discovered by outsiders—a subset of what anthropologists term "indigenous resource management" (Alcorn 1997). Because of their foreign assistance focus, outsiders often see community forestry as formal forestry projects, and fail to see the self-generated forest management that people do on their own without projects. The lifeblood that sustains self-generated CF is local innovation in communities exercising de facto tenure rights.

Self-generated CF exists in all regions of the world, but tends to be less visible in Asia and Africa than in Latin America due to the regions' colonial histories.

This historical difference also underpins the differences in the ways that REDD+ programs are incorporating CF in Asia and Africa versus in Latin America. Prior to European colonization, communities living in the world's natural forests followed long-held indigenous scripts[6] for managing enriched[7] natural forests (Alcorn and Royo 2014; Colfer et al. 2014). Agroforestry crops were integrated into natural forests as part of rotational agricultural systems; these, in turn, produced the great diversity of products that were traded through vast, pre-colonial trading networks in the Americas, Asia and Africa.

After the catastrophic population crash as a result of European colonization, natural forests re-established themselves over large areas, and indigenous CF continued in those areas under indigenous control. Indigenous forest management systems were also adopted by settlers in remote areas, such as the *ribereños* and *caboclos* in South America, because these systems require low labor inputs and are well adapted to the local conditions (Alcorn 1989). However, in Asia and Africa, during the centuries of colonial domination, the *taungya* system (derived from the German *Waldfeldbau* system) replaced traditional CF in many places by forcing the local population to plant and tend plantations (Nair 1993). Local people were allowed to plant crops between the newly planted trees for 2 years, providing short-term subsistence in return for their labor to establish plantations. This system included colonial laws that formally converted customary community forests into state land so that forests could be harvested for the benefit of the colonial or neocolonial state. *Taungya* legitimized forest colonization and the political violence used in postcolonial counterinsurgency (Peluso and Vandergeest 2011). The *taungya* system created the large forest plantations that persist in many parts of Asia and Africa (Nair 1993). As *taungya* was not imposed in Latin America, in contrast to Asia and Africa, only 1 percent of Latin American forests are plantations today (IADB 2012). Today 43 percent of Latin American forests are under community or indigenous control with some type of legal recognition, and most of the remaining forest area is claimed by indigenous, Afro-descendent, and other rural communities who live in forests without legal recognition. In Asia and Africa, however, communities have limited or no formal forest tenure rights despite recent actions to expand their tenure rights (e.g., the 2006 Forest Rights Act in India, and the 2012 Indonesia constitutional court decision that customary *adat* forests do not belong to the state). Outside of China, only 10 percent of Asian forests are legally under community control (RRI 2012). In Africa, some 2 percent of forests are legally under community control, although 1.4 billion ha in sub-Saharan Africa are under communities' customary control (Wiley 2012). Nonetheless, indigenous CF systems persist throughout Asia and Africa.

Self-generated CF arguably has the best record of maintaining forests while adjusting to changing conditions (Porter-Bolland et al. 2011). Over the past 40 years, communities around the world have modified their self-generated community forestry systems as they adjust to increasing interventions of markets, new economic opportunities, labor migration and external projects.

Lessons from the failures of imposed (or designed) CF began to arise in the 1980s. Yet, 20 years later, a critical World Bank review (Mansuri and Rao 2004), found that attempts to apply rural development lessons from the 1980s were hampered by continued top-down tendencies, lack of evaluations, and the complexities of local situations. The manner and degree to which externally developed projects have affected community forest governance depend on how self-generated systems have responded to such interventions, and this is poorly understood (van Laerhoven and Barnes 2014). At the same time, self-generated CF has, without external project assistance, adapted to produce new products for markets, such as coffee and cacao, by incorporating commercially valuable species into existing CF systems (Scriven 2012).

Policy reforms have had major impacts on community forests. For example, in the 1980s, community forestry was formally recognized by Latin American national policy reforms (Wentzel 1998), beginning in Mexico with forest law reform, and then in Brazil with creation of "extractive reserves" for communities in response to Chico Mendez and the rubbertappers' movement in the early 1990s. At the same time in Asia, colonial *taungya* became social forestry, as the state claimed national forest dominion, and negotiated with communities to provide labor. This colonial practice has been legitimized in legal jargon as "forest villages" and "forest leases" (Peluso 1994; Poffenberger 1999). Under the seemingly benign social forestry rubric, the CF now being promoted in Southeast Asian REDD+ (ASEAN 2014) maintains the state's claims over community forest and management decision-making rights, continuing the tradition of treating communities as labor.

The global human rights context of CF shifted in 2008, when the UN Declaration on the Rights of Indigenous Peoples mandated that all nations and donors recognize and support indigenous rights. While the implementation of UNDRIP has lagged in Asia and Africa, the jurisprudence of the Inter-American Commission on Human Rights provides guidance applicable to resolving national governments' claims to forests in other regions (IACHR 2009).

The Labor Investment Dimension of Community Forestry

A second key dimension of CF is viewed through the labor lens. CF ranges from low labor intensity to high labor intensity—with low intensity having the least impact on natural forest. Rotational (swidden) agriculture that favors useful tree species (Alcorn 1989) is the main CF practice at the low end of the intensity spectrum. This type of CF tends to be found in remote areas with low population densities, where indigenous territorial rights have been recognized or demanded, and where forests are highly valued for their game, non-timber forest products and ecological services, as well as for cultural values. Medium labor intensity CF is found in less remote areas, where forests are enriched by favoring or planting useful species in standing forest or regenerating forest, and may be logged under timber harvest management plans, as in the well-documented Mexican cases (Bray 2013).

High labor intensity CF tends to be found in areas with greater population density, in indigenous areas with limited land, in areas that have been colonized by new migrants and are in conflict, and/or in areas where markets for high value forest products lead farmers to favor forest gardens over clearing for agriculture (as in Dayak forests of West Kalimantan). High intensity CF also includes reforestation and commercial agroforestry.

Methods

This chapter presents key lessons from recent literature and from regional meta-analyses commissioned by USAID (US Agency for International Development) in 2012, to encourage the development of REDD+ interventions and policies that integrate social considerations into climate change program design and implementation (Alcorn 2014; Blomley 2013; Fisher 2014).

The CF literature generally offers ex ante advice for designing interventions and policy rather than evidence-based recommendations (Helming et al. 2011). It is dominated by descriptive academic studies that seldom assess their representativeness before generalizing findings. Very few project evaluations are published or used as baselines for follow-up evaluations. Disciplinary divides also mark the literature, as foresters seldom interact with agronomists who understand traditional agrarian mosaics of forests and fields, while biodiversity specialists and anthropologists seldom collaborate. Surprisingly few researchers have focused long-term on a specific site with the intent of following all the factors relevant to the cause-and-effect relationships behind an observable situation.

Assertions about the early impacts of REDD+ projects were not grounded in rigorous evaluations and lacked counterfactuals[8] (Caplow et al. 2011). Research by the International Forestry Resources and Institutions (IFRI) program (Ostrom 2011), established by Nobel laureate Elinor Ostrom, is an exception. The Center for International Forestry Research (CIFOR), World Agroforestry Center (ICRAF) and Bioversity International have begun long-term monitoring of a few key sites with community forestry (CIFOR 2012), and promise to deliver new insights useful for REDD+.

Highlights from the Community Forestry Literature of Interest for REDD+

This section highlights lessons for REDD+ gleaned from the CF literature.

Payments for Environmental Services as a Model for REDD+

REDD+ has been described as the world's largest experiment in payments for environmental services (PES) (Corbera 2012). A meta-analysis of 301 studies of 40 different PES schemes in Latin American watersheds (Martin-Ortega et al.

2012, 2013) found that payments to "sellers" (communities with watershed forests) are 60 percent higher than what buyers of environmental services (such as hydropower companies and water users) are willing to pay for those services. Significant subsidies pay the high costs of the intermediary institutions (usually NGOs) that promote and mediate PES payments. Similar to the pricing of watershed services, the market price of carbon is far below what would be an acceptable payment to communities who agree to maintain forests for their carbon rather than put their forests to an alternative use (such as agriculture or logging) that would be banned under a carbon payment agreement. REDD+ will likely require considerable subsidies from public or donor funds to be successful.

The Effects of Forest Management Plans and Regulations on Community Forestry

A forestry technocracy that demands expensive forest management plans can impede communities from benefiting from CF (Giri and Ojha 2011). In an unusual departure from the global norm, Mexico's CF program (PROCYMAF, initiated with World Bank funding in 1997) enabled government forest agencies to transition from their usual role as anti-logging enforcers to a role of providing technical advice to communities for forest management plans for logging (Sarukhan and Merino 2007).

Another technocratic requirement that negatively affects CF revolves around the licensing regulation for non-timber products. For example, in Peru, short-term Brazil nut licenses do not enable collectors to protect the forests in which they are found, but in Bolivia, collective land rights have been titled to Brazil nut communities (Alcorn et al. 2006). However, in Bolivia, Peru and Brazil (the three countries where Brazil nut trees grow), the financial benefits to communities from managing forests with Brazil nuts are limited, because that market is controlled by a monopoly that keeps the price artificially low (Laird et al. 2012). In addition, state-licensed logging concessions overlap with the forests with Brazil nut trees, and illegal logging prevails beyond the borders of concessions. State regulation of both logging and nontimber products is not supporting CF.

Forest certification is another technocratic solution that creates new costs and labor demands on communities that are already meeting the goals of certification on their own. Certification aims to improve community forest management, but a review by Blackman and Rivera (2010) found little evidence that certification changed producer behavior or forest outcomes. The minimal difference in market price for certified products, the competition with falsely certified timber and non-timber forest products (NTFPs), and the higher annual costs for getting certified by outside experts, do not provide sufficient financial rewards for the costs of changes to meet certification requirements. The additional costs of monitoring forests for certifying REDD+ compliance likewise discourages community interest in REDD+.

Community Forestry and Reforestation Limits for REDD+

Community-driven reforestation is generally limited to small patches, often in steep areas where deforestation for farming led to soil erosion (Alcorn et al. 2010)—places where forests are not regenerating naturally. However the World Bank's Clean Development Mechanism (CDM) climate change reforestation funding protocols cannot monitor small patch reforestation;[9] CDM is only apt for industrial scale reforestation. CDM reforestation projects can also conflict with communities living in an area targeted for industrial reforestation when those communities are forced to relocate or invest their labor in *taungya* systems. In Brazil, 90 percent of reforestation involves large industrial plantations of exotic pines and eucalyptus (Mancano Fernandes et al. 2012). Further study is needed to compare the costs and outcomes of large reforestation projects versus self-generated reforestation, where community nurseries use native seeds gathered from nearby forests (Alcorn et al. 2010) or use "assisted natural regeneration"[10] (Shono et al. 2007; Smith and Scherr 2003).

There are few rigorous studies of the success of reforestation implemented by governments or NGOs. Assessment of one local NGO-assisted reforestation effort in the Upper Parapeti River in Bolivia, where 32 communities planted native trees, found a range of 5 percent to 90 percent tree survival 2 years after planting, depending on the quality of care given by the community (Alcorn et al. 2010). Almost all communities achieved above 80 percent tree survival, a remarkably high rate compared to plantation forestry. Low survival occurred in the few communities that did not invest labor in caring for the saplings. Trees need to be cared for during the first 5 years, and in communities where responsibility is taken for doing that, the trees survive. This corroborates the expert opinion, and the colonial *taungya* experiences in Asia and Africa, that labor investment is key for reforestation; successful reforestation depends on commitment to care for the newly planted trees until they are established.

Reforestation programs for mitigating climate change can paradoxically incentivize deforestation for timber harvest and deforestation in order to create space for reforestation or oil palm plantations, or to establish legal title (Scriven and Malhi 2012).

Women have played key roles in reforestation through their agroforestry in traditional swidden forest garden systems (Perreault 2005) and in frontier areas that have been logged (Kelly 2009), yet women are poorly represented in designed CF projects (Schmink and Arteaga Gómez-Garcia 2015).

Key Factors for Community Forestry and REDD+ Success

The USAID-funded CF regional meta-analyses confirm earlier conclusions from the Padgett et al. review (2006) of 69 case studies worldwide. All four conclude that key factors influencing CF success include secure resource tenure, congruence between the boundaries of socio-cultural/political institutions and biophysical boundaries of the resources, effective enforcement, expectation of benefits, and shared interests between authorities and communities.

Risks and Resilience of Community Forestry in a Changing Environment

Change is a constant factor to which CF must adapt. For example, in Panama, the Embera Waunaan are moving from low intensity CF into higher intensity CF in their 430,000-ha territory. With funding and assistance from external donors, the Embera Waunaan are expanding forest management and commercial logging plans in 178,000 ha of their extensive Darien forests (Congreso General Embera Wounaan 2013). The country, the region and the Embera-Waunaan people are adapting to new challenges and opportunities introduced by the projected completion of the highway connecting Central America with Colombia.

Other challenges threatening community forests worldwide include armed conflict, illegal trade and extractive industries. Where communities and their forests are not protected by tenurial security, CF is also under global threat by deforestation to establish agro-industrial plantations, especially from oil palm plantations, which have also been approved for REDD+ funding (Pusaka et al. 2013; RFUK 2013).

Conclusions and Discussion

This section summarizes and discusses conclusions in relation to REDD+ by responding to questions typically posed by development practitioners and REDD+ project designers.

What Are the Impacts of Supporting Community Forestry for REDD+?

Meta-analyses show that gains from providing support to CF include: improvement in forest and carbon management; access and management rights for indigenous and local people; secure global and local benefits; implementation of the UN Declaration on the Rights of Indigenous Peoples (UNDRIP); and improvements in forest governance and greater social capital. Useful support from donors, governments and NGOs ranges from technical assistance to legal and policy reforms that reinforce the legitimacy of diverse, self-generated community forestry systems (Alcorn 2014; Blomley 2013; Fisher 2014). Additional evidence (Stevens et al. 2014) shows lower rates of deforestation in areas under self-generated CF, making a compelling case that securing community forest rights can secure forests.

What Have We Learned about Threats to Community Forestry and REDD+?

Large coordinated infrastructure plans, extractive industry, illegal activities, oil palm and other forms of agro-industrial expansion are threatening CF. Coherence between development policies to support community forestry and REDD+ is urgently needed. Actions to support CF for REDD+ will need to counter the strong threats that are currently undermining CF. Illegal activities not only threaten forests but also the local social fabric and local institutions that protect forests.

What Have We Learned about Community Empowerment for REDD+?

Forest communities must have legitimate and effective rights to make and apply rules governing forest use and access. Policy and legal reforms that strengthen collective tenure rights are the most effective intervention—*if* implementation makes paper rights into effective rights (Alcorn 2014). Mexico's standout success in building REDD+ on CF (Bray et al. 2010; Bray 2013) is based on two elements that consolidated community control over forests. First, the Mexican revolution legitimized community decision-making assemblies of household heads—assemblies that are generally typical of customary community governance bodies in many parts of the world despite influences by colonial impositions of chiefs and new officers required by national laws. Subsequently national forest laws recognized Mexican communities as the legitimate forest managers who can apply rules and control access.

While the tenure rights of indigenous peoples may be legally recognized and documented in the Amazon (RAISG 2012) and other areas of the world (RRI 2014), indigenous communities are often unable to realize their tenurial rights because law enforcement is lacking, or administrative bureaucracies have not finalized the paperwork for titles to be fully registered, as in Peru (Che Piu and Menton 2013). Realizing those rights is a key challenge for REDD+ (UN-REDD 2014a).

What Have We Learned about Governance and Stakeholder Engagement in Community Forestry for REDD+?

Self-generated forest governance institutions that have the greatest legitimacy to local people are most likely to produce community forestry success for REDD+. Law and policy can be written to recognize existing self-generated institutions. The rights of local communities to make decisions about forest management in turn bolsters the legitimacy and effectiveness of those institutions. REDD+

could be guided by Ostrom's law (Fennell 2011) regarding local forest management institutions—if it works in practice, it can work in theory and policy. REDD+ would be wise to avoid creating parallel processes that weaken self-generated institutions, and instead use rights-based approaches, mapping, life plans and community-based development plans generated by inclusive processes controlled by communities.

Ignoring the presence and institutions of people living in forests creates conflicts and will undermine REDD+ success. There is a very real need to survey populations of people living in state forests so that forest dwellers are acknowledged and included in solutions. The enlightened Philippines Indigenous Peoples Rights Act (1997) was drafted after survey of population in the state forests opened the path for recognition of indigenous forest rights (Alcorn 2011). Empowered, diverse forest communities find ways to coexist and negotiate agreements about forest rights and access that greatly reduce forest conflicts (Alcorn et al. 2010).

What Have We Learned about REDD+ Benefits and Incentives for Communities?

Benefits that communities already derive from CF are both intangible and tangible. These benefits are economic, environmental, spiritual and cultural. Analyses of Mexican community forest enterprises have found that, with supportive policies and technical support, most are producing financial benefits that enable communities to invest in other enterprises or in needed services for the community. Employment opportunities are often allocated according to household needs, with income distributed directly to households after the community enterprises' operational costs are covered (Cubbage et al. 2013).

The largest carbon deal in the world (6.2 million ha, $120 million in carbon sales) was carried out by Coastal First Nations in the Great Bear Forest, Canada. The carbon sales profits are split equally between the province of British Columbia and the Coastal First Nations communities, after the First Nations' costs of forest management are paid (Shawn Burns, personal communication, 2011). This deal was based on Coastal First Nations' ongoing CF. Benefits from similar carbon sales could stabilize forests under CF, yet the international framework for carbon rights does not recognize community rights to carbon, and a recent study (Alemeida et al. 2014) shows that only 2 of 23 countries eligible to access support from UN-REDD and FCPF (Forest Carbon Partnership Facility) have national legislation defining carbon tenure.

In discovered CF, communities figure out their own workable practices that respond to their own assessment of opportunities, costs and benefits. Assisted natural regeneration is the forest management tool most often used in discovered CF and ideal for REDD+. It is less labor-intensive, less costly, more effective, and provides more benefits than plantation reforestation.

Strengthened community tenure offers a strong incentive for self-generated forest management institutions and reduces risk and costs for REDD+. Incentives or subsidies established without taking into account the community's own assessment of the costs and benefits are not sustainable. National and local objectives often differ, so they need to be reconciled for successful outcomes. REDD+ benefits are best viewed as supplemental to the existing CF benefits valued by communities.

What Have We Learned about the Role of the State and Capacity Building for REDD+?

The state plays a critically important role for REDD+ by supporting and defending formal, complete recognition of tenure rights—the necessary basis for communities to control access to and benefit from forests while providing services to the larger society—including carbon sequestration. Planning, monitoring and evaluation, and adaptive management are skills that contribute to successful forest management, particularly in cases where timber is being harvested. The state and NGOs can play key roles by providing technical assistance on request by communities, including assistance with institutional design that enhances governance capacities for two-way accountability[11] for decisions. Aiming for good governance is an essential goal for both communities and government. Local communities also benefit from technical assistance and training in bookkeeping skills for better accountability and trust of their leadership where forests generate revenue to communities.

In sum, technical and softer skills for both government and communities are needed for successful REDD+. Government agencies benefit from re-orientation to better communicate with and serve forest communities and indigenous peoples.

What Have We Learned about Scaling-Up Community Forestry for REDD+?

Proven CF spreads itself. The fact that discovered CF exists, persists and scales up without external support indicates that it is meeting felt needs and is adapted to changing local conditions. Reforming policies, laws and regulations, and their enforcement are key to scaling up. Relevant reforms include forest tenure and management laws, as well as regulations that support continued forest maintenance.

For REDD+ to work, the many conflicts over forest rights must be acknowledged and addressed at scale. REDD+ will be stronger if it builds on ground-truthed information about forests and the self-generated institutions that are de facto governing choices about forests. Nurturing diversity is essential; one size does not fit all (Nagendra and Ostrom 2012).

What Have We Learned about Sustainability of Community Forestry for REDD+?

CF supports ecological sustainability at local and global levels. CF sustainability in turn depends on supportive policies, especially tenure policies and policies that legitimize local self-generated institutions that govern forest decisions locally. A meta-analysis of 40 cases (de Jong et al. 2010, 310) succinctly describes the opportunities and challenges for REDD+:

> The community forestry development establishment has indeed observed the self-generated forestry models, but has failed to interpret those adequately to propose forestry development models that are acceptable to local forestry protagonists. Externally proposed forestry development models do not last unless they are rooted in the local social structures, economies, and value systems. If they don't meet these conditions, they become ephemeral and constitute a drain on national and international resources. Even though linking to export markets appears to be necessary in order to achieve some significant economic benefits, they are not the best departure points from which to design new models, unless those models have truly been adjusted to local realities. And the only ones who can truly judge whether or not that is the case, are the local producers, and nobody else.

What Have We Learned about Designing and Implementing Social Soundness, and Its Lessons for REDD+?

Social soundness, in international development jargon, refers to the comprehensive consideration and incorporation of social issues into REDD+ program design and implementation (FCMC 2013). REDD+ social soundness tools include the development and application of social safeguards and standards, components of REDD+ standards that address social concerns, social impact assessment approaches, benefit distribution, and free, prior, and informed consent (Mackenzie 2012). Yet we have learned that these valuable social soundness tools have limited impact unless attention is also paid to reforming the key overarching policies that set the rules of the REDD+ game—including those that legitimize or undermine community tenure rights and governance systems. This lesson was recently highlighted in the first 5-year evaluation of the UN-REDD program (UN-REDD 2014a), and additional attention to community forest tenure has subsequently been incorporated into the UN-REDD 5-year strategy for 2016–2020 (UN-REDD 2014b).

Notes

1. The global climate change negotiations in Paris (2015) for a new international agreement to curb greenhouse gases are drawing donor investment into other climate gas reduction mechanisms beyond REDD+, even as REDD+ shifts into the

implementation stage (Norman et al. 2014). Growing international financial commitments to the Green Climate Fund are refocusing funding attention onto adaptation and green growth. The carbon market and price per ton continue to shrink. Whether REDD+ becomes a shrinking sidebar among efforts to affect climate change or not, community forestry will continue to play a key role in projects aimed to maintain the world's forests and community well-being.

2. See http://usaidlandtenure.net/commentary/2014/09/incentives-adopt-climate-smart-agriculture for an overview of other climate mitigation programs announced at the New York Climate Summit, September 2014; and http://reddplussafeguards.com/?p=1110 on the German-Norwegian-Peruvian Climate and Forest Partnership welcomed by indigenous organizations. A similar Letter of Intent for a Norway-Indonesia REDD+ partnership was signed earlier.
3. For example, Kalame et al. (2011) support *taungya* social forestry for REDD in Ghana because it "satisfies the farmer's quest for arable land" (p. 519), "provid[ing] food, employment and 'energy security' [firewood] for farmers" (pp. 524–525).
4. Reforestation refers to the planting of tree plantations (generally non-native species) in areas where natural forest has been severely disturbed or removed and cannot be naturally regenerated. Afforestation refers to planting trees in areas that were not naturally forested or have long been under agricultural use.
5. Seymour's label "discovered" implicitly acknowledges that self-generated CF is invisible to outside agents until they discover/acknowledge its existence.
6. Script is a cognitive psychology term that refers to behaviors appropriate to situations that an individual absorbs through cultural influence and association with society; in this case it includes the ritual and practical steps necessary for maintaining complex socio-ecological systems such as milpa, chacra and other swidden, rotating agricultural systems (Alcorn and Toledo 1998).
7. People enrich forests by planting or protecting certain species instead of others; an enriched forest may appear natural but the species composition has been altered by people. For example, the rustic fincas of Central America are natural forests that include introduced coffee trees in their shade; Indonesian and Malaysian Dayak enriched rainforests have higher densities of durian, introduced rubber and other economically valuable trees. Enriched forests appear, and perform ecologically, as natural forests.
8. Counterfactuals are what-if scenarios used to test the strength of a prediction of an outcome, rather than accepting conditions as constant for predicting causal relationships. Counterfactual reasoning allows for alternative possibilities and outcomes.
9. A community-forestry CDM project for the Pico Bonito area in Honduras was cancelled because communities did not follow the large-scale industrial plantation models required for FCPF carbon monitoring; instead they only reforested patches where the community deemed that reforestation was necessary (cf., CDM evaluation by Rainforest Alliance 2011).
10. Natural regeneration is the process by which forests naturally regrow after old trees fall or wind knocks down a gap in forest. Sun-loving species move in along with those species of trees with stumps that sprout naturally, and seeds of shade-loving species typical of mature forests slowly sprout after those seeds are introduced by bats, birds or other animals who carry the fruit from nearby mature forests. Assisted natural regeneration refers to specific practices such as thinning out unwanted weedy saplings or bending down grasses that suppress stump sprouts and/or by preventing fires, so that re-sprouting trees are no longer suppressed or burned back, and grow up to shade out the grasses.
11. Two-way accountability means that not only the community is accountable to donor/buyer for maintaining the forest, but the donor/buyer/government is accountable to the community for making payments, providing promised benefits, and complying with other promises to communities.

References

Alcorn, Janis B. 1989. Process as resource. In *Natural resource management by indigenous and folk societies in Amazonia*, D. A. Posey and W. Balee, eds., 63–77. Bronx, NY: New York Botanical Garden.

Alcorn, Janis B. 1997. Indigenous resource management systems. In *Beyond fences: Seeking social sustainability in conservation*, Volume 2, G. Borrini-Feyerabend, ed., 8–13. Gland, Switzerland: IUCN.

Alcorn, Janis B. 2011. *Tenure and indigenous peoples*. Property Rights and Resource Governance Briefing Paper #13, for USAID. Burlington, VT: TetraTech.

Alcorn, Janis B. 2014. *Lessons learned from community forestry in Latin America and their relevance for REDD+*. Washington, DC: Forest Carbon, Markets and Communities (FCMC).

Alcorn, Janis B., Carol Carlo, David Rothschild, Julio Rojas, Alaka Wali and Alejo Zarzycki. 2006. Heritage, poverty, and landscape-scale conservation—An alternate perspective from the Amazonian frontier. *IUCN Policy Matters* 14:272–285.

Alcorn, Janis B. and Antoinette G. Royo. 2014. Best REDD scenario, reducing climate change in alliance with shifting communities and indigenous peoples in Southeast Asia. In *Shifting cultivation and environmental change*, Malcolm F. Cairns, ed., 286–306. London: Routledge.

Alcorn, Janis B. and V. M. Toledo. 1998. Resilient resource management in Mexico's forest ecosystems: The role of property rights. In *Linking social and ecological systems for resilience and sustainability*, F. Berkes and C. Folke, eds., 216–249. Cambridge: Cambridge University Press.

Alcorn, Janis B., Alejo Zarzycki and Luis Maria de la Cruz. 2010. Poverty, governance and conservation in the Gran Chaco of South America. *Biodiversity* 11:39–44.

Almeida, Fernanda, Alexandre Corriveau-Bourque, Jenna DiPaolo Colley, Arvind Khare, Tony La Viña, Jenny Springer and Andy White. 2014. *Status of forest carbon rights and implications for communities, the carbon trade, and REDD+ investments*. Manila and Washington, DC: Rights and Resources Initiative and Ateneo University.

Association of Southeast Asian Nations (ASEAN). 2014. ASEAN joint statement on climate change delivered by Hla Maung Thein, Deputy Director General, Ministry of Environmental Conservation and Forestry, the Republic of the Union of Myanmar, on behalf of ASEAN member states at the Joint High-Level Segment 20th session of the conference of the Parties to UNFCCC and 10th session of the Conference of the Parties serving of the meeting of parties to the Kyoto Protocol. http://www.asean.org/images/pdf/2014_upload/ASEAN%20Joint%20Statement%20on%20Climate%20Change%202014.pdf (accessed January 3, 2014).

Blackman, A. and J. Rivera. 2010. *The evidence base for environmental and socioeconomic impacts of "sustainable" certification*. Stockholm: Environment for Development.

Blomley, Tom. 2013. *Lessons learned from community forestry in Africa and their relevance for REDD+*. Washington, DC: Forest Carbon, Markets and Communities (FCMC).

Bray, David. 2013. From Mexico, global lessons for forest governance. *Solutions* 4(3):51–59.

Bray, David, Deborah Barry, Sergio Madrid, Letica Merino and Ivan Zuñiga. 2010. *Sustainable forest management as a strategy to combat climate change: Lessons from Mexican communities*. Mexico City and Washington, DC: Consejo Civil Mexicano para la Silvicultura Sostenible and Rights and Resources Initiative.

Caplow, S. P. Jagger, K. Lawlor and E. Sills. 2011. Evaluating land use and livelihood impacts of early carbon projects: Lessons for learning about REDD+. *Environmental Science and Policy* 14:152–167.

Center for International Forestry Research (CIFOR). 2012. *Learning lessons from smallholder and community forest mosaics to support the sustainable livelihoods in the Amazon.* Bogor, Indonesia: CIFOR.

Che Piu, Hugo and Mary Menton. 2013. *Contexto de REDD+ en Perú.* Documentos Ocasionales 90. Bogor, Indonesia: CIFOR.

Chishakwe, N., L. Murray and M. Chambwera. 2012. *Building climate change adaptation on community experiences: Lessons from community-based natural resource management in southern Africa.* London: International Institute for Environment and Development (IIED).

Colfer, Carol, Janis B. Alcorn and Diane Russell. 2014. Swiddens and fallows: Reflection on the global and local values of "slash and burn". In *Shifting cultivation and environmental change*, Malcolm F. Cairns, ed., 62–86. London: Routledge.

Congreso General Embera Wounaan. 2013. A Forest for Life, Forest Management in Embera Wounaan. Video produced by Fundacion Prisma, San Salvador. https://www.youtube.com/watch?v=6naIDg4dPrk.

Corbera, Esteve. 2012. Problematizing REDD+ as an experiment in payment for environmental services. *Current Opinion in Environmental Sustainability* 4:1–8.

Cubbage, Frederick, Robert Davis, Diana Rodriguez Paredes, Gregory Frey, Ramon Mollenhauer, Yoanna Kraus Elsin, Ignacio Antonio Gonzalez Hernandez, Humberto Albarran Hurtado, Ana Mercedes Salazar Cruz and Diana Nacibe Chemor Salas. 2013. *Competividad y acceso a mercados de empresas forestales comunitarias en México.* Mexico City and Washington, DC: CONAFOR and World Bank.

de Jong, C. Cornejo, P. Pacheco, B. Pokorny, D. Stoian, C. Sabogal and B. Louman. 2010. Opportunities and challenges for community forestry: Lessons from Latin America. In *Forests and society—Responding to global drivers of change*, G. Mery, P. Kaila, G. Galloway, R. I. Alfaro, M. Kanninen, M. Lobovikov and J. Varjo, eds., 299–314. Vienna: IUFRO.

Fennell, Lee Ann. 2011. Ostrom's Law: Property rights in the commons. *International Journal of the Commons* 5:9–27.

Fisher, Robert J. 2014. *Lessons learned from community forestry in Asia and their relevance for REDD+.* Washington, DC: Forest Carbon, Markets and Communities (FCMC).

Forest Carbon, Markets and Communities (FCMC). 2013. *Social dimensions of REDD+: Issues brief.* Washington, DC: Forest Carbon, Markets and Communities (FCMC).

Giri, Kalpana and Hemant R. Ojha. 2011. *How does techno-bureaucracy impede livelihood innovations in community forestry?* Discussion Paper. Kathmandu: Forest Action Nepal.

Helming, K., K. Diehl, H. Bach, O. Dilly, B. König, T. Kuhlman, M. Perez-Soba, S. Sieber, P. Tabbush, K. Tscherning, D. Wascher and H. Wiggering. 2011. Ex ante impact assessment of policies affecting land use, Part A: Analytical framework. *Ecology and Society* 16(1):27.

Inter-American Commission on Human Rights (IACHR). 2009. *Indigenous and tribal peoples rights over their ancestral lands and natural resources; Norms and jurisprudence of the Inter-American Human Rights System.* OEA/Ser.L/V/II Doc.56/. Washington, DC: Organization of American States.

Inter-American Development Bank (IADB). 2012. *The climate and development challenge for Latin America and the Caribbean: Options for climate resilient low carbon development.* Washington, DC: IADB.

Kalame, Fobissie B., Robert Aidoo, Johnson Nkem, Oluyede C. Ajajie, Markku Kanninen, Olavi Luukkanen and Monica Idinoba. 2011. Modified taungya system in Ghana: A win-win practice for forestry and adaptation to climate change. *Environmental Science & Policy* 14:519–530.

Kelly, J. J. 2009. *Reassessing forest transition theory: Gender, land tenure insecurity and forest cover change in rural El Salvador.* PhD dissertation, New Brunswick, NJ: Rutgers University.

Laird, Sarah A., Rebecca J. McLain and Rachel P. Wynberg. 2012. *Wild product governance: Finding policies that work for non-timber forest products.* London: Routledge.

Mackenzie, Catherine. 2012. *REDD+ social safeguards and standards review.* Washington, DC: Forest Carbon, Markets and Communities (FCMC).

Mancano Fernandes, B., C. A. Welch and E. C. Goncalves. 2012. *Land governance in Brazil: A geo-historical review.* Land Governance in the 21st Century: Framing the Debate Series, No. 2. Rome: International Land Coalition (ILC).

Mansuri, G. and V. Rao. 2004. Community-based and-driven development: A critical review. *World Bank Research Observer* 19(1):1–39.

Martin-Ortega, Julia, Elena Ojea and Camille Roux. 2012. *Payments for water ecosystem service in Latin America: Evidence from reported experience.* BC3 Working Paper Series. Bilbao: Basque Centre for Climate Change (BC3).

Martin-Ortega, Julia, Elena Ojea and Camille Roux. 2013. Payments for water ecosystem services in Latin America: A literature review and conceptual model. *Ecosystem Services* 6:122–130.

Murombedzi, James C. 2010. Agrarian social change and post-colonial natural resource management interventions in southern Africa's "communal tenure" regimes. In *Community rights, conservation and contested land: The politics of natural resource governance in Africa*, Fred Nelson ed., 32–51. London: Routledge.

Nagendra, Harini and Elinor Ostrom. 2012. Polycentric governance of multifunctional forested landscapes. *International Journal of the Commons* 6(2):104–133.

Nair, P.K.R. 1993. *An introduction to agroforestry.* Boston: ICRAF and KLGWER.

Norman, Marigold, Alice Caravani, Smita Nakhooda, Charlene Watson and Liana Schlatek. 2014. *Climate finance thematic briefing: REDD+ and finance.* ODI Climate Funds Update. London and Washington, DC: Overseas Development Institute and Heinrich Böll Stiftung North America. Böll Stiftung.

Ostrom, Elinor. 2011. Why do we need to protect institutional diversity? *European Political Science* 11:128–147.

Pagdee, A., Y. S. Kim and P. J. Daugherty. 2006. What makes community forestry management successful: A meta-study from community forests throughout the world. *Society and Natural Resources* 19:33–52.

Peluso, Nancy Lee. 1994. *Rich forests, poor people.* Berkeley: University of California.

Peluso, Nancy Lee and Peter Vandergeest. 2011. Political ecologies of war and forests: Counterinsurgencies and the making of national natures. *Annals of the Association of American Geographers* 101(3):587–608.

Perreault, T. 2005. Why chacras (swidden gardens) persist: Biodiversity, food security, and cultural identity in the Ecuadorian Amazon. *Human Organization* 64(4):327–339.

Poffenberger, Mark, ed. 1999. *Communities and forest management in Southeast Asia.* Working Group on Community Involvement in Forest Management. Gland, Switzerland: IUCN.

Porter-Bolland, Luciana, Edward A. Ellis, Manuel R. Guariguata, Isabel Ruiz-Mallén, Simoneta Negreta-Yankelevich and Victoria Reyes-García. 2011. Community managed forests and forest protected areas: An assessment of effectiveness across the tropics. *Forest Ecology and Management* 268:6–17.

Pusaka, Forest Peoples Programme (FPP), and Sawit Watch. 2013. *A sweetness like unto death: Voices of the indigenous Malind, Merauke, Papua.* Moreton-in-Marsh: Forest Peoples Programme (FPP).

Rainforest Alliance Smartwood Program. 2011. *Validation assessment report for: Pico Bonito-Ecologic USA, LLC in San Marcos and La Libertad, Honduras*. Richmond: Rainforest Alliance.

Rainforest Foundation UK (RFUK). 2013. *Seeds of destruction: Expansion of industrial oil palm in the Congo Basin, potential impacts on forests and people*. London: RFUK.

Red Amazónica de Información Socioambiental Georreferenciada (RAISG). 2012. *Amazonía bajo presión*. Sao Paulo: RAISG.

Rights and Resources Initiative (RRI). 2012. *Turning point*. Washington, DC: RRI.

Rights and Resources Initiative (RRI). 2014. *Lots of Words, Little Action: Annual Review of Rights and Resources 2013–2014*. Washington, DC, Rights and Resources Initiative.

Sarukhan, Jose and Leticia Merino. 2007. *Challenges to sustainable forest management and stewardship in Mexico*. Mexico City: Instituto de Investigaciones Sociales and Instituto de Ecología, Universidad Nacional Autónoma de México (UNAM).

Schmink, Marianne and Marliz Arteaga Gómez-Garcia. 2015. *Under the canopy: Gender and forests in Amazonia*. CIFOR Occasional Paper No. 121. Bogor, Indonesia: CIFOR.

Scriven, J.N.H. 2012. Preparing for REDD: Forest governance challenges in Peru's Central Selva. *Journal of Sustainable Forestry* 31:421–444.

Scriven, J.N.H. and Y. Malhi. 2012. Smallholder REDD+ strategies at the forest-farm frontier: A comparative analysis of options from the Peruvian Amazon. *Carbon Management* 3(3):265–281.

Seymour, Frances. 1994. Are successful community based conservation projects designed or discovered? In *Natural connections*, D. Western and M. Wright, eds., 472–496. Washington, DC: Island Press.

Shono, K., E. A. Cadaweng and Patrick B. Durst. 2007. Application of assisted natural regeneration to restore degraded tropical forestlands. *Restoration Ecology* 15:620–626.

Smith, J. and S. J. Scherr. 2003. Capturing the value of forest carbon for local livelihoods. *World Development* 31(12):2143–2160.

Stevens, Caleb, Robert Winterbottom, Jenny Springer and Katie Reytar. 2014. *Securing rights, combatting climate change: How strengthening community forest rights mitigates climate change*. Washington, DC: World Resources Institute and Rights and Resources Initiative.

UN-REDD. 2014a. External evaluation of the United Nations collaborative programme on reducing emissions from deforestation and forest degradation in developing countries (the UN-REDD programme), validated during the July 2014 UN-REDD policy board meeting in Lima, Peru. http://www.un-redd.org/Portals/15/documents/UN-REDD%20Evaluation%20Final%20Report%20Volume1%20June2014%20EN.pdf (accessed January 3, 2015).

UN-REDD. 2014b. *Developing the UN-REDD programme 2016–2020 strategy*, Discussion Paper. http://www.un-redd.org/Portals/15/documents/Discussion%20Paper%20Developing%20the%20UN-REDD%20Programme%202016–2020%20Strategy.pdf (accessed January 3, 2015).

van Laerhoven, Frank and Clare Barnes. 2014. Communities and commons: The role of community development in sustaining the commons. *Community Development Journal* 49(51):118–132.

Wentzel, S. 1998. Social forestry in Latin America—A first overview of the issues. *Wald-Info* 2, Bonn: GTZ.

Western, David and R. Michael Wright, eds. 1994. *Natural connections*. Washington, DC: Island Press.

Wiley, Liz Alden. 2012. Rights to Resources in Crisis: Reviewing the Fate of Customary Tenure in Africa, Brief #1. Washington D.C., Rights and Resources Initiative.

SECTION V
Alternative Configurations of Community and Governance

16
EMPOWERING FOREST DEPENDENT COMMUNITIES

The Role of REDD+ and PES Projects

Mark Poffenberger

Background

Forest dependent communities around the world have acted as de facto natural resource managers for millennia, while the nationalization of forests in many countries has taken place largely in the past two centuries. Many national forest management institutions now recognize the need for community participation in management, with governments around the world, including Indonesia, Philippines, Nepal and Brazil, increasingly giving legal recognition of community forest rights. A recent study found that over the past 20 years, over 200 million ha of public forests has been transferred to community managers in 60 countries. The study also found that in a 10-country survey of 80 community forestry (CF) groups, decentralization of public forest management generated improved community livelihood and carbon storage benefits. The authors concluded that "decentralization . . . is not only about forest governance—it is equally about development and climate policies" (Chhatre and Agrawal 2009, 17,669).

Over the past 30 years, Asia has experienced growing interest in recognizing the resource rights of forest dependent people on state forest lands. In India, over 100,000 communities have recognized Forest Protection Committees (FPCs) managing over one-third of India's forests. Yet progress in linking community forestry to national REDD+ strategies and programs has been slow to emerge in India and other South and Southeast Asian countries, with few examples of certified projects that are generating funds from carbon offsets or other environmental services. This process has been impeded by the glacial pace of national REDD+ policy and program development, as well as a lack of support for subnational community-based projects. This chapter provides an example of a community-oriented REDD+ project that illustrates both opportunities for developing community carbon projects, as well as the constraints faced by project developers and participating communities.

The Khasi Hills Community REDD+ Project

The Khasi Hills is located in the state of Meghalaya in Northeast India. The project area is situated on an upland plateau dissected by rivers that drop quickly to the Bangladesh floodplain. The Khasi people who have historically inhabited the region have a tradition of forest management and sacred grove protection, and have expressed concern over the rapid destruction of communal forests. At the request of the community leaders in Mawphlang village, Community Forestry International (CFI) began supporting a small pilot project in 2005 to explore how changes in resource management behavior could be incentivized through payments for environmental services. Based on the positive impacts of the pilot project, the formal REDD+ project design process was initiated in 2010. The community REDD+ project brought together 10 indigenous Khasi tribal kingdoms that possess legal rights to the 27,000-ha Umiam sub-watershed under the Sixth Schedule of the Indian Constitution (Community Forestry International 2013). In 2011, the 62 communities that make up the 10 indigenous governments (*hima*) formed a sub-watershed federation (*synjuk*) to create one of India's first community-based REDD+ initiatives. The REDD+ project supports the participating communities, coordinated through their own federation, to control drivers of deforestation, including forest fires, mining, fuelwood collection and forestland conversion. By conserving and restoring forest cover and hydrological functions, the federation seeks to help facilitate a transition to more sustainable livelihood activities and agricultural systems that are climate-resilient. The communities also believe that forest loss is changing the local micro-climate, resulting in longer, hotter dry seasons with corresponding declines in water availability. The project has been approved by the Khasi Hills Autonomous District Council, with support from the chief secretary of the state of Meghalaya. The project seeks to restore and connect natural forest fragments and sacred groves to create a wildlife corridor along the Umiam River, while improving the sustained productivity of utilization forests.

The project evolved from growing concerns among the participating Khasi communities and leaders regarding the rapid degradation and loss of their community forests. In the East Khasi Hills District, between 2000 and 2006, forest loss exceeded 5 percent per year, contributing to deteriorating surface and ground water supplies, erosion and sedimentation problems, and perceived changes in the micro-climate. Approximately 39 percent of forest lands in the project area are severely degraded as a result of unsustainable fuelwood harvesting, grazing and fire, as well as by quarrying and timber extraction (CFI 2013). Many of these drivers are being mitigated through community actions that include improved fuelwood harvesting rules based on rotation, adoption of fuel-efficient wood stoves and liquid natural gas cook stoves, changes in animal husbandry systems and fire control. Other drivers of deforestation, such as surface mining of coal and limestone, are controlled through the action of *hima* governments that hold the authority to oversee leases on community lands in their jurisdiction under the Sixth Schedule of the Indian Constitution. Agreements to limit mining and quarrying leases by the 10 *hima* under the umbrella of the *synjuk* (federation) are helping to reduce the impact of these drivers in the future, as are sedimentation ponds that are now required at mine sites.

Under the REDD+ project framework, the federation or *synjuk* plan is implementing a 30-year climate adaptation strategy for their upper watershed. The project is designed to establish an initial 10-year income stream to support the federation. Based on initial projections of the impact of community-based activities to avoid deforestation and forest degradation (avoided deforestation), as well as through forest restoration (sequestration), an additional 20,000 to 30,000 tCO_2 credits will be generated each year, yielding a gross income of US$100,000 to US$150,000 annually to finance the *synjuk* management institution and livelihood activities for participating communities. Credits come from the 9,200 ha of dense forest under REDD+ and 5,900 ha from open forests that are being restored generating sequestration benefits.

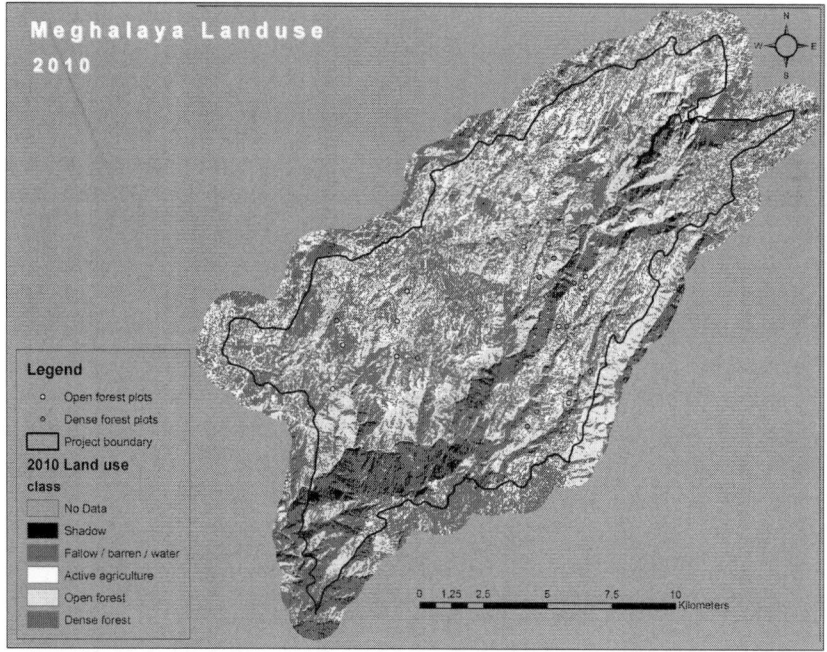

FIGURE 16.1 Map of Khasi Hills project area.

Source: CFI (2011).

To qualify for REDD+ credits, Plan Vivo, like other project certifiers, requires the project to prove that carbon offsets are additional and would not occur in the "without project" scenario. The Khasi Hills project model estimates that the rate of deforestation will be reduced gradually as mitigation activities are implemented, eventually halting forest loss over the project period. A risk buffer of 20 percent is maintained by Plan Vivo to ensure against unforeseen losses of forest cover through fire or other events, with an additional 5 percent deducted for "leakage." The project received REDD+ registration under Plan Vivo standards in March 2013, requiring a performance-based approach to project design and implementation. Key variables that must be monitored include carbon stocks, forest conditions and other environmental indicators, including changes in biodiversity and hydrology.

Socio-economic performance indicators are also being monitored by the community, including changes in household income, microfinance account balances and repayment rates, participation in alternative income-generating activities, energy use and activities of newly established farmers' clubs, such as adoption of sustainable farming practices. The project was validated by a Nepali third-party nongovernmental organization (NGO) in 2013 as part of the Plan Vivo certification process. The process took 18 months from the start of the issuance of the Project Identification Note (PIN), through the development, review and approval of the Project Design Document (PDD). The final step was the third-party validation, which was completed in March 2013.

The project is one of the first REDD+ initiatives in Asia to be developed by indigenous tribal governments on communal and clan land. If successful, the project has potential for broad-based replication among northeast India's 240 ethnolinguistic tribal communities. The federation has established agreements with companies that focus on brokering carbon credits generated by community forestry to corporate social responsibility (CSR) buyers in the private sector. Project brokers include C-Level of London, Zero Mission of Stockholm, WE Forest of Brussels, COTAP of San Francisco and EVI of New Delhi. The brokers seek to establish long term relationships with companies that will buy a fixed quantity of credits each year over the next 5–10 years.

The success of any community-based resource management system depends on the community itself and its commitment to sustaining the land, forest and water it depends upon. The 10 indigenous governments in the Umiam sub-watershed have agreed to federate and protect and restore their forests, because of their own sense of an urgent need to halt deforestation and restore important ecosystems that are central to their history and culture. REDD+ provides both a planning framework for setting goals, monitoring outcomes, and receiving financial support. REDD, payment for environmental services (PES) or any other project support will facilitate this process, but it is simply a means, not an end. What is perhaps more significant is that important socio-cultural institutions in Khasi society that have been largely bypassed by national and state government in the past are now emerging as key elements in a grassroots attempt to protect and restore local forests that possess valuable biological and cultural diversity. Communal governance structures, like the village council (*dorbar*) and indigenous government (*hima*), which rely on democratic processes to enable consensus-based decision-making, are being re-empowered through this project. This process strengthens traditional land tenure rights by focusing attention on communal forest resources whose management has been neglected in recent decades.

While the Khasi experience with REDD+ project development as a financing mechanism is still in an early stage, the project was certified under Plan Vivo (Edinburgh) standards in March 2013. In June 2013, 21,805 tons of CO_2e certificates were issued and entered in the Markit Registry. The Markit Registry acts as a clearing house for project carbon, allowing credits to be transferred or retired by buyers. Two carbon brokers, including U and We (Stockholm) and C-Level (London), entered into agreements to market the federation's carbon credits. From May 2013 to October 2015, the federation sold 27,777 tons of CO_2, valued at US$164,947 at an average price of US$5.94 per tCO_2. In addition, the federation recently signed a contract with We Forest, a Belgian NGO, to reforest 1,500 ha of degraded land for approximately $249,000. This represents the first 2 years of a 10-year agreement to regenerate 5,000 ha of degraded land. The federation is utilizing funds from carbon sales and reforestation activities for mitigation activities and to provide funds for community development and enterprise activities. In 2015, 52 village development grants were made to participating communities with two-thirds of the communities utilizing them to improve drinking water systems, often with their own co-funding. In addition, 35 small livelihood grants were provided to predominantly women-run micro-finance groups, which were used for establishing piggeries, poultry raising, and other economic initiatives.

The design and development of the Khasi Hills Community REDD+ Project catalyzed a series of dialogues, actions and consensual decisions that are improving forest governance, development processes and the environment:

- Formation of a federation by neighboring kingdoms to coordinate plans and activities raised their visibility to local government programs, allowing direction of government projects to priority development needs and capacity building activities;
- Resource mapping and planning, which created dialogue among community members about the need for conservation and restoration;
- Creation of Local Working Committees at the micro-watershed level. These provide technical support to local clusters of three to five villages in creating long-term resource management plans and implementing activities such as forest monitoring, fire control, fuelwood production and so forth;
- Support to women-led, micro-finance groups in the forms of tree nursery contracts and grants for alternative livelihood activities such as piggeries, ecotourism businesses and so forth.

Discussion

Enabling Factors

A number of factors allowed the Khasi Hills Community REDD+ Project to be successfully designed, developed and certified in a relatively short period of time. These included an effective operational strategy; a strong consensus among indigenous leaders and members to act; legal community control of the forest lands; and necessary financial and technical support from partners. The REDD+ project design built upon on a 4-year demonstration pilot project initiated by CFI, which had proven the effectiveness of forest protection and PES mechanisms operating in two communities. This pilot project generated interest among the neighboring communities and indigenous governments, leading to the creation of the federation and a commitment to implement conservation and restoration activities. Since the participating communities owned the forest land, they did not need to formally coordinate or seek the approval of local or national governments, though both were informed. This avoided potential delays in seeking inputs on project design and approvals for certification and marketing transactions. Finally, the federation received both technical and financial support for 6 years from CFI for the design; monitoring, reporting and verification (MRV); certification and marketing aspects of the project. From the beginning, there was an agreement that the federation would take full responsibility for the project by the end of 2014.

Indigenous Institutions and Apex Bodies

In many Asian countries, there has been a tendency for national and local governments to bypass indigenous institutions and leaders to interact with state-sponsored leaders and organizations. This often limits the effectiveness of government resource management projects in areas where forests and water use is controlled by traditional organizations. Recognizing and empowering indigenous resource managers with financial and technical support can enhance the effectiveness of conservation and restoration investments. As in the case with the Khasi federation, indigenous communities may be further empowered politically when they establish apex bodies such as federations and associations to coordinate resource management activities at a landscape or watershed level. This was enabled by the homogeneity of the Khasi community and the traditional network of communications that existed prior to the project.

Building on Indigenous Management Systems

In the case of the Khasi Hills, traditional systems of forest management were present and somewhat effective in controlling drivers of degradation and deforestation. Yet local pressures on forests have been increasing for decades, driven by rapid population growth, while new external market pressures are entering into the area. Indigenous governments (*hima*) were increasingly leasing community lands to commercial quarrying and mining operations in order to generate funds. Though the income was modest, the environmental impacts on the forest were substantial. To control deforestation, the communities recognized that indigenous management systems needed to be updated to address the underlying causes of deforestation and forest degradation. The REDD+ project required the establishment of baseline assessments and the monitoring of changes in the environment. The project also required the development of long-term conservation and restoration management plans, with clear implementation duties allocated to traditional government institutions as well as to newly created, micro-watershed management groups (Local Working Committees). The process of designing this REDD+ project catalyzed a broader dialogue among the participating communities regarding the environmental problems in the area and their solutions. As a consequence, the evolving management system was founded on Khasi forest conservation and sustainable use values and institutions, with new planning, monitoring and reporting systems added and coordinated through the federation.

Reducing Project Risk

Like the Khasi Hills Community REDD+ Project, REDD projects generally face a variety of risks at the field level, as they typically are situated in areas with high rates of deforestation, which they must slow substantially to generate carbon offset credits. In the case of the project area, the design team reduced the risk of project failure by selecting project sites where communities and traditional governments were aware of the environmental problems and had already begun to initiate forest conservation activities. In some cases, community forest protection may be part of an indigenous system, such as in the Khasi Hills, or related to a more recent grassroots movement, while in others it may be a component of traditional forest conservation or development projects.

Social appraisals are important as part of the project development process, illuminating local attitudes, motivations and goals for forest resources, as well as the extent and intensity of deforestation drivers operating in the area. The project developers sought to identify social capital in the form of effective community leaders, supportive local government officials, concerned community members and supportive local NGOs. The project also conducted environmental appraisals in the degraded forests to assess the biotic capital, as a way of reflecting regenerative potential that can be released through "social fencing."

Through the project, 62 participating communities have identified degraded forests with the biological conditions (soil, tree seed sources, volunteer saplings etc.) that will allow for rapid regeneration, and have closed those areas to grazing and fuelwood collection, while also protecting them from forest fire. This form of social fencing is based on a community-wide agreement to implement the management plan and associated activities. The Khasi Hills possess substantial social capital in the form of active and effective indigenous organizations and leaders, traditional conservation values, as well as a growing number of literate, educated and concerned youth. The biological capital includes high rainfall, moderately good soils and neighboring forest seed sources. This type of context might be viewed as the low-hanging fruit for REDD+ projects that should be given priority by governments and development organizations.

Partnerships for Project Design

Project developers are caught between the requirements of many REDD/REDD+ certification standards, such as the Voluntary Carbon Standard, Gold Standard and Plan Vivo, which require sophisticated project designs, carbon modeling scenarios and long term monitoring and reporting. These, in turn, require skill sets that poor, rural communities rarely possess. Launching a hybrid, forest conservation and community development process, while attending to the manifold requirements involved in monetizing and marketing carbon, requires a diverse design team. Rarely are communities and their local NGO project developers endowed with a strong capacity in all of these areas. In the case of the Khasi Hills project, this required the creation of various forms of institutional collaboration that bring diverse kinds of expertise together. This enhances design team capacity, provides a larger network of contacts in the community, public and private sector, and facilitates communication between key stakeholders. Initially facilitated by CFI, the development of the needed institutional linkages, including dealings with certifiers, brokers and carbon registries, is now increasingly managed by the federation, though CFI and other institutions continue to provide some technical support.

Assisted Natural Regeneration (ANR)

While developing forest carbon projects can create opportunities for communities to restore degraded forests, the costs of forest restoration through planting may be high, far exceeding carbon revenue flows, especially given the current low prices for forest carbon in international markets. In the case of the Khasi Hills project, ANR was identified as one of the most cost-effective approaches to restoration, relying on natural ecological regenerative capacities, while largely requiring community skills and manpower to control local drivers of deforestation and forest degradation. Payment to community groups to control illegal logging, reduce the impact of forest fires and conduct basic silvicultural activities (such as thinning, shoot-cutting, weeding and enrichment planting) is accelerating the regeneration of natural forests in the project area and increasing the productivity of non-timber forest products (NTFPs).

NTFPs will be important to increase the value of local forests for communities. Wintergreen oil is one product that could be produced from the *Gaulteria spp.* plants that are increasingly abundant as the forest recovers. The federation will need to seek out technical guidance in developing distillation systems and marketing for this product, which they believe will create alternative employment for families currently engaged in charcoal production.

Reducing Financial Risk

A major source of project risk relates to the failure to secure project funding and the disruption of funding flows during the course of the project. The Khasi Hills Community REDD+ Project was developed under a 12-month grant from a private foundation in the UK, with a grant of 100,000 British pounds. The project required 18 months to become certified and an additional 4 months until its first carbon sale, putting considerable financial pressure on the nascent federation team. Insufficient and uneven funding during the early phases of project development caused concerns to communities and NGO support groups, especially as funding for the transaction costs of validation and certification had not been covered in project budgets. While some projects rely on the presale of carbon to private investors to cover costs until certification and subsequent validation of credits can be completed, discounts may be considerable for this type of "ex-ante" sale. At several points in the development of the project, cash flow caused a major problem that was only resolved with the increasing number of sales of credits.

Once the Khasi Hills project was certified, finding buyers on the voluntary markets for forest carbon was challenging. Buyers are difficult to identify, may have unreal expectations regarding profits and risks, and may not value social and environmental benefits. Market prices for forest carbon are difficult to ascertain and appear to vary widely depending on the motives of the buyer. Project designers in developing countries may have little experience dealing with private investors, carbon brokers, certifiers and validating organizations, further limiting accessibility and raising transaction costs in terms of time and money.

In the case of the Khasi Hills, it took time for the project developers to create partnerships with carbon brokers, buyers and NGOs that support PES-type projects. In part, this has been hampered by a weak global market for forest carbon. At the same time, the value of the Plan Vivo certification has helped ensure prices of $6 to $9 per ton between June 2013 and June 2014, though downward pressure on carbon offset prices globally have pushed the price for bulk purchases to $5 per ton in 2015. Corporate social responsibility (CSR) buyers are attracted by the high social and environmental benefits that are associated with the project, as well as the forest carbon. Thirteen sales made by the project up to October 2015 generated $164,947, which has constituted the vast majority of funding for the federation staff and office, meetings, tree nurseries and small grants program.

As the project matures, publicity through the Internet and short films is boosting its visibility, and should attract additional PES buyers. The long term reforestation contract with We Forest has helped stabilize the long-term funding strategy, with a committed repeat buyer now engaging in the project that is based on individual trees grown as the monetary unit for PES.

Reducing Political Risk

Community REDD+ projects like the Khasi Hills must also address a range of issues related to their interactions with local and national governments. Since many countries, including India, are still designing their national REDD+ policies and programs, small subnational activities may face an uncertain relationship with government agencies. The Khasi project managers have struggled with the tax implications of receiving funds from carbon sales. Attempting to register as a non-profit organization required 3 years of audited accounts and interactions with a distant bureaucratic approval process. While the state government has not inhibited project progress, there has been little in the way of any direct support from the state or central government. The Khasi Hills project team, with their small innovative REDD+ initiative, were exposed to reviews, evaluations and critiques by external researchers, journalists and competing NGOs that had the potential to undermine the initiative and discourage the leadership.

Start-Up Financing for Community-Based Projects

A major issue for small REDD+ community projects is how to finance them until payment for environmental services begin to flow. Start-up costs are substantial in preparing the project design document, initiating planning activities and beginning mitigation actions. In many countries, there are few grant funds available to support this work. Some policy analysts suggest that national REDD programs could provide facilities for cost-effective carbon stock accounting, develop approved methodologies, mobilize private funding, and generally create an enabling environment for subnational REDD projects (see Streck et al. 2010, 15), which could be funded through a centralized mechanism to control local drivers of deforestation. In the Khasi Hills project experience, the absence of direct government involvement in the project has helped reduce potential delays from bureaucratic approvals and reporting requirements. In this case, the partnerships between communities, indigenous governments, NGOs and the private sector has worked effectively, though cash flow issues were a problem while the project awaited carbon sales to take place. Further national government restrictions on NGOs' ability to receive foreign funds has also created problems for the project, despite strong support from the state government.

Role of National and Global REDD Institutions

The establishment of an international fund to provide start-up financing for project design and early implementation is also a possibility. A community forest carbon mutual fund could provide initial funding to projects within their respective portfolio. This could also reduce risk by creating a group of diverse projects reflecting varied approaches and locations. In Indonesia, CFI is currently assisting a consortium of NGO project developers to establish a project development pipeline, certified by Plan Vivo, that could potentially sell credits to the national Ministry of Environment and Forests. This would create the potential for a national portfolio of community REDD+ and PES projects

Reliance on a top-down, national bureaucratic model for REDD management, funded by multilateral and bilateral organizations alone, may not facilitate local initiatives, nor would an exclusive reliance on private markets. For REDD project coverage to expand from the bottom up, project development processes may require a number of components, including responsive national certification systems, possibly managed by civil society organizations that understand local socio-cultural and environmental conditions. Also indicated may be sustainable financing to encourage project developers to build local capacities. National REDD institutions will need to play a supportive role to enable the emergence of thousands of local projects and to release the latent human and ecological capital already present in many countries. Creating catalytic and synergistic linkages between global REDD policies and programs, national forest sector reform strategies, and grassroots community movements will likely be key for a successful and truly sustainable REDD effort.

Conclusions

Bringing the Khasi Hills Community REDD+ project to scale by uniting the communities within a federation was key to creating a successful project. Reliance on local, dynamic leaders, together with the efforts of CFI and other partners to build their technical capacities to clear the certification and marketing hurdles, established a local community-based management team that is gradually reducing dependence on external actors. By working with the Plan Vivo system, the project identified a community-oriented set of project standards that reflected the goals and capacities of the Khasi communities. Buyers have been excited by the opportunity to invest in a multi-benefit project that generates carbon offsets, as well as a range of other environmental and socio-economic values. For that reason, carbon prices for high quality community projects with good stories have remained at $5 to $10 per tCO_2, while those from industrial REDD+ projects have plummeted during the 2013–2015 period.

Perhaps equally important, the process of going through the project development cycle has united the Khasi communities in an effort to collectively address both environmental and development issues, empowering their traditional institutions in the process. The project helped build new skills in planning, monitoring and reporting systems, as well as injecting financial resources over which the community has direct control, unlike government schemes where decision-making and management is largely in the hands of departmental staff.

As the global and national rules governing REDD are developed, it is important that emerging field experiences, like those from the Khasi Hills, inform the thinking of policy-makers. If REDD+ is to play a substantial role in reducing GHG emissions from forests, the policy environment must be designed to enable the meaningful involvement of forest-dependent communities. Early REDD+ project experiences can identify constraints and opportunities that can guide the formulation of an enabling environment for project development and widespread replication.

The historic transition in forest governance systems in Asia that has been characterized by the emergence of national community forestry legal frameworks and laws in India, Nepal, Vietnam, Thailand, Indonesia and the Philippines can now be linked to global climate change initiatives empowered under the UN Framework Convention on Climate Change (UNFCCC). Unfortunately, the potential synergy generated by linking community forestry activities to REDD+ projects has not been achieved in the Asia region. This has in part been due to the UNFCCC emphasis on supporting an unending series of international and national REDD+ policy dialogues, rather than creating an enabling environment for local community-based forest carbon projects through the provision of technical and financial support, as well as encouraging the development of a market specifically for community forest carbon. This has slowed the development of community-oriented project standards, MRV systems and certification mechanisms, as well as an efficient carbon market for environmental services provided by forest communities.

The author concludes that to enable subnational, community REDD+ projects to grow in number and impact, REDD project design and certification processes must be simplified with more accessible sources of flexible financing. Some of the emerging international carbon standards have set the measurement bar so high that only project design teams with access to sophisticated carbon measurement and modeling methodologies and data can hope to see their projects certified. In contrast, the Khasi Hills project adopted the community-friendly Plan Vivo standards and placed the onus of design-making decisions on the community, though even that process is requiring sophisticated technical specifications and detailed project design documentation that is reviewed by a technical committee and field validated.

There is an urgent need for additional learning regarding operational processes for developing community REDD+ and other payment for environmental services projects. With more proof of concept examples like that from the Khasi

Hills, policy-makers, planners and private sector investors will be both better informed regarding the keys to creating an enabling project environment, and encouraged to build a global REDD+ strategy from the bottom up.

References

Chhatre, Ashwini and Arun Agrawal. 2009. Trade-offs and synergies between carbon storage and livelihood benefits from forest commons. *Proceedings of the National Academy of Sciences* 106, 42:17667–17670.

Community Forestry International (CFI). 2011. *Management Plan for the Khasi Hills Community REDD+ Project: Project Design Document for The Khasi Hills Community REDD+ Project*. Antioch, CA: CFI.

Community Forestry International (CFI). 2013. *Project Design Document: Khasi Hills Community REDD+ Project*. Antioch, CA: CFI.

Streck, Charlotte, Robert O'Sullivan, Toby Janson-Smith and Richard G. Tarasofsky. 2010. *Climate Change and Forests, Emerging Policy and Market Opportunities*. Washington, DC: Climate Focus.

17
CLIMATE MITIGATION BASED IN ADAPTATION

El Salvador's Restoration of Mangrove Ecosystems, 2011–2013

Fiona Wilmot

Political ecologists have suggested that carbon forestry, as a payment for environmental services (PES) approach to environmental governance, deepens socioeconomic inequalities in rural communities rather than reducing vulnerability, and that the burden of implementation falls disproportionately on local actors. In some instances it restricts access to communal resources, thereby limiting livelihoods, and in others leads to outright dispossession of land and forest resources through state privatization, or involves coercive reforestation schemes (Beymer-Farris 2013; Lansing 2012; Shapiro-Garza 2013). Institutional scholars have expressed concern about recentralization of forest governance through REDD+ and have framed decentralization, often associated with structural adjustment policies in developing countries, in a positive light as a desirable increase in rights and responsibilities for local communities (Phelps, Webb, and Agrawal 2010).

Yet governance is not manifested uniformly in all places, and political cycles are shorter than the time frame of most carbon forestry schemes (Friess 2013). Changes in government can bring about significant changes in relationships between state and non-state actors, affecting the tension between government and society (Hajer and Versteeg 2005). Further, electorates in a growing number of Latin American states have brought nominally left-leaning administrations to power, notably Venezuela (2000), Brazil (2002), Argentina (2003), Bolivia (2005), Chile (2006), Ecuador (2006) and El Salvador (2009), with a mixture of policies aimed at addressing the social concerns generated by earlier neoliberal regimes (Yates and Bakker 2014).

The new administration of President Salvador Sánchez Cerén in El Salvador (who took office in June 2014), in a rhetorically similar vein to the early Morales and Correa administrations in Bolivia and Ecuador, has promoted the concept of well-being (*buen vivir*), based on principles of human coexistence with nature rooted in indigenous philosophies, in tandem with democratic pluriculturism (Sánchez Cerén 2013). While providing partisan continuity with the preceding administration of President Mauricio Funes discussed here, the Sánchez Cerén administration is an unknown quantity. However, it is likely to be even more committed to social justice in rural areas than its predecessor, so that the case study of a changing paradigm in environmental governance described in this chapter is unlikely to be overturned in the immediate future. The academic literature privileges case studies concerning PES established under neoliberal regimes (McAfee 2012), which now seems less representative of the full range of institutional and ideological frameworks emerging in Latin America (Yates and Bakker 2014). It is in need of refreshment.

This chapter examines the new environmental governance matrix that developed in 2011 in El Salvador under the first left-leaning administration of the Frente Farabundo Martí para la Liberación Nacional (FMLN) in that nation. It lays out the financial patchwork that underpinned the turnover to FMLN, and the networks of resource-reliant communities, development practitioners, aid agency staffers and politicians that brought it into being. Serendipitously, a congruence between the programs of the newly engaged state and mangrove-dependent communities also aligned with international environmental policies through a mangrove restoration project during the UN International Year of the Forest in 2011.

This alignment coincided with my first visit to El Salvador as a PhD student to refine a topic on mangrove restoration. It was captivating. As a result, during 2011–2013, I analyzed mangrove restoration as a pilot carbon forest project for my doctoral research in El Salvador's Lower Lempa River floodplain (Wilmot 2014). I applied a mixed-methods approach of "key actor" interviews ($n = 30$), household surveys ($n = 39$), roundtable discussions ($n = 3$) and participant observation during workshops and meetings (> 15). Household surveys were conducted with participants involved in mangrove restoration from hamlets in the Lower Lempa. Key actors included international mangrove experts, staff from aid agencies and nongovernmental organizations (NGOs) working in El Salvador, Environment Ministry staffers, and Salvadoran science educators. Roundtable discussions were held with the key local NGO, international institutions, and community-based mangrove guards from the Lower Lempa. Mangrove restoration workshops were held in Mérida, Mexico, and the Lower Lempa, El Salvador, in 2011. Germane to this chapter, I attended two policy-setting meetings convened by Herman Rosa Chávez, the Environment Minister, in 2011, and attended climate change forums in 2012 and 2013 in San Salvador, hosted by the Minister. Findings indicate that strong local institutions and community need, combined with an informed and sympathetic Ministry of Environment, contributed in the short term to the realization of social equity goals promoted by the UNFCCC for REDD-Readiness, including reduced vulnerability to environmental hazards.

Background

El Salvador is the smallest mainland Latin American country, covering a territory of 21,040 km^2. Bounded by Guatemala and Honduras, it has a Pacific coastline of 300 km along the seismically active Cocos Plate. Since colonial times, human settlement and agricultural activities have occupied the narrow coastal plain and extended up the slopes of 22 volcanoes, some currently active, producing an ecological matrix of great diversity (Hecht, Morrison, and Padoch 2014; Perfecto, Vandermeer, and Wright 2009). Indigo, coffee, cotton, cattle and sugar have successively fueled the export-oriented economy (Williams 1986). After the civil war (1980–1992), migrant human labor (mostly to the United States) provided significant foreign revenue until the financial crisis of 2008 curtailed remittances from abroad. The substantial informal economy at home has more recently been infiltrated by the international drug cartels through the street gangs (*maras*), an unwelcome development that has transformed the daily lives of Salvadorians in innumerable ways, and warped government policy, business investment and international politics (Boyce 1996; Farah 2012; Lauria-Santiago 1999; Luciak 2001). During the civil war, right-wing governments adopted neoliberal policies, which were retained through three successive Alianza Republicana Nacionalista (ARENA) administrations until the left-leaning FMLN party (formed from a coalition of five guerrilla groups) came into office, for the first time, through the ballot box in 2009. Structural adjustment policies and membership in the Central American Free Trade Agreement (CAFTA) since 2004 left the incoming FMLN administration of President Mauricio Funes considerably reduced in capacity to confront the myriad economic, internal security and environmental challenges it had inherited (Seelke 2012).

The environmental challenges concern us here. One-third of the population of 6 million Salvadorans living in the country are subsistence, rain-fed maize farmers. Nearly 3 million urbanites live in the greater metropolitan area of the capital, San Salvador (Seelke 2012). Almost everyone is vulnerable to environmental hazards of some sort—whether living in informal settlements on unconsolidated volcanic debris in the metropolitan area, on deforested mountain slopes, along heavily polluted rivers or in the floodplains of the river mouths (Wisner 2001). All Salvadorans are affected by the recently erratic climate, with increasing occurrence and intensity of major rain events punctuating drought years, inflicting significant material losses. Tropical Depression 12E in October 2011 caused $800 million in damage, followed by a devastating drought that brought almost total ruin to the maize crop the following year (MARN 2012). The rural and urban poor, as everywhere, are more vulnerable to disasters than landowners and the urban elite.

This case study concerns communities of ex-combatants and refugees from both sides of the civil war, who resettled in the mangrove-rich area that became the Biosphere Reserve of Xiriualtique-Jiquilisco (henceforth the Reserve) in 2007, located in the flood plain of the Lower Lempa River (Figure 17.1).

Mangrove Restoration in El Salvador   **289**

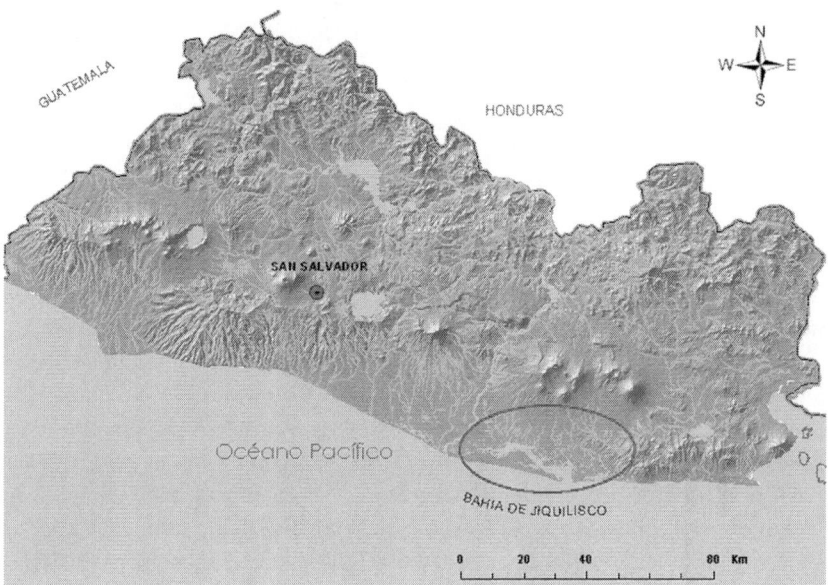

FIGURE 17.1   Project area, Lower Lempa River and Jiquilisco Bay, El Salvador.

Before the civil war, this historically sparsely populated area had been appropriated by cotton- and rice-producing companies, called *latifundias*, where laborers lived in designated and controlled areas. During the conflict, the *latifundias* ceased to function, and the violence emptied the countryside through outmigration to urban areas and the search for refuge abroad (Hecht and Saatchi 2007). The *latifundia* lands were later expropriated under the agrarian reforms of the early 1990s, and the former land owners turned their attention from agriculture to finance and business in the capital city (Hecht et al. 2014). Only a handful of communities persisted around the bay and along the coast during the war, joined intermittently by groups of FMLN fighters in training camps and guerrilla field hospitals seeking refuge in the mangroves. Under the land reform provisions of the 1992 peace accords, peasant cooperatives were permitted to purchase some of the *latifundia* lands, and returning exile communities were allowed to attach themselves to existing cooperatives (Reyes 2012). Infrastructure in the form of roads and drainage canals had decayed during the war years, and the newcomers, mostly originally from the highlands, had little preparation for living in the flood-prone lowlands. The government provided nothing in the Lower Lempa, even for ex-military families, to adapt to living in the mangroves (Dávila 2011).

The areas into which these settlers moved are dominated by mangroves, salt-tolerant, woody trees that thrive in intertidal zones of the humid tropics, flourishing particularly well in river deltas. Because they act as custodians of the carbon-rich, anoxic sediments beneath them, mangrove forests were introduced at the UNFCCC's COP 16 at Cancun, Mexico, in 2010 as carbon forests under the rubric "Blue Carbon" (Pendleton et al. 2012). The Environment Minister of El Salvador, Herman Rosa Chávez, a faithful attendee of the COPs since the FMLN government took elected office in 2009, took advantage of this development to incorporate El Salvador's mangroves into his regional climate change proposal for mitigation based in adaptation, a precursor to REDD-Readiness (PRISMA 2010, 2013).

The most extensive tract of Salvadoran mangroves occurs along the Lower Lempa River and around Jiquilisco Bay. Here the hasty and unplanned settlement of ex-military combatants by the ARENA government in the days leading to the signing of the peace accords in 1992, along with the cooperatives and groups of returned exiles already in the Lower Lempa, contributed to heavy resource exploitation of the area. Neoliberal, right-wing governments ignored the area after the peace accords of 1992, creating a political void that fostered a local environmental governance regime involving social movements, international NGOs and debt-for-nature financing from the United States through the Fund of the Initiative of the Americas, El Salvador (FIAES). The ARENA government materialized in the Lower Lempa to establish the Reserve in 2007, as part of its contribution to the Mesoamerican conservation corridor, without significant consultation with local actors (Díaz 2011). It left management of the Reserve to local organizations. Under the FMLN several years later, two environmental governance initiatives emerged from the national environment ministry (Ministerio del Ambiente y Recursos Naturales, or MARN) involving the mangroves of the Lower Lempa at the core of the Reserve and the people who lived in the buffer zone of the Reserve: the Program to Restore Ecosystems and Landscapes (PREP) on a watershed basis; and a policy for climate change mitigation based in adaptation (MARN 2012). Both PREP and the policy directed by MARN contribute to the goal of REDD-Readiness by fulfilling required performance measures involving stakeholder consultation and preparation, in addition to enhancement of forest carbon stocks in Salvadoran territory—the "+" of REDD+. Preparatory funding of €15 million was anticipated from the German Agency for International Cooperation (GIZ) after the REDD-Readiness proposal was submitted in May 2012 to the World Bank's Forest Carbon Partnership Facility (FCPF) (Quezada 2012).

Biophysical situation, colonial history, Cold War geopolitics, participation in the global economy and climatic extremes converged to shape the environmental governance regime and the bricolage of livelihood strategies that the members of the Lower Lempa communities elaborated for their survival. Unexpected interventions by new actors during the years 2011–2013 enabled the development of new governance arrangements in El Salvador. Let us consider first the actors and then their interventions.

Restoration Governance

Neither a carbon market nor funding streams and technology transfers through the mechanisms debated at the UNFCCC meetings, particularly COP Cancun in 2010, materialized in El Salvador through 2014. Instead, a small number of state and non-state actors have been key players in the regulatory, decision-making and knowledge-circulating networks in the political economy of ecosystem restoration for climate mitigation in El Salvador.

The Salvadoran state is represented at the UNFCCC by the Minister of the Environment and Natural Resources (MARN), a post held by Herman Rosa Chávez, an electrical engineer turned environmental advocate, during the FMLN administration of President Funes (2009–2014). Under Rosa Chávez's stewardship, MARN led intra-governmental initiatives to cope with climate mitigation and adaptation through the national Committee on Climate Change, and promoted intergovernmental cooperation on climate and conservation efforts at the regional level through SICA (Sistema de la Integración de Centro América) and CCAD (Comisión Centroamericana de Ambiente y Desarollo). Rosa Chávez also had ultimate responsibility for the nation's mangroves.

MARN was created by presidential decree in 1997, and came into effect with the first National Environmental Law (MARN 1998). Major agrarian reform and institutional reconstruction took place at that time, placing 35,500 ha of mangroves under the jurisdiction of the state, with restrictions on access and use, in addition to the establishment of national protected areas (NPAs). Under the right-wing ARENA government of 2007, MARN created the Biosphere Reserve of Xiriualtique-Jiquilisco in 2007 in the Lower Lempa. The Reserve buffer zone was mapped over the communities of ex-combatants and returned exiles in their mosaics of fields and pastures, which abut the extensive tracts of mangroves that form the core zone. It thus formalized its stewardship role of the Lower Lempa and Jiquilisco Bay. In the locally extinct language, Náhuat, Jiquilisco means "the place of the stars."

Non-state actors include the aid agency, FIAES; a local non-profit organization, Mangle; and an international NGO, EcoViva. A Salvadoran environmental think tank, PRISMA, is always present in any decision-making that involves climate change in the region, including initiatives for adaptation and mitigation. Rosa Chávez directed PRISMA for nearly two decades before his appointment as Minister of the Environment, and carried over his concerns for degraded ecosystems and rural poverty into office. A perennial concern of the organization has been ensuring that PES schemes, including REDD, are not turned into "instruments of social exclusion," but rather, through a participatory landscape perspective, build a shared societal vision of equitable use and stewardship of environmental services (Rosa et al. 2004). Although not directly involved in this particular mangrove restoration project, PRISMA actively supports landscape restoration for REDD-Readiness as a national goal (PRISMA 2010).

The Coordinadora del Bajo Lempa (henceforth the Coordinadora) is a social movement representing nearly 100 communities in the coastal plain of the Lempa River and around Jiquilisco Bay. A few visionary individuals organized the Coordinadora in 1996 to resolve flooding and infrastructure problems that became apparent at the end of formal hostilities, after peacemakers resettled ex-combatants and returning exiles in the flood plains of the Lempa. It became a legal entity after Hurricane Mitch devastated Central America in 1998. The founders of the Coordinadora had honed their community-level problem-solving skills while in exile in the rainforests of Panamá; they refined them in their new homes in the Lempa mangroves and scrublands. All communities are represented in the Coordinadora by elected committees, with a high degree of participation by women—a legacy of the wartime structure of the FMLN, which included women as full-time members, albeit mostly in supporting roles, such as radio operators, message-bearers, nurses and provisioners, and not as field combatants (Reyes 2012; Silber 2011; Wood 2003).

The Coordinadora, in response to neoliberal retrenchment policies, established its own extension agency, Asociación Mangle (henceforth Mangle).[1] Mangle has an elected board, which deliberates requests from local community groups, solves livelihood problems, reviews proposals and disburses funds. The board sets the agenda for a small, ingenious, self-sacrificing, permanent staff that works on potable water, food security, climate adaptation, housing, sanitation, disaster preparedness, youth education and mangrove management. As discussed in the next section, it also employs an accountant and keeps a tight rein on procurement and disbursements.

EcoViva, a US-based NGO, has given material, technical and moral support to Mangle for 18 years. Originally a faith-based organization that came to the region after the civil war, it has provided direct funding for projects, brought volunteer tourism into the local communities, and fostered alliances between Mangle and other NGOs, including the Mangrove Action Project (MAP), also based in the United States. EcoViva brought MAP to El Salvador in 2011 to lead a workshop on ecological mangrove restoration (EMR), a practice that has subsequently become part of the national program for ecosystem restoration (PREP).

Finance for Restoration

Funding sources for mangrove restoration are various. One source includes debt-for-nature swap agreements made between the governments of El Salvador and the United States at the conclusion of the war and administered through FIAES, the debt-for-nature aid organization funded in part by the US Agency for International Development (USAID). Staffed by Salvadorans, FIAES has strong institutional ties with Mangle and the communities of the Lower Lempa, which it has supported through many projects, including mangrove reforestation. Most of the FIAES restoration projects were hand-planted silviculture schemes that, for ecological reasons, have not resulted in significant success (Weller 2012). In the case of MARN, the Ministry has received funding from the World Bank, through the Global Environmental Facility (GEF), to systematize land tenure in the protected areas under the Program for the Administration of Conservation and Protected Areas (PACAP). PACAP also contributed toward the restoration at El Llorón (described later in this chapter).

An important source of financial support in the Lower Lempa, greater than migrant remittances, comes from non-governmental donors (Reyes 2012; Wisner 2001). EcoViva, mentioned earlier, is a reliable and dependable source of funding and technical and policy advice for conservation, sustainability and innovation in the Reserve. Financing for the training workshop on mangrove restoration arranged by EcoViva in the Lower Lempa during July 2011, and consequently the pilot restoration project at El Llorón, came from FIAES, World Bank GEF monies through MARN, EarthWatch, the Mangrove Action Project and smaller grants in-kind from other non-profit organizations. As mentioned earlier, financing for mangrove restoration has been a patchwork from various sources, not a single stream administered by the state, as the Minister for the Environment hoped for through REDD+.

Mangle is able to execute environmental work in the Lower Lempa for three important reasons. First, it has the legal capacity in El Salvador to undertake contracts with nationally based funding agencies, such as FIAES, as well as with USAID, the University of El Salvador, and the government, principally MARN and the Agriculture Ministry.[2] Second, it provides a centralized location to manage complex funding arrangements, as was the case for payments to local contractors for mangrove restoration. Through EcoViva, it has the institutional capacity to handle private donations and payments made to US banks for volunteer and educational tourism programs. Third, and most importantly, it has a reputation for scrupulous accounting and even-handedness in the region, as not only does the board review financial transactions in considerable detail, but Mangle employs an accountant and procurement officer and thus is able to handle finances in a legible manner for external scrutiny. Mangle is therefore in a unique position in the region to act as the financial conduit for future carbon forestry projects under the auspices of the UNFCCC (Quezada 2012). While the German government, through its overseas aid fund GIZ, pledged €15 million to support El Salvador's REDD-Readiness preparations, no funding from this source had been applied to ecosystem restoration in the study area by the time the research conducted for this chapter concluded in 2013.

REDD as a Source of Funding for Community-Based Restoration

The landscape between the Lempa River and around Jiquilisco Bay is a mosaic of sugarcane fields, cashew orchards, cattle pasture, woodlots, shrimp ponds, salt pans, Lilliputian maize and vegetable plots. All surround small hamlets, which are connected by roadways only passable seasonally. Land ownership is private in the buffer zone of the Reserve, whether by peasants or larger owners and cooperatives, and was systematically documented by PACAP, the World Bank GEF program.[3] Mangroves in the core zone of the Reserve belong to the state, which permits limited harvesting of resources, such as shrimp, crabs, fish, clams and dead wood by local residents, regulated by the local committees of the Coordinadora. Edging the creeks draining into larger channels and the bay are mangroves, mostly the red mangrove *Rhizophora mangle* and the taller *R. racemosa*, although on slightly higher ground the black mangrove *Avicennia germinans* provides burrowing habitat for the prized mangrove crab *Ucides occidentalis*. Before the Reserve was created, eight mangrove-dependent communities had joined forces to create a resource management plan called the Plan Local para Extracción Sostenible (PLES). PLES was intended to address overharvesting for local consumption, the poaching of crabs by non-participating communities, and illegal extraction of mangrove lumber and fuelwood by outsiders before establishment of the Reserve. In the words of one of the PLES leaders:

> I came up with the idea of promoting resource guards with some friends who were interested in the idea, to volunteer as guards. We began to patrol the zone, then we went to Mangle for training and the equipment you need in the mangroves, in the creeks, in the communities. From there we were capable of doing the work of resource guards. Seven years have passed since then. We don't have the same problems we had at the beginning. But we still mount guard at night, two people from the community, keeping watch.
>
> *(PLES interview 2012)*

Mangle did not shirk from providing the PLES communities with resources, or access to international NGOs with deeper pockets, but at the same time held views on where the responsibility should lie:

> We believe the government should, first of all, arrange conditions, management plans, lines of coordination, supervision, monitoring and concessions as well as administration. And the government should pay for the programs. It's political.
>
> *(Mangle interview 2011)*

Yet Salvadorans understand that, in the brutally honest words of an environmental governance review: "Many nation states no longer have the resources to manage their environments" (Lemos and Agrawal 2006, 302). One of the resource guards put it in his own words: "When they put together the national budget, the Ministry of the Environment is the one that gets the least resources. It's the system" (Guard interview 2012).

A spokesperson for FIAES elaborated in an interview on alternative sources for funding conservation and environmental management, a suggestion well in line with Rosa Chávez's publicly expressed views at the COPs:

> The case that concerns us today, which is the Bay of Jiquilisco, is a RAMSAR site, and a Biosphere Reserve. Because of that I believe that it has all the international qualifications to capture external financing to strengthen its conservation processes. Mangroves are at the heart of ecosystem restoration in our national territory.
>
> *(FIAES interview 2011)*

There was thus a locally perceived need for environmental management, a sense of the state's obligation, an understanding of government limitations, and a possible mechanism for addressing financial shortfalls. The Society for Ecological Restoration concurs with this argumentation: "REDD+ . . . has the potential to provide the developing nations with significant funding for forest restoration activities that contribute to climate change mitigation, sustainable management, and carbon-stock enhancement" (Alexander et al. 2011, 683).

Intervention Moments in the Place of the Stars

The stars aligned for the entry of carbon offsetting to the Reserve during 2011. Early in the year, the PLES communities put forward a request to Mangle and FIAES for technical, legal and financial help to restore a large area of degraded mangroves in order to bring them back to productivity. In San Salvador, on July 8, 2011, the Environment Minister Herman Rosa Chávez addressed a forum that EcoViva and FIAES had convened entitled "Mangrove Restoration: The Challenge for Adapting to Climate Change," preceding a weeklong workshop led by MAP on restoration grounded in hydrology and ecological principles, Ecological Mangrove Restoration (EMR). Rosa Chávez took the opportunity to link the national ecosystem restoration program, PREP, with climate change, through the UNFCCC's proposal for mitigation based in adaptation.

It was a happy coincidence that the minister was present. The day before, his assistant had scheduled him for a press photograph, planting mangrove seedlings in a swamp outside the city to celebrate the UN International Year of the Forest. He had agreed on one condition: "Only if it doesn't get my shoes dirty." Unable to meet his impossible demand, the resourceful assistant proffered a flyer for the forum. Rosa Chávez immediately seized the opportunity to make the moment his own. Less than 24 hours later, he inaugurated the forum and approved EMR as part of the new national PREP policy, with the Lower Lempa as the pilot project. Thus, he set in motion the opportunity for new governance arrangements, and the possibility of new landscapes—all in impeccable wingtips.

The Restoration of El Llorón—Mitigation Based in Adaptation

Nobody quite knows why 70 ha of mangroves died at the head of the creek known locally as El Llorón ("the one who weeps"), but most suggestions implicate extreme weather events, including droughts interspersed with heavy rains at the upper landward limit of mangrove habitat.[4] In July 2011, the area was covered in stagnant brackish water supporting a mass of belly-up fish, indicative of a sudden change in water chemistry. The stumps of dead trees broke the surface of this unappealing scene. The restoration workshop participants agreed that it was a perfect place to experiment with EMR, based on principles endorsed by the Society for Ecological Restoration (Alexander et al. 2011). With ministerial support, international funding, technical expertise and local labor, the request made by the PLES communities to restore mangroves could be met. An opportunity to test ecological restoration, with full community participation and as a mitigation strategy under the guidelines proposed for Blue Carbon at COP 16 in Cancun, had presented itself (Pendleton et al. 2012).

The work of restoring tidal flow, and thus enabling natural regeneration, was carried out by 30 of the poorest members of the community from a cluster of hamlets with about 600 households nearest to El Llorón. Hiring preference was given to female-headed households and day laborers with families to support. All but one of the workers persevered for the 2 months of grueling work of hand-clearing dead mangroves and digging out clogged ditches in the tropical heat. The rate offered by FIAES and MARN, paid through Mangle to a local subcontractor, was US$10 per person per day, above the local rate for a cane-cutter of US$6. Incentives went beyond the pay, as articulated by several women I surveyed, who stated that the prospect of enhanced livelihoods for future generations from clam, shrimp, crab and fuelwood harvesting from restored mangroves was as important to them as cash in hand. Other important factors were flood control and provision of ecosystem services—"one hectare of mangroves produces a lot of oxygen," commented one survey respondent.

Lawlor, Weinthal and Olander (2010) propose "citizen access to grievance mechanisms" as a necessary institutional condition for protecting rural livelihoods. My surveys offered respondents the opportunity to talk about potentially coercive work practices, but they never took advantage of it. Instead, most of the respondents said they would willingly participate in such projects again, citing the camaraderie of the work, the financial reward, and the experience of working for a project outside their immediate surroundings. Some made recommendations for practical improvements for future projects. Although the participants remained relatively poor, there was no indication that at any time they felt exploited or in danger of having their livelihoods curtailed by mangrove restoration work, as appears to be the case in the Rufiji Delta of Tanzania, in the REDD project reported by Beymer-Farris and Bassett (2012).

Some scholarly critiques of ecosystem service markets and the new carbon economy, in which REDD falls squarely, suggest that individuals and communities will end up being exploited by PES and carbon offset projects (Bumpus and Liverman 2011; McAfee 2012; Newell, Boykoff, and Boyd 2012; Shapiro-Garza 2013), but this did not seem to be the case in the El Llorón restoration and carbon project. The inhabitants of the Lower Lempa, through Mangle's interventions, had already absorbed the cost of creating the Reserve, and accommodated their bricolage of livelihood pursuits to the restrictions on extraction imposed by MARN (specifically, removal of whole trees) in the buffer zone of the Reserve in 2007. Asked if they felt that climate mitigation was a valid reason for ecosystem restoration in the Lower Lempa, even though they knew that their contribution to the global climate problem was minuscule, the response from the people I surveyed was almost uniformly positive (36 out of 39).

Conclusion

In 2011, El Salvador's national ecosystem restoration program (PREP), as well as local communities' need for legal recognition, financial support and technical advice for local mangrove management plans, coincided with the promotion of "Blue Carbon" (mangrove forests as carbon mitigation) at the 2010 UNFCCC COP 16. An association representing the mangrove communities (Mangle), a US-Salvadoran Foundation (FIAES), and an international NGO (EcoViva) assembled the restoration network that brought the tides back into the mangroves at El Llorón, thus promoting regeneration and enhancement of national forest carbon, a REDD+ goal. The Environment Minister provided the catalyst to re-engage the government in rural El Salvador when he quite opportunistically attached himself and his theoretical policy to a very concrete program being contemplated by non-state actors.

Having contributed little more than its legitimating authority to the program, the Salvadoran state then brought ecosystem restoration into its pro-REDD, climate finance appeal at the Doha COP of 2012. Environment Minister Rosa Chávez addressed the High Level Segment thus on December 6, 2012:

> We are not remaining passive. We know what to do and we are forging ahead with our limited resources, but we need to do much more if we are going to have a chance in advancing towards climate resilience and low carbon development. And that will only be possible if our developed country partners also take seriously, and without excuses and further delay, their commitments, particularly those regarding climate finance. The coming costs will simply become catastrophic for us all. We have to act and we have to act now.
>
> (Rosa Chávez 2012)

Neoliberal governments in El Salvador from 1989–2009 had not engaged with the needs of the rural populations of El Salvador, nor expressed any interest in rural affairs other than through national legislation, such as the poorly enforced National Environment Act of 1998. Non-state actors, such as the locally developed extension agency, Asociación Mangle, national aid agency FIAES and US-based NGO EcoViva, occupied the space left by the absence of government in the face of need. The personal and intellectual characteristics of the FMLN Minister of the Environment, Herman Rosa Chávez, between 2009 and 2014 expanded the agency of the state for one electoral cycle, so that stars aligned, social moments were appropriated, and landscapes evoked that might not otherwise have happened. Environmental governance reached beyond the self-organization and local interests of the Lower Lempa communities, exceeded the troubled US-El Salvador nexus, expanded eco-philanthropy and entered the wider world of the carbon fix.

To conclude, lessons learned from the Salvadoran example described in this chapter concur broadly with the institutional conditions and policies outlined by Lawlor et al. (2010) for REDD or other carbon forestry projects to succeed. First, the safeguards provided by Mangle were in place to avoid elite capture of funding by central government. Second, the FMLN government has clearly operational, pro-poor rural development policies, which occur under other left-leaning governments in Latin America that subscribe to the concept of *buen vivir*. Third, the people of the Lower Lempa, with their background in the FMLN guerrilla group, clearly sought outside assistance for addressing the social equity imbalances of female participation in paid work, resource access and climate vulnerability reduction. It was not a matter of institutions fulfilling an eligibility requirement for funding by paying lip service to "stakeholder participation."

Within El Salvador, mangrove communities to the east in Jiquilisco Bay of El Llorón have adopted the community-based EMR restoration method used by Mangle without carbon financing. Communities in the west, in the Jaltepeque Estuary complex, have approached Mangle for advice on mangrove management. The Mangrove Action Project that promotes EMR held a similar workshop in Honduras in September 2014, and has a long history of working in Thailand and southeast Asia. The community-based, ecological and contextually sensitive principles of EMR align well with the reform processes originating in social movements that are reconfiguring and democratizing neoliberal national regimes in Latin America.

Notes

1. Mangle (Spanish for "red mangrove") is named for the forests in the Lower Lempa that provided shelter to FMLN guerrillas during the war. The founder, Arístides Valencia, is a former guerrilla fighter who represents the Department of Usulután, covering the Lower Lempa and Jiquilisco, in the National Assembly. The second representative for Usulután, Estela Hernández, is a former director of Mangle's board, and the daughter of one of the current members, a colonel in the army during the war.

2. As a legally registered non-profit organization in El Salvador, Mangle may accept money for its projects from any source, which, according to the policy director of EcoViva, technically includes entities such as the World Bank, although it is unlikely for political reasons (Nathan Weller, personal communication, 2014).
3. Land tenure in the buffer zone of the Reserve is secure under the agrarian reforms, although the practice of securing loans for seed and fertilizer against land has bankrupted individuals and cooperatives in years of weather extremes, and thus allowed wealthier individuals to accumulate larger tracts. Land speculation has taken place by outsiders, against the day when the ocean front is opened up for mass tourism by infrastructure improvements.
4. One suggestion implicated local honey gatherers, who were rumored to have set fire to the trees during a dry period to facilitate honey collection and make wood available for construction, but according to the mangrove expert who led the restoration workshop in 2011, the trees were most likely dead already from non-human causes (Jim Enright, personal communication, 2011).

References

Alexander, Sasha, Cara R. Nelson, James Aronson, David Lamb, An Cliquet, Kevin L. Erwin, C. Max Finlayson, Rudolf S. de Groot, Jim A. Harris, Eric S. Higgs, Richard J. Hobbs, Roy R. Robin Lewis III, Dennis Martinez and Carolina Murcia. 2011. Opportunities and challenges for ecological restoration within REDD+. *Restoration Ecology* 19, 6:683–689.

Beymer-Farris, Betsy A. 2013. Producing biodiversity in Tanzania's forests? A combined political ecology and ecological resilience approach to "sustainably utilized landscapes." In *Land Change Science, Political Ecology, and Sustainability*, edited by Christian Brannstrom and Jacqueline M. Vadjunec, 84–106. London: Routledge.

Beymer-Farris, Betsy A. and Thomas J. Bassett. 2012. The REDD menace: Resurgent protectionism in Tanzania's mangrove forests. *Global Environmental Change* 22:332–341.

Boyce, James K. ed. 1996. *Economic Policy for Building Peace: The Lessons of El Salvador.* Boulder: Lynne Rienner.

Bumpus, Adam and Diana Liverman. 2011. Carbon colonialism? Offsets, greenhouse gas reductions, and sustainable development. In *Global Political Ecology*, Richard Peet, Paul Robbins and Michael J. Watts, eds. 203–224. London: Routledge.

Dávila, Medina and María Inés. 2011. *La Coordinadora del Bajo Lempa como Agente del Desarrollo Local Sostenible. Estudio de Caso.* Master's thesis, Faculty of General Studies, University of El Salvador.

Díaz, Oscar. 2011. *La evolución del rol territorial de la Bahía de Jiquilisco.* San Salvador, El Salvador: Fundación Programa Salvadoreño de Investigación para el Desarollo y Medio Ambiente (PRISMA).

Farah, Douglas. 2012. Central American gangs: Changes in nature and new partners. *Journal of International Affairs* 66:53–67.

Friess, Daniel. 2013. Tropical wetlands and REDD+: Three unique scientific challenges for policy. *International Journal of Rural Law and Policy*, 68–73.

Hajer, Maarten and Wytske Versteeg. 2005. Performing governance through networks. *European Political Science* 4:340–347.

Hecht, Susanna B., Kathleen D. Morrison and Christine Padoch. 2014. *The Social Lives of Forests: Past, Present and Future of Woodland Resurgence.* Chicago: University of Chicago Press.

Hecht, Susanna and Sassan Saatchi. 2007. Globalization and forest resurgence: Changes in forest cover in El Salvador. *BioScience* 57, 8:663–672.

Lansing, David M. 2012. Performing carbon's materiality: The production of carbon offsets and the framing of exchange. *Environment and Planning A* 44:204–220.

Lauria-Santiago, Aldo A. 1999. *An Agrarian Republic: Commercial Agriculture and the Politics of Peasant Communities in El Salvador, 1823–1914.* Pittsburgh: University of Pittsburgh Press.

Lawlor, Kathleen, Erika Weinthal and Lydia Olander. 2010. Institutions and poicies to protect rural livelihoods in REDD+ regimes. *Global Environmental Politics* 10, 4:1–11.

Lemos, Maria Carmen and Arun Agrawal. 2006. Environmental Governance. *Annual Review of Environment and Resources* 31:297–325.

Luciak, Ilja A. 2001. *After the Revolution: Gender and Democracy in El Salvador, Nicaragua and Guatemala.* Baltimore: Johns Hopkins University Press.

McAfee, Kathleen. 2012. The contradictory logic of global ecosystem services markets. *Development and Change* 43, 1:105–131.

Ministerio del Ambiente y Recursos Naturales (MARN). 1998. *Ley del Medio Ambiente.* San Salvador, El Salvador: MARN.

Ministerio del Ambiente y Recursos Naturales (MARN). 2012. Presidente Mauricio Funes presenta la Política Nacional del Medio Ambiente. Ministerio de Medio Ambiente y Recursos Naturales. http://www.marn.gob.sv (accessed June 5, 2012).

Newell, Peter, Maxwell Boykoff and Emily Boyd. 2012. *The New Carbon Economy: Constitution, Governance and Contestation.* Chichester, Sussex: Wiley-Blackwell.

Pendleton, Linwood, Daniel C. Donato, Brian C. Murray, Stephen Crooks, W. Aaron Jenkins, Samantha Sifleet, Christopher Craft, James W. Fourqurean, J. Boone Kauffman, Núria Màrba, Patrick Megonigal, Emily Pidgeon, Dorothee Herr, David Gordon and Alexis Baldera. 2012. Estimating global "Blue Carbon" emissions from conversion and degradation of vegetated coastal ecosystems. *PloS One* 7, 9:e43542, doi:10.1371/journal.pone.0043542.

Perfecto, Ivette, John Vandermeer and Angus Wright. 2009. *Nature's Matrix: Linking Agriculture, Conservation and Food Sovereignty.* London: Earthscan.

Phelps, Jacob, Edward L. Webb and Arun Agrawal. 2010. Does REDD+ threaten to recentralize forest governance? *Science* 328:312–313.

Programa Salvadoreño de Investigación para el Desarollo y Medio Ambiente. 2010. *Designing a REDD+ Program that Benefits Forestry Communities in Mesoamerica: Sythesis Report.* San Salvador, El Salvador: PRISMA.

Programa Salvadoreño de Investigación para el Desarollo y Medio Ambiente. 2013. *Mesoamerica at the Forefront of Community Forest Rights: Lessons for Making REDD Work.* San Salvador, El Salvador: PRISMA.

Quezada Díaz, Jorge Ernesto. 2012. *Readiness Preparation Proposal for El Salvador: Version 6 Working Draft. May 31, 2012.* San Salvador: MARN.

Reyes Granados, José Nohé. 2012. *Llenos de Vida.* San Salvador: Author.

Rosa, Herman, Deborah Barry, Susan Kandel and Leopoldo Dimas. 2004. Compensation for environmental services and rural communities: Lessons from the Americas. In *Political Economy Research Institute.* Amherst: University of Massachusetts.

Rosa Chávez, Herman. 2012. Doha Climate Conference: http://www.youtube.com (accessed January 2013).

Sánchez Cerén, Salvador. 2013. *El Buen Vivir en El Salvador: Construyendo un Nuevo Paradigma.* San Salvador, El Salvador: Impresos Los Planes.

Seelke, Clare Ribando. 2012. *El Salvador: Political and Economic Conditions and U.S. Relations.* Washington, DC: Congressional Research Service.

Shapiro-Garza, Elizabeth. 2013. Contesting the market-based nature of Mexico's national payments for ecosystem services program: Four sites of articulation and hybridization. *Geoforum* 46:5–15.

Silber, Irina Carlota. 2011. *Everyday Revolutionaries: Gender, Violence, and Disillusionment in Postwar El Salvador*. New Brunswick: Rutgers University Press.

Weller, Nathan D. 2012. *Restauración Ecológica de Manglares en el Cauce El Llorón de la Bahía de Jiquilisco: hace un manejo comunitario de los bosques de manglar*. San Salvador, El Salvador: Asociación Mangle, EcoViva.

Williams, Robert G. 1986. *Export Agriculture and the Crisis in Central America*. Chapel Hill, London: University of North Carolina Press.

Wilmot, Fiona C. 2014. *Making Mangroves: Ecologies of Mangrove Restoration, El Salvador, 2011–2013*. Unpublished PhD dissertation, Department of Geography, Texas A&M University.

Wisner, Ben. 2001. Risk and the neoliberal state: Why post-Mitch lessons didn't reduce El Salvador's earthquake losses. *Disasters* 25, 3:251–268.

Wood, Elisabeth Jean. 2003. *Insurgent Collective Action and Civil War in El Salvador*. Cambridge: Cambridge University Press.

Yates, Julian S. and Karen Bakker. 2014. Debating the "post-neoliberal turn" in Latin America. *Progress in Human Geography* 38, 1:62–90.

18
A CRITICAL REFLECTION ON SOCIAL EQUITY IN UGANDAN CARBON FORESTRY

Adrian Nel

Introduction

This chapter seeks to critically reflect on social equity in carbon forestry using the case of Uganda and a range of projects within it. It takes as a starting point the assertion that carbon forestry projects should not be evaluated in isolation, without reference to both the comparative experiences of the different carbon project types, or without an understanding of the social conditions of forestry governance in any particular country. In keeping with this sentiment, Roth and Dressler (2012) argue for empirical analysis that accounts for the particularities of 'place,' in this case Uganda, that shape market-oriented conservation in practice and expose the 'messiness' of such ventures.

The insights here draw from 6 months of fieldwork conducted in 2012 in Uganda toward a PhD in human geography. I identified and explored the ways that three REDD+ projects, three afforestation/reforestation Clean Development Mechanism (A/R CDM) projects, and three voluntary carbon market (VCM) forestry offset projects (or prospective projects) emerged as part of a new transnational 'assemblage' of forestry governance in Uganda. The focus of the research was more explicitly on governance changes, and thus the reflections here cannot substitute for longitudinal studies of individual projects. I took this approach for the reason that, while generalizations with regard to 'the state of carbon forestry' are challenging—as with much qualitative research (Bryman 2012)—they are well worth pursuing for the insights they provide.

The common thread identified among these projects is the provision of carbon funding for forestry-related activities, yet the projects have continuities and discontinuities with each other, not least in the ways they draw differing combinations of state and non-state actors into particular projects, and thereby into the broader environmental governance sphere in Uganda. Indeed, while REDD+ is the current buzzword in carbon forestry, it is, according to Poffenberger (2009), only an extension of earlier A/R-oriented initiatives under the Kyoto Protocol's CDM. In Uganda, older A/R CDM projects use afforestation/reforestation methodologies as components of industrial-scale activities, such as forestry plantations on leased Central Forest Reserve lands (CFRs, along with some national parks, constitute a form of 'protected forest estate' held by the national government). This is largely because 'natural' carbon from standing forests is excluded from the CDM. The projects thus provide supplementary carbon funding to companies leasing CFR lands.

In contrast, REDD projects tend to take place on private lands in Uganda, and are largely donor-funded at this stage. 'Private land' here is a term adopted in the new governance dispensation, first used in the National Forestry and Tree Planting Act of 2003 (though not applied to non-forestry land use in Uganda) to relate to forests on land of varying tenure types[1] outside the state-owned 'protected forest estate.'

The third category of projects is that of voluntary offsetting projects, which fund local tree planting initiatives involving small-scale farmers on so-called private land, or the reforestation of 'degraded' CFRs. These are administered by nongovernmental organizations (NGOs) or private business carbon providers who sell credits in the so-called voluntary carbon market (VCM). This project type in Uganda arose to tap into corporate 'greening' or corporate social responsibility (CSR) activities, through which they are predominantly funded. At times the distinctions between the project types blur, with for instance a number of REDD projects attempting to draw funds from the VCM market.

In the first section of this chapter, I briefly introduce the changing context of forestry governance and forest territory control in Uganda. I argue that the 'environmentality'[2] (Agrawal 2005) underpinning conservation policy and practice is increasingly turning toward market-based interventions (even though market-oriented conservation exhibits continuities with coercive conservation) to reconcile the growing conflicts between environmental conservation and rural livelihood needs. This background provides the context for second section of this chapter, where I describe the three categories of Ugandan projects themselves, establishing the contrasts and differences among them. In the third section, I focus on three key themes that emerged from the research: those of asymmetric benefits, expulsions, and false promises. In the concluding section, I reflect on what carbon forestry might mean for equity and climate mitigation in Uganda.

Context—From Territory to Flow

This section sets out the context for my comparison of carbon forestry projects in Uganda, focusing on the way power and social space are ordered in the Ugandan landscape. I look at the ways in which forestry actors and state functionaries have sequentially endeavored to territorialize and reterritorialize the exploitation and conservation of forest resources in the colonial and postcolonial period. By 'territorialization' I mean activities related to the establishment and defense of forest territories, particularly by the state (Vandergeest and Peluso 1995). Just as in Uganda as a whole, where the problems of violence, patronage and corruption cannot simply be reduced to moral deficiencies, problems in forestry governance stem from the social conditions of forestry where, despite regular reformulations and revisions, it remains unclear whether forestry policies, laws (and I would add territories) are acceptable to the local people and appropriate to the local situation (Turyahabwe and Banana 2008).

In Uganda, a shift in forestry governance has occurred through the application of market mechanisms and institutions to the governance of the environment; an arrangement administered through 'coalition' networks while still enrolling the state for its territorial powers (Sikor et al. 2013). It also includes the establishment of payment for environmental services (PES) initiatives, as well as the formation of voluntary certification schemes and standards, both for projects themselves and for wood products more broadly. The Forest Stewardship Council standard is the exemplar of the latter. According to Sikor et al. (2013), such changes in general forms of land-based governance are to a large extent reflected in the increasing shift from territorial governance to neoliberal, flow-based arrangements of particular resources or goods such as carbon.[3]

In Uganda, the shift takes shape more specifically in the changing emphasis, policy focus, and allocation of resources from territorial forms of governance (focused on individual forest territories and the maintenance of the 'forest estate') to flow-based governance realized in the control of biomass (including both timber and carbon) and of the populations of trees and people that interact with it, and in the facilitation of public-private partnerships in planting and carbon forestry (Nel forthcoming). In the forestry sector, as in others in the country, the use of foreign investment for development has been institutionalized and championed at both the national and international levels, and in both policies and processes. Plantation forestry and private sector investment have been identified as key to protecting natural forests, based on the assumption that if enough commercial timber is produced, then natural forests will not be degraded. This assumption is highly contested, however, and even the Director of National Forests at the National Forestry Authority (NFA) itself questions its validity. Nevertheless, the bias toward commercial forestry has been enshrined within the National Forestry Policy (2001) and the National Forestry and Tree Planting Act (2003)—both of which conform with Uganda's Poverty Eradication Action Plan (PEAP) of 2005.

There are strong social implications of such changes to consider: 'flow-based arrangements' can be dominated by powerful actors, and have the potential to generate new forms of social exclusion, political struggle, inequity and ecological simplification (Sikor et al. 2013, 2). Just as environmentalities are multiple, they can be dysfunctional and dissonant when they interact, with sovereign and neoliberal approaches, for instance, overlapping, coexisting or at times contesting each other.[4] As uncertainty exists over which actors sit at those scales, in particular at the powerful yet uncertain scale of the national, contestations and conflicts can be exacerbated (Li 2007), and can lead to displacement (or as Sassen 2013 puts it, expulsion), marginalization and 'false promises' toward residents (Heynen et al. 2007). Here 'global public goods' or 'services' relevant to the international community, such as carbon sequestration, are considered of greater value than the local livelihoods and the traditional ownership claims of local communities.

In Uganda, then, the focus on market environmentalism could be said to extend the *deterritorialization* of the protected forest estate in Uganda, with significant social and environmental implications (Nel forthcoming). A major problem concerns the maintenance of forest territories that are contested, an issue that is exacerbated when funding is shifted away from local governance capacity building to the facilitation of public-private partnerships, and to NGO activities and carbon forestry (Nel forthcoming). I turn to the projects themselves to show how this can happen in practice.

Projects Studied in Uganda

The typologies of the nine projects examined in this chapter are set out in Table 18.1.

TABLE 18.1 Project types

Acronym	Project Type	Brief Description of Case Study Projects
REDD+	Reducing the Effects of Deforestation and Forest Degradation, conservation and enhancement of forest carbon stocks, and sustainable management of forests.	Predominantly international NGO implemented conservation projects, with donor funds or finance from multilateral institutions, including the World Bank through its BioCarbon Fund, Norwegian Agency for Development Co-operation (NORAD), the African Development Bank, the Global Environment Facility (GEF), and the Congo Basin Forest Fund. The projects are not actively selling carbon credits, though they are in varying stages of 'readiness.'
A/R CDM	Afforestation/Reforestation Clean Development Mechanism	World Bank–funded projects, implemented by forestry companies seeking to reforest demarcated protected areas with exotic species. The A/R CDM projects are far less hesitant in their rollout, as carbon offsets are funded through the World Bank.
VCM	Voluntary Carbon Market	Private company carbon providers implement projects, using payments from voluntary offsets to work with private forest owners on reforestation.

Seven of the studied projects are in the west of the country, and two in the east. Six projects are in protected areas (PAs); the remaining three projects, located outside of PAs, all have relationships with the PA system (see Figure 18.1). One seeks to protect forest corridors between CFRs; another is located in 'buffer zones' of national parks and local forest reserves; and a third is in two community forests adjacent to a national park.

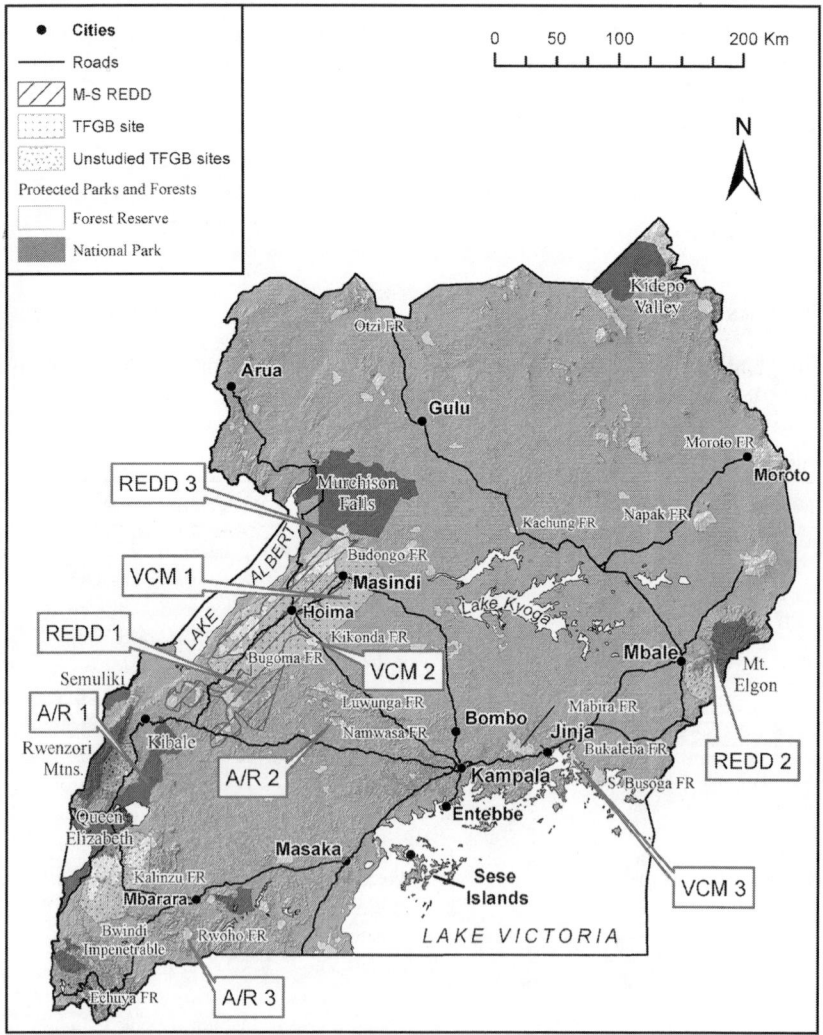

FIGURE 18.1 Uganda project map.

Source: Brice Gijsbertson, UKZN Geography.

TABLE 18.2 Project summaries

Code	Type	Project name	Methodology/ Standard	Land cover type/ tenure	Authority	Implementing Actors	Funding Sources other than offset sales
A/R 1		FACE Kibale reforestation	A/R CDM (Kibale)	Degraded national park	UWA	FACE the Future, Dutch electricity board, SGS	FACE Netherlands, Nedbank
A/R 2	CDM	New Forests Reforestation	A/R CDM	Degraded CFR	NFA	New Forests, NFA, HSBC, SPGS, SGS	New Forests, IFC of World Bank, HSBC, SPGS
A/R 3		Nile Basin Reforestaion	A/R CDM	Harvested plantation CFR	NFA	World Bank	World Bank, SPGS, CDM
R 1		NARCG REDD	REDD VCS	Rivernine forest corridors on private land	NFA, NEMA	Ngo consortium – WCS, JGI, CSWCT, WWF, Nature Harness, UNEP	Darwin initiative, GEF, WWF, WCS, UNEP, NORAD, Tullow Oil
R 2	REDD	MERECEP REDD	none	Degraded National Park	UWA	IUCN, LVBC, World Agroforestry Centre	NORAD
R 3		Ecotrust Community Forests	VCS	Tropical high forests, woodland	DLG	Ecotrust	VCM through My Climate
V1		Ecotrust TFGB Hoima Masindi	Plan Vivo, Rainforest Alliance	Degraded 'private land'	DLG	Ecotrust, ECCM, ICRAF	USAID, DFID
V2	VCM	Kikonda Reforestaiton	Carbon Fix	Degraded woodland CFR	NFA	Global Woods	SPGS, DANIDA, TUV SUD verified, CCBA
V3		Bukaleba Reforestaion	VCS (failed A/R CDM)	Degraded CFR	NFA	Green Resources, former tree farms.	SPGS, Noard

Equity Evaluation in the Context of Deterritorialization

Drawing from field research on the projects, this section describes three themes: (1) asymmetrical benefits, (2) expulsion and marginalization and (3) false promises, which relate to the question of equity in Ugandan carbon forestry.

Asymmetrical Benefits

While there are certainly benefits and costs that can accrue from carbon forestry implementation, interpreting benefits through an equity lens can show how these can be asymmetrical in the ways that they arise and be skewed against the rural poor.

Concerning the A/R CDM and voluntary projects in protected areas, particularly those involving forestry companies (Green Resources AS, New Forest Company and Global Woods), instances of asymmetrical benefits are somewhat easy to point out. Given a focus on private sector tree planting in forestry policy since the 2003 National Forestry and Tree Planting Act, as well as a government-mandated investment promotion as set out in PEAP (Lyons and Westoby 2014), private sector planters enjoy various privileges, including tax breaks, cheap land leases, grant support from the donor-funded Small Production Grant Scheme (SPGS), and at times, political support.[5] Though it is claimed these incentives are justified, given the narrow profit margins in the timber market, the benefits accruing to communities are meager, and these projects have significant social externalities for peoples living on or adjacent to the project sites (see the next section). In these projects, there are no formal revenue-sharing arrangements with communities, but companies do initiate 'sustainable development' projects for communities from the carbon revenues.

The Green Resources Bukaleba (VCM-3) project, for instance, aims to direct 10 percent of carbon revenues into local infrastructure, roads, schools and training (Green Resources 2010). However, communities complain of the long distances to the small primary school and clinic along poor roads, and do not have access to government services or NGO support (Interview, Bukaleba, June 2012), because of their designation by the National Forestry Authority and the company as encroachers on the contested Bukaleba CFR (the vast majority of CFR lands in Uganda are contested). 'Encroacher' is a term utilized by the NFA that relates to residents on de jure protected forest territories, even though they might have de facto access rights to the area.

Similarly, in the Forests Absorbing Carbon Emissions (FACE, now known as Face the Future) Kibaale (A/R-1) project, a 10 percent share of the carbon revenues received by the Uganda Wildlife Authority (UWA) is distributed to targeted parishes adjacent to the National Park (this is combined with 20 percent of the gate takings from the park as a contribution from UWA). However, the revenues are not delivered to communities in the many parishes and villages but to local district councils, which decide on benefit projects for whole area, not necessarily for the individual, marginal and underrepresented minority groups adjacent to the forest. While these initiatives may provide these limited development projects, employment in the plantation projects is left as the major local benefit. However, the jobs are limited in number (Nakakaawa 2011, 277), and often go to migrant labor hired through contractors, under poor working conditions, as at the Rwoho (A/R-3), Kikonda[6] (VCM-2) and Bukaleba[7] (VCM-3) projects (CDI 2012; Nakakaawa 2011; Nel and Hill 2013).

As McDermott et al. (2009) describes, preventing elite capture of benefits in project interventions requires pro-poor activities within participant groups, which is far more complex than merely insisting on equal participation, and such pro-poor activities or benefits are weakly developed in both protected area projects and those on so-called private or communal land among the study projects. In protected area projects, there have been stalled community planting activities at the Rwoho Nile Basin (A/R-3), and Kikonda (VCM-2) projects (see the next section). In projects outside of protected areas, there has been a tendency of failure to engage the rural poor. From observations and interview responses, market-based payments seem to favor small subsets of the relatively better off 'community members' on private lands, who hold suitable land, collateral and financial knowledge to be a safe bet for the projects to invest in.

This was the case with the Trees for Global Benefit (TFGB, VCM-1) project, where benefits were skewed to rural elites (Fischer 2011; et al. 2010), which in turn has led to possibilities for land concentration and conflict, as those who are able to do so secure more land (Carter 2009; Peskett et al. 2011). This asymmetry follows the trend—which Nakakaawa argues applies to the TFGB, GR Bukaleba (VCM-3) and Face the Future (A/R-1) projects, and which Peskett et al. (2011) identify in the Nile Basin reforestation project[8]—that less powerful actors tend to be marginalized, while carbon payments tend to reinforce pre-existing inequalities, alienating resource users that have traditionally accessed resources through local informal arrangements (Nakakaawa 2011, 2). Though the Mount Elgon Regional Ecosystem Conservation Programme (MERECP) project (REDD-2) shows some distinct local benefits (Mwayafu and Kimbowa 2011),[9] these are dependent on limited project funds, community initiatives that were already in place, good relations with UWA, sufficient collateral for loans and evidence of suitable financial management capacities—criteria that many communities are unable to meet and thus fall through the gaps.

Finally it is also clear that such asymmetries can undermine projects themselves. The "spectacular failure" of the Mount Elgon A/R CDM project (Cavanagh and Benjaminsen 2013) provides an example. After communities experienced little benefit from the project, they destroyed 4,000 of the 10,000 ha planted. Furthermore, following criticisms of unfair negotiations that characterized the projects, the carbon rights had to be returned to UWA, which is the legal landowner (Cavanagh and Benjaminsen 2013; Nakakaawa 2011). To conclude, then, while it seems the A/R CDM projects are more susceptible to asymmetric benefits, the REDD and VCM projects also evidence asymmetric tendencies.

Expulsion and Marginalization

In contrast to mere asymmetries of benefits, the findings discussed earlier also present instances of more pernicious negative effects, including marginalization, dispossession and expulsion from protected forest estates as well as projects outside of protected areas.

We also see that there are inherent risks and dangers in the neoliberal environmentality that seeks to enforce or promote private land tenure registration through carbon forestry. We may think there is nothing problematic with land titling on so-called private lands, but where unanswered questions relating to absentee landlords and long term tenants arise, even attempts to define property rights might be insufficient to secure community carbon rights. As it stands, carbon rights are tied to land ownership in Uganda, and in this regard, attempts at enclosing new carbon spaces and the effective conversion of communal land and use rights to individualized private control can mean dispossession of vulnerable communities and the accumulation of resources by elites (Mamdani 2013). In the TFGB (VCM-1) project, for instance, Nakakaawa chronicles how some landless people were displaced by the project, as powerful actors took advantage of contradictions in existing policies, laws and institutions to further their own interests and invest in the sequestration process at the expense of tenants or registered landowners (Nakakaawa 2011, 3). Similarly, concerning the M-S project (REDD-1), carbon finance and project implementation has brought absentee landlords into conflict with 'tenants' who have been there for years[10] (particularly in what are known as the lost counties of Kibaale).

The story is somewhat more complex on the CFR and national park territories that make up the protected forest estate, and can only be properly appreciated in the context of an ongoing deterritorialization, or weakening of protected areas as bounded entities (Nel forthcoming). Here we have seen the emergence of a hybrid form of neoliberal and sovereign environmentalities, both unable to provide the resources to bolster protected area territoriality, and it is the awkward alignments of the two, that take place differently in different places, that partially determine the equity outcomes of carbon forestry projects.

In one instance of this variegation, and in contrast to the disciplinary form of control through incarceration or compensation that would be expected in a purely sovereign environmentality (underpinned by conceptions of political citizenship), non-conforming individuals in the neoliberal model can be marked for expulsion as they are not seen as legitimate stakeholders. The violent eviction of long-term encroachers from the New Forests project (A/R-2) provides an example: an Oxfam and Uganda Land Alliance report in September 2011 controversially alleged that up to 22,500 people (the number is more likely 15,000) were evicted from their land and 'left destitute' in Kiboga and Mubende districts by the NFA, in what was characterized as a 'land grab'[11] (Grainger and Geary 2011).

However, the case studies show that the way marginalization and expulsions happen in practice is differentiated and mediated by local geography and power relationships in different places. In these 'zones of awkward engagement' (Tsing 2005), where interventions meet local contexts, expulsions can be both contested and accommodated. In contrast to the direct violence of the New Forests case (A/R-2), where there was political support for evictions, a form of expulsion that is not directly but rather structurally violent, and that could be termed 'slow violence' (Nixon 2011), can arise. Here, rural populations can be caught in the middle. In the Bukaleba project (VCM-3), evictions were not politically sanctioned, yet communities inside and immediately adjacent to the project area are made invisible while experiencing continued long-term marginalization.[12] Such cases of expulsion or marginalization are seen as the norm, made mundane and acceptable within the longer horizon over which the territoriality of the forest estate is reproduced across time and space.

In another example, grazing at the Rwoho Nile Basin project (A/R-3), which used to take place on the approximately 50 percent of deforested land now used for tree planting, has been criminalized. By its own admission, however, the NFA's efforts to police the plantations have not worked and have given rise to conflicts with communities that have protested against the 'denial of access' to forest resources by local communities (Nel and Sharife 2012).

Beyond the extremes of slow violence and direct expulsions, there are a number of ways in which marginalization can be resisted, avoided and accommodated. In the FACE projects at both Kibaale Central Forest Reserve (A/R-1) and Mount Elgon National Park, there were turbulent histories of evictions in the areas where the projects were initiated,[13] but only the former project has survived to this day, as the latter was undermined because of local resistance. At the Global Woods Kikonda project (VCM-2), management is changing from a self-styled militaristic approach to one of "learn between the lines of the law and context of the investment" (Global Woods Community Officer, Interview, Kikonda, October 2012). In essence, they are trying to facilitate plantation development by working with communities and within the already blurred lines between legality and illegality. In some cases this involves turning a blind eye to illegal activities of NFA rangers, who allow cropping or cattle grazing for a fee in the parts of the reserve to be cleared for tree planting, which can then lead to conflictive community relations.[14] Yet this could be said to support local corruption and further encroachment, with local communities' precarious positions no better off.

Finally there can also be, as in the case of MERECEP project (REDD-2), what Lewis (2009) would call a 'progressive space of neoliberalism,' where new neoliberal spaces engender positive outcomes. In this instance, although the project is very limited in scope, it nevertheless serves to mediate and mitigate conflict over the boundaries of the Mount Elgon National Park.[15]

False Promises

Here, I detail how disappointment at the false promises of carbon forestry arises, not only through scarce materialization of purported benefits, but also through the lack of the community 'participation' that carbon forestry ostensibly represents. Community development components of projects, such as a halted tree outgrower scheme at the Kikonda (VCM-3) project,[16] or a Collaborative Forest Management initiative in Rwoho[17] (A/R-3), have had very limited success, and led residents to express frustration and disappointment. At Rwoho, the executive of the community group Rwoho Environmental Conservation and Protection Association (RECPA) lamented the cooling of community interest after the project backslid because of both internal challenges on the part of the project, and a lack of planting capacity, funding, and benefits for the community (Interview, Baguma Ancelet, RECPA, Rwoho, November 2012).

In fact, the failure of project funds to materialize from donors and carbon markets can be excruciatingly frustrating for project implementers, who face a disparity between the ability of the project to deliver and the expectations generated by it.[18] This happened with the M-S project (REDD-1), where officials from the Jane Goodall Institute (JGI) and Wildlife Conservation Society (WCS) expressed their frustration and disappointment at the lack of progress. As the director of the JGI put it, "once you introduce REDD to communities, they see dollars immediately, which then don't come" (Interview, Entebbe, October 2012). Further, non-materialization of benefits can create tensions that would not otherwise have been there. In the Ecotrust Community Forest project (REDD-3), the lack of funds, as well as perceptions of unevenness in the distribution of benefits, are creating tensions in local villages.[19]

When disappointment surfaces, further complications can arise, as they have at Ecotrust's TFGB (VCM-1) sites, where not all farmers are allocated the same buyers and thus receive different carbon payments, which they declare are low.[20] Such conditions can undermine the project's implementation. It is evident from my interviews with community members that there is little understanding of what carbon is, and little direct information about the actual funding networks involved and the limitations and constraints experienced by projects. This leads communities to a simplistic, apolitical understanding of carbon forestry as a vehicle to get rich or to get paid. Anderson and Zerriffi (2010) thus question whether there is sufficient knowledge of the project on the part of beneficiaries to make informed decisions, which can exacerbate difficulties between participants at different levels.

Furthermore, even if the carbon payments do not cover the true opportunity costs of alternative land use,[21] as seems to be the case in the Ecotrust TFGB projects (VCM-1), people will continue to participate for as long as is convenient because of a dire need for cash and lack of alternatives (Nakakaawa 2011, 71). The implication seems to be that carbon forestry is used instrumentally by individuals and communities, as well as by local government. It is a strategy for short-term benefit in areas where NGOs are active, where 'facilitations' are requested from the intervention proponents, who communities know may only be there for a relatively short period of time. The problem with this, however, according to Anderson and Zerriffi (2010), is that it might foster an attitude of 'no pay, no care,' whereby insufficient benefits actually undermine environmental stewardship.

Finally, instead of the effective participation of communities in all aspects of the project and the safeguarding of community interests, I suggest that community engagement within carbon forestry is more about disciplining landowners to behave in ways that are consistent with capitalist notions of rational self-interest, and about unevenly disciplining and re-enforcing exclusionary arrangements on protected areas. In the protected area projects, there was only limited consultation and transparency,[22] and limited or non-existent involvement and participation of local resource users in project design and implementation. The only way for project-adjacent people to be taken note of is through the formation of Community-Based Organizations external to protected areas, or as rational, private forest owners conforming to a specific eco-governmentality, with relevant, recognizable, and implementable stakes in the forestry governance. In top-down interventions, however, the rural poor[23] and encroachers are excluded from such representation. There are two sets of practices and principles of participation instituted in Uganda, in accordance with UN-REDD prescriptions, that are meant to address and ensure substantive community involvement: free, prior, and informed consent (FPIC) on private lands, and social safeguards that should come into play on protected areas where expulsions are occurring. However, it is clear that in the complex context of expulsion and marginalization described in the previous section, these fall short of the mark.

Conclusion

In this critical reading of carbon forestry in Uganda, these interventions, underpinned by a logic of market environmentalism, seem to be falling short in their equity commitments. This appears to have as much to do with the social conditions of forestry in Uganda as it has with the faltering neoliberal environmentality and emergent instances of asymmetric benefits, marginalization and exclusion, and false promises that characterize the projects. Here, the notion of a dissonance (Erb 2012, 11) is a fitting characterization of carbon forestry in the context of deterritorialization—where intention and outcome in interventions conflict in the new, hybrid form of governance, and can exacerbate conflicts and environmental damage.

This view helps to contextualize the asymmetries of benefit, false promises, and expulsions (through both direct and slow violence) associated with carbon forestry in Uganda, with the expulsions occurring where neoliberal and sovereign environmentalities overlap on protected areas. At its worst, the geography of dissonance makes visible how it becomes necessary to trade off the well-being of poor, marginal, environmentally problematic communities and 'excess populations' on protected area territories (in coterminous spaces of sequestration and dispossession) against the generation of the kinds of resources and 'value' that make new forms of landscape management (as reflected in carbon forestry), if not possible, then potentially realizable (Sullivan 2008). Less perversely, carbon forestry can represent an asymmetrical form of benefit, where the accumulation of benefits among narrow segments of the population (see Matulis 2013 for a PES parallel in Costa Rica) through carbon forestry alters land access as part of the larger process of rendering surplus value and facilitating capitalist expansion (Robertson 2012, 397). Carbon forestry does not preclude progressive outcomes, which I certainly see as forthcoming in the prospective MERECP project (REDD-2), and there are certainly some projects that could be judged to be more or less equitable than others, with A/R CDM projects in the latter category (see also Bond et al. 2012). In the final analysis, I suggest that carbon forestry must be appreciated in the context of the contested nature of protected forest territories, which has been exacerbated by the shift toward market environmentalism (Nel forthcoming). Moreover, while dynamics are different on and off protected areas, it is clear that the problems in carbon forestry are related to the contested social context of forestry in Uganda, and any evaluation of equity in the countries' projects must be appraised with that in mind.

Notes

1. The four tenure types in Uganda include Mailo (a Buganda term derived from the square mile), freehold, leasehold and communal tenure. Based on 2012 interviews with forestry officials, 65% of Uganda forest stocks are on private and community lands, while 35% is in the permanent forest estate.
2. Relating to ecological rationalities and technologies of governance in relation to what is delineated as 'the environment' in specific contexts.
3. State-based territorial governance equates to a sovereign environmentality—manifesting in the 'fortress conservation approach' by the state with regard to protected areas—which is aimed at the governance of a territory through compelling subjects' obedience to a sovereign will, under direct threat of punishment (Foucault 2008, 312). Here, land regulation is the sole purview of central governance, while land use planning is administered by local governments. This form of governance is to an extent overlapped and superseded by the commodifying, flow-based neoliberal environmentality underpinning carbon forestry, where monetary incentives are offered to stimulate flows of carbon and induce conservation-friendly behavior (Fletcher 2010).
4. Erb posits a concept of dissonance to suggest that, because of variations in responsibilities and styles of governance across different scales, dissonance occurs, even if it is not intentional, creating uncertainty for communities about their rights over

different scales (local, national and global), and resulting in disjointed policy implementation (2012, 11).
5. Interviewees alleged that the New Forests Company (NFC), for instance, had the political support of the Museveni government, which was willing to support the removal of encroachers and enforce the agreement with investors that stipulated a uncontested lease of the CFR land. The NFC and the Uganda Tree Growers Association (UTGA) officials had met with the Presidential Round Table on Investment in 2009, and interviewees insist that it would have required connections of political patronage to achieve sanction of the evictions, given there was a (selective) presidential ban against evictions standing since 2006. Further, despite claims to non-involvement by the NFC in the 'government-led' evictions, an NFA official, conceded the 'facilitation' of the evictions, at least in part, by the NFC (Interview, September 2012).
6. Migrant contract workers from various parts of the country, including Arua, Mbale and Mityana (there were no local workers among the contract workers I met) highlighted that the contracts are temporary and with middlemen who employ the migrants for 200,000 Ugandan shillings per month (about US$78), depending on outputs (GW contract worker, Interview, Kikonda, October 2012). There have also allegedly been incidences of rape of local women by these workers and forest rangers (CDI 2012).
7. The company claims it employs 550 casual workers under contract (of whom 150 are women) and 63 permanent or semi-permanent staff (of whom 23 are women) (Nakakaawa 2011, 277). However, temporary employees cite low pay and poor working conditions. Employment of the latter category of workers has dwindled now that the bulk of clearing and planting has been done. In the worst case, a Nakalaanga village resident I interviewed cited low (and irregular) payments of 50,000 Ugandan shillings a month (US$23), unpaid over 10 months.
8. Peskett et al. (2011, 224) further suggest that the association's project membership is strongly biased toward wealthier community members (generally male) and professionals (allegedly even including Janet Museveni, the wife of the president), because of need to meet membership fees and annual subscriptions.
9. For Mwayafu and Kimbowa (2011), the project's strength is that it is different from merely giving the community money in the form of grants, payments or donations determined by project developers, which may end up being misused. Instead, benefits are given through a 'community revolving fund' in which all decisions about management and payback are made by the community. Now in its second loan cycle, the revolving fund has increased from 20 to 24 million Ugandan shillings (US$9,000) to support the purchase of livestock for the members (Mwayafu and Kimbowa 2011).
10. While the project stipulates it requires titles of customary tenure, the stance of the project seems to be that it will simply exclude cases where customary title is revealed to be contested. The potential for land conflict in the area is high, because of an influx of migrants due to tea and oil developments and land scarcity elsewhere. Historical contestations in Kibaale and in Hoima are prevalent, and a District Land Board official estimates that nascent land conflicts exist in at least 20% of the cases that pass his desk (Interview, Hoima, September 2012).
11. The New Forests Company, however, describes itself as a 'sustainable and socially responsible' company contributing jobs, revenue, carbon finance and the timber products the country needs as it develops, which they argue would otherwise be logged from natural forests (NFC 2011). In contrast, villager testimonies suggest that no consultations or resettlement plans were undertaken prior to eviction, and that people were violently forced from their land (Institute for Security Studies 2011). Oxfam's call for transparency and fairness was certainly justified, as Human Rights Network's (HURINET) Peter Magelah reported, with a large army and police

presence, a lack of community consultation, the use of violence, the near arrest of Hurinet officers and the arrogance and lack of compassion of New Forests' officers during the process (Interview, Kampala, August 2012).

12. The communities in and around the reserve consider their livelihood situations to be extremely difficult, with few opportunities outside agriculture, declining fish stocks and rising fuel costs (Lohmann 2006, 237–247). Lyons and Westoby note that the agricultural chemical regime of the plantation, followed as part of the Forest Management Plan, has killed both food crops and livestock, according to many villagers (2014, 1). Residents decry the destruction of their crops, planted where they are able; between tree stands, on vacant land, on hillsides above the tree line (considered unfit for tree planting), and below the infertile ridge tops, all of which are considered designated reserve land. Some interviewees in Lwanika and Budhala commented on the deteriorating conditions at the villages in comparison to 1999, when a report by Makerere University academics documented their livelihoods (UFRIC 1999).

13. The projects at Kibaale National Park and at Mount Elgon National Park include histories of human rights abuses and conflict, which the company was associated with but not solely responsible for. In 1992, 30,000 forest dwellers and peasant settlers were expelled without warning from a strip of land between the Kibaale Forest Reserve and the Queen Elizabeth National Park. They lost most of their livestock and belongings so that a wildlife corridor could be created between the reserve and the national park. The expulsions took place under the Kibaale Forest and Game Corridor Programme, part of the World Bank's Forestry Rehabilitation Project, which was co-financed by the European Community (Global Justice Ecology Project 2011). There is further history of local conflict related to land use, boundary disputes and ethnic violence (Dor et al. 2013). At the Mount Elgon site, similar evictions took place in 1993, in which there were no consultations or compensation; in 2002, UWA evicted 550 families from Mount Elgon and destroyed their houses and crops (Guide for Africa 2011).

14. In one example at Kikonda, an NFA patrolman and his deputy have allegedly been renting areas to cattle keepers and accepting payments from residents and encroachers to plant crops and graze cattle on the CFR land that constitutes these long-term expansion areas. This has caused further social conflict between cattle keepers and local agriculturalists farming in the reserve. One encroacher, who had been in the area 13 years, perceived that the official was employed by Global Woods and alleged he had been charged 50,000 Ugandan shillings (US$20) per acre for one season. He was never receipted and does not know where the money went, yet he has been arrested three times over the last 5 years for charges of harassing cattle (Nkulinza Vincant, Interview, Kikonda, October 2012).

15. The members of the community show pride in their achievements and attest that they now protect the park from encroachment by others, and work to sensitize the local community about its benefits.

16. From 2005 there was to be a collaboration with 300 community members, including both individual households and institutions such as the church and school, through a group called the Kikonda Community Forestry Association (KiCoFa). However, this fell away in late 2009 due to the combined effect of a lack of registered land titles, held by only 4% of the community (Eklof 2012), and the realization that planting in buffer zones might offer usage rights to communities in the buffer zone. The company claimed that the NFA was not willing to have the land sub-licensed to farmers (Peskett et al. 2011).

17. The project proponents also found the CDM excessively bureaucratic, and with faltering NFA capacity, poorly implemented CFM arrangements (EMPAFORM 2006) and the pullout of donor funding that had supported the NFA, the rate of planting at Rwoho was disappointing for all parties. While expected to plant at least 20% of the project area (or about 400 ha), only 70 of the 250 members of the Rwoho

Environmental Conservation and Protection Association (RECPA) joined the project (Institute for Security Studies (ISS) 2011, 70).
18. At one community meeting of the M-S project (REDD-1), for example, members of the Private Forest Owners Association (PFOA) that had been haphazardly developed for the project were asked to record their objectives. They emphatically stated "we want to be rich!" and aimed to establish eco-tourism, both of which are in reality highly unlikely (Project workshop observations, Hoima, August 2012).
19. At Ongo, for instance, it is clear there is no single community, but different groups and interests that do not always overlap. In an informal focus group in Wunini village, individuals attested that, whereas the community forest bylaws govern the whole community (which benefits from water services, firewood, poles and reeds), the Community Based Organization (CBO) executive was perceived to 'eat the money' without paying a percentage of the revenue from pole and timber harvesting to the community. They said that in 3 years they had not received benefits from the forest, and further claimed that parts of the forest were destroyed and cleared for agriculture in anger as a result (Interview, Ongo, October 2012).
20. The contribution of carbon payments to total income of farmers was estimated at only 12% by Nakakaawa (2011, 278). And according to Mwayafu (2016), the vast majority of people (93.8%) thought that improvements in project benefits should be made, with just under 50% calling for payment increases and alternative income generating activities, and 45% of non-participants stating they didn't join because of the low profitability of trees compared to other land uses.
21. For instance, 'marginal land' in the TFGB project could generate US$70 per ha per year if rented, generating higher returns in the short term than under the carbon project. If rented for a sugarcane outgrowers scheme, the potential returns are much higher, and many farmers in Masindi, in Bwijinga sub-county in particular, are converting to that practice.
22. With the exception of a few examples of a limited social baseline survey in the Kikonda project, and workshops in the Nile basin project.
23. Gilbert Wathum, of Unique Forestry in Uganda, argues that

> it is not for lack of desire for access or the potential for planting within community managed agreements but that its it is a problem of power—communities are wanting to plant, but the power gap is so wide . . . they might have a policy, but when it comes to private investors it is by will alone will that they are showing participation, and it not clearly defined, it is not mandatory in practice. (Gilbert Wathum speaking of Collaborative Forestry Management Arrangements, Kampala July 2012)

References

Agrawal, A., 2005. *Environmentality: Technologies of Government and the Making of Subjects*, Durham: Duke University Press.

Anderson, E. & Zerriffi, H., 2010. Can planting trees bring co-benefits? Smallholder tree planting for development and carbon mitigation. *Climatic Change*, 115(3–4), pp. 741–757.

Bond, P., Sharife, K., Allen, F., Amisi, B., Brunner, K., Castel-Branco, R., Dorsey, D., Gambirazzio, G., Hathaway, T., & Nel, A., 2012. *The CDM cannot deliver the money to Africa*, Durban: Ejlot.

Bryman, A., 2012. *Social Research Methods*, Oxford: Oxford University Press.

Carter, S., 2009. *Socio-Economic Benefits in Plan Vivo Projects: Trees for Global Benefits, Uganda*, Plan Vivo Foundation and ECOTRUST. http://planvivo.org.34spreview.

com/wp-content/uploads/TFGB_socio_economic_study_SarahCarter_2009.pdf (accessed March 17, 2013).
Cavanagh, J.C., & Benjaminsen, T. A., 2013. Virtual nature, violent accumulation: A critical political ecology of carbon market failure at Mt. Elgon, Uganda. *Geoforum*, 56, pp. 55–65.
CDI. 2012. *Jungle Fever*, Kampala: Climate and Development Initiatives.
Dor, T.F., Heskam, A.B., Madison, I.B., & Reichel, K.D., 2013. *Missing the Poorest for the Trees? REDD+ and the Links between Forestry, Resilience and Peacebuilding*, London: LSE and International Alert.
Eklof, G., 2012. *REDD Plus or REDD 'Light'?: Biodiversity, Communities and Forest Carbon Certification*, Stockholm: Swedish Society for Nature Conservation.
EMPAFORM, 2006. *Participatory Forest Management Initiatives in Uganda: Key Implementation Concerns and Recommendations for Policy Actions*, EMPAFORM Policy Briefing Paper No. 1. Kampala: EMPAFORM.
Erb, M., 2012. The dissonance of conservation: Environmentalities and the environmentalisms of the poor in Eastern Indonesia. *Raffles Bulletin of Zoology*, 25, pp. 11–23.
Fischer, J., 2011. *Payments for Ecosystem Services in Forests: Analysing Innovations, Policy Debates and Practical Implementation*. Doctor of Philosophy. School of International Development University of East Anglia.
Fletcher, R., 2010. Neoliberal environmentality: Towards a poststructuralist political ecology of the conservation debate. *Conservation and Society*, 8(3), p. 171.
Foucault, M., 2008. *The Birth of Biopolitics*, Basingstoke: Palgrave Macmillan.
Grainger, M. & Geary, K., 2011. *The New Forests Company and Its Uganda Plantations*, Washington, DC: Oxfam International.
Green Resources. 2010. *Bukaleba Forest Project Idea Note*. http://www.greenresources.no/Portals/0/Carbon/PIN%20Bukelaba_27_04_2010.pdf.
Heynen, N., McCarthy, J., Prudham, S., & Robbins, P., 2007. *Neoliberal Environments: False Promises and Unnatural Consequences*, Abingdon: Routledge.
Institute for Security Studies. 2011. *Carbon Trading in Africa-A Critical Review*, ISS Monography no. 184. The Hague: ISS.
Lewis, N., 2009. Progressive spaces of neoliberalism? *Asia Pacific Viewpoint*, 50(2), pp. 113–119.
Li, T.M., 2007. Practices of assemblage and community forest management. *Economy and Society*, 36(2), pp. 263–293.
Lohmann, L., 2006. Carbon trading. *Development Dialogue*, 48, pp. 31–218.
Lyons, K. & Westoby, P., 2014. Carbon colonialism and the new land grab: Plantation forestry in Uganda and its livelihood impacts. *Journal of Rural Studies*, 36, pp. 13–21.
Mamdani, M., 2013. *The Contemporary Ugandan Discourse on Customary Tenure: Some Theoretical Considerations*. Kampala: Makerere Institute of Social Research.
Matulis, B.S., 2013. The narrowing gap between vision and execution: Neoliberalization of PES in Costa Rica. *Geoforum*, 44, pp. 253–260.
McDermott, C., Cashore, B. & Kanowski, P., 2009. Setting the bar: An international comparison of public and private forest policy specifications and implications for explaining policy trends. *Journal of Integrative Environmental Sciences*, 6(3), pp. 217–237.
Mwayafu, Mujasi David. 2016. Masters thesis in Forestry. Contribution of the Trees for Global Benefits Project to Rural Communities in Masindi District, Uganda. Kampala: Makerere University College of Agricultural and Environmental Sciences.
Mwayafu, D. & Kimbowa, R., 2011. *Issues and Options for Benefit Sharing in REDD+ in East Africa: A Case Study of Mount Elgon Regional Conservation Programme*, Kampala: Uganda Coalition for Sustainable Development.

Nakakaawa, C. A., 2011. *Forest Carbon Sequestration: Contribution of the Private, Public and Civil Societies to Poverty Alleviation and Management of Forest Resources in Uganda*, Oslo: Norwegian University of Life Sciences (Norway).

Nakakaawa, C. A., Vedeld, P. O. & Aune, J. B., 2010. Spatial and temporal land use and carbon stock changes in Uganda: Implications for a future REDD strategy. *Mitigation and Adaptation Strategies for Global Change*, 16(1), pp. 25–62.

Nel, A., 2015. The neoliberalisation of forestry governance, market environmentalism and re-territorialisation in Uganda. *Third World Quarterly*, 36(12), pp. 2294–2315.

Nel, Adrian & Douglas Hill. 2013. Constructing walls of carbon–the complexities of community, carbon sequestration and protected areas in Uganda. *Journal of Contemporary African Studies*, 31(3), pp. 421–440.

Nel, A. & Sharife, K., 2012. East African trees and the green resource curse. In Bond, P. (ed.), *The CDM in Africa*, pp. 62–76. Durban, South Africa: Centre for Civil Society.

New Forests Company, 2011. New Forests Company: Response to Oxfam. http://www.newforests.net/index.php/responsibility/response-to-oxfam (accessed December 13, 2011).

Nixon, R., 2011. *Slow Violence*, Cambridge, MA: Harvard University Press.

Peskett, L., Schreckenberg, K. & Brown, J., 2011. Institutional approaches for carbon financing in the forest sector: Learning lessons for REDD+ from forest carbon projects in Uganda. *Environmental Science & Policy*, 14(2), pp. 216–229.

Poffenberger, M., 2009. Cambodia's forests and climate change: Mitigating drivers of deforestation. *Natural Resources Forum*, 33(4), pp. 285–296.

Robertson, M., 2012. Measurement and alienation: Making a world of ecosystem services. *Transactions of the Institute of British Geographers*, 37(3), pp. 386–401.

Roth, R. J. & Dressler, W., (2012). Market-oriented conservation governance: The particularities of place. *Geoforum*, 43(3), pp. 363–366.

Sassen, S., 2013. Land grabs today: Feeding the disassembling of national territory. *Globalizations*, 10(1), pp. 25–46.

Sikor, T., Auld, G., Bebbington, A. J., Benjaminsen, T. A., Gentry, B. S., Hunsberger, C., Izac, A.-M., Margulis, M. E., Plieninger, T., Schroeder, H. & Upton, C., 2013. Global land governance: From territory to flow? *Current Opinion in Environmental Sustainability*, 5(5), pp. 522–527.

Sullivan, S., 2008. Conservation and Food Sovereignty Workshop Presentation. http://www.youtube.com/watch?v=fuX0-KACTJg (accessed June 18, 2013).

Tsing, A. L., 2005. *Friction: An Ethnography of Global Connection*, Oxford: Princeton University Press.

Turyahabwe, N. & Banana, A., 2008. An overview of history and development of forest policy and legislation in Uganda. *International Forestry Review*, 10(4), pp. 641–656.

UFRIC, 1999. *A Site Report Prepared for Presentation to the Local People of Lwanika, Budhala and Forest Department Office, Iganga District*, Uganda: Uganda Forestry Resources and Institutions Centre (UFRIC) Research Note Number 3.

Vandergeest, P. & Peluso, N. L., 1995. Territorialization and state power in Thailand. *Theory and Society*, 24(3), pp. 385–426.

INDEX

Page references in bold indicate tables. Page references in italics refer to figures.

A/R *see* afforestation
accountability 5n1, 9, 14, 51–2, 82, 89, 95, 99, 101, 205, 264, 266n11
Acid Rain Treaty 5, **6**
additionality 7, 30, 51, 84n2
afforestation 192, 266n4; afforestation and reforestation (A/R) 5, 66, 78, 302–3, **305**
agro-forestry 18, 78, 153
AIDESEP *see* Inter-Ethnic Association for the Development of the Peruvian Amazon
Alliance of Small Island States 63–4
Amazon, Amazonia 16, 64, 220–35
Amerindians 14, 94–5, 100, 102
ANR *see* assisted natural regeneration
AOSIS *see* Alliance of Small Island States
Argentina 286
Asia 17, 39, 44, 67, 94, 140, 143, 255–7, 260, 273, 276, 278, 284, 298
assisted natural regeneration 260, 263, 266n10, 280–1
auditing, audit 14; limitations 91–103
Australia 63, 65, 155–7
avoided deforestation 5, 64, 66, 71n14, 76–8, 81–2, 157, 184, 204, 209, 213, 216n1, 220, 226–7, 230, 275

Bali *see* United Nations Framework Convention on Climate Change: COP-13 (Bali)
baseline data 7, 52, 66, 114, 191, 196, 204–5, 245, 258, 279, 317n22

BDS *see* benefit distribution systems
Belém Letter 3
benefit distribution systems 190, 194, 196; *see also* Reducing Emissions from Deforestation and Forest Degradation: benefit distribution
benefit(s): asymmetrical 18, 308–10, 313–14; benefit sharing 5, 12, 173, 188, 206, 208–9, 212, 214n5; capacity-building 7, 10, 65, 189, 225; and incentives for communities 5, 18, 76–8, 81, 99, 145, 152, 162–3, 165, 174–5, 178–9, 207–9, 225, 230–3, 235, 241, 255, 259, 261, 263–4, 266n11, 312–13, 317n19–20; multi-benefit 283; social and environmental benefits 48, 53–4, 66, 120, 141, 190, 234, 242, 248–9, 255, 281; *see also* co-benefits; livelihoods: benefits; Reducing Emissions from Deforestation and Forest Degradation: community benefits
biodiversity 15–16, 27–9, 38, 40, 47, 49, 66, 92, 138–41, 184, 203–6, 220, 228, 246
Biosphere Reserve of Xiriualtique-Jiquilisco (El Salvador) 288, 291
blue carbon *see* mangroves
Bolivia **6**, 65, 259–60, 286
Bonn Agreements 64
Brazil **6**, 16, 40, 42, 52, 54, 64, 92, 129, 171–2, 177, 220–35, 257, 259–60, 273, 286

322 Index

CAFTA *see* Central American Free Trade Agreement
California 15–16, 40, 42, 119–131, 171–80
California Air Resources Board 121
California Global Warming Solutions Act 119, 129, 171–80
Cambodia 207
Cancun *see* United Nations Framework Convention on Climate Change
Cancun Statement on REDD 187
cap-and-trade **6**; allocation 124–6, 129; in California 15, 42, 119–131, 171–2, 178–9; critiques 129; rationales 41–2
capitalism 13, 25, 28, 30–1, 37, 83, 91, 111, 117, 146, 212, 313–14; capitalist market 107, 146–7
CARB *see* California Air Resources Board
carbon accounting 5, 155, 188; critique xx, 14, 66–7, 70, 111–13; function of 107–108, 117; inclusion and exclusion 109; optimization of 113–15; subjective nature of 14–15, 107–17
carbon broker 76, 116, 157, 159, 277, 281; *see also* carbon investor
carbon conversion factor 109
carbon credits 2, 7–8, 10, 12, 17, 29, 39–40, 43, 47, 52, 55n3, 55n13, 77, 116, 157–8, 171, 209, 213, 221–2, 225–7, 231–4, 245, 276–7, **305**
carbon emission equivalents 109
carbon finance 76, 208, 310
carbon footprint 43, 107–8, 111, 113–14, 117, 140–1
carbon forestry xix–xxi, 3–4, 8, 12, 286, 293, 298, 302–14
carbon funding 8, 302–3
carbon investor 159 *see also* carbon broker
carbon market xix–xxi, 1, 5, **6**, 82–3, 206, 291; benefits 37–9, 42–3, 52, 143, 171, 231; critiques 9–10, 40, 46–9, 51–2, 142, 155, 161, 196, 255, 312; cultural construct 2, 9, 13, 108, 116–17; definition 2, 41–2, 107; environmental consequences 9–10, 50–2; impacts on greenhouse gas emissions 13, 38, 42, 49–50, 146–7; opportunity costs 47–8, 146, 163; rule-making 14–15, 119–30; *see also* carbon markets, compliance; markets; voluntary carbon market
carbon markets, compliance 9, 15, 41, 43; compliance mechanisms 179; costs 122, 130; institutions 121
carbon neutrality 92, 143–4, 156

carbon offsets xix, 5, 10, 14, 18, 39, 43–4, 74–84, 108, 115, 117, 140, 158, 173, 175, 225, 248, 250, 273, 276, 283, **305**; as a cost-containment mechanism 128; cost-effectiveness 3, 8, 184, 231; fungibility of 128
carbon prices xix, 10, 41–3, 121, 189, 283; carbon valuation 196; definition 38; low carbon prices 12, 130, 226, 259; volatility 122
carbon producer 16
carbon rights 10–12, 155–6, 209, 231–4, 255, 263, 309–10
carbon sequestration i, 45, 47–8, 50, 55n7–8, 61–3, 66, 70, 83, 142, 155–6, 220–235, 239, 242, 249–50, 264, 275, 305
carbon sink 1–4, 5, 13, 39, 47–8, 50, 60–70, 159; as a development opportunity 65
carbon standards 280, 284; data collection 113; Social Carbon 10; technical support 81; *see also* Climate, Community, and Biodiversity Standard; Gold Standard; Plan Vivo; Verified Carbon Standard
carbon stocks xix, **6**, 19n1, 74, 191, 221–3, 234, 242, 245, 249, 276, 290, **305**
carbon storage 3, 53, 207, 220, 230, 232, 239, 273
Caribbean 94, 203
CBC *see* community-based conservation
CBD *see* Convention on Biological Diversity
CBNRM *see* community-based natural resource management
CCBA *see* Climate, Community, and Biodiversity Alliance
CCBS *see* Climate, Community, and Biodiversity Standard
CDM *see* Clean Development Mechanism
Center for International Forestry Research **6**, 78, 97, 258
Central American Free Trade Agreement 288
Central Forest Reserve 303, 306, **307**, 308, 310–11
certification 92, 173, **176**, 205–7, 215, 225, 242, 259, 276, 278, 281, 283–4, 304; socially responsible 10; standards 10, 249, 280
CF *see* community forestry
CFI *see* Community Forestry International
CFR *see* Central Forest Reserve
Cherangany Hills (Kenya) 142

Chiapas (Mexico) 16, 40, 54, 129, 171–2, 177–9
Chile 286
China 52, 64, 256
CIFOR *see* Center for International Forestry Research
Cinta Larga 17, 221, 226–8, 230–1, 233
clan(s) 17, 161–3, 224, 276
Clean Development Mechanism xix, 5, 18, 60–70, 74–5, 78–80, 141, 302–3, **305**, **307**, 314; critiques 5, 13–14, 42–3, 51, 67, 84, 260, 308–10; precedents **6**, 14; rationale 37–8
climate change xix–xx, 1–3, 137–147; adaptation 75, 83, 142, 159, 239, 255, 275; greenhouse gases xix, 1, **6**, 37, 39, 49–50, 60, 62, 80, 115, 119, 122, 125, 129, 140–1, 175, 204, 220, 255; international climate negotiations 4, **6**, 7, 37, 171, 234, 255; mitigation xix–xx, 1, 4, **6**, 7–8, 13, 18, 29–30, 38, 40, 43, 45, 50, 61, 70, 75–6, 80, 83, 93–4, 138, 140, 142–3, 159, 171–2, 174, 180, 213, 220, 249–50, 254, 276–7, 282, 286–98, 303; North/South xix–xx, 2–3, **6**, 8, 11, 13, 38–9, 40, 42, 46–8, 50, 53, 63, 69, 95–6, 110–11, 115–16, 129, 154, 173–4, 188, 220–1, 255; resilience 77, 80, 83–4, 254, 261, 297; responsibility for emissions xx, 41, 67, 144, 174, **176**; risk 4; *see also* climate governance
climate governance 7, 38, 76, 146–7, 172; *see also* environmental governance
Climate, Community, and Biodiversity Alliance 157, 189, 215, 242, 247, 249, **307**
Climate, Community, and Biodiversity Standard 10, 206, 225, 247, 249; *see also* Gold Standard
CO$_2$e *see* carbon emission equivalents
Coalition for Rainforest Nations 152
co-benefits 3, 11–12, 74, 100, 184
Cochabamba Accord of 2010 3, **6**
COICA *see* Council of Indigenous Organizations of the Amazon Basin
Colombia 65, 213, 261
colonialism, colonial legacies 11, 15, 46, 64, 139, 151–4, 161, 240, 255–7, 260, 262, 288, 290, 304
co-management 240–2, 247
command-and-control 45, 139
commodification xx, 2, 9, 40

commoditization 203
community forestry 17–18, 53, 77, 191, 254–65, 273–4, 276, 284
Community Forestry International 274, 278, 280, 283
community participation 11, 273, 296 *see also* participation
community-based conservation 139, 255
community-based natural resource management 77, 241, 255
community-based organizations 313
complexity 27, 50, 67, 69, 75–6, 80, 114, **176**, 203, 205
Conference of the Parties *see* United Nations Framework Convention on Climate Change
conflicts of interest 51, 155
Congo Basin Fund 92
conservation xx, 1–2, **6**, 7–8, 14–16, 18, 27–8, 30–1, 38–40, 42–4, 46–54, 63–6, 77–8, 80, 83, 95, 110, 137–47, *152*, 156, 159, 161–5, 173–5, **176**, 184, 191–2, 204–8, 211–12, 214–15, 220–1, 223–5, 228–32, 234–5, 238, 241–2, 244, 246, 250, 255, 277–80, 290–1, 293, 295, 302–4, **305**
Convention on Biological Diversity 28, 33, 195
COONAPIP *see* Panama's National Coordinator of Indigenous Peoples
Coordinadora del Bajo Lempa (El Salvador) 292
COP *see* United Nations Framework Convention on Climate Change
corporate social responsibility 107, 110–11, 210, 276, 281, 303
corruption 51, 76, 94, 101, 152, 154–5, 203, 209, 304, 311
Corruption Perception Index 94
Costa Rica 7, 9, 15, 65, 75, 78, 137–47, 152, 221, 314; National Climate Change Strategy (NCCS) 142–3
Council of Indigenous Organizations of the Amazon Basin 76, **81**, 83
CPI *see* Corruption Perception Index
CSR *see* corporate social responsibility
customary land tenure *see* land tenure
customary landowners *see* land tenure

debt-for-nature 290, 292
decentralization **81**, 153, 190, 273, 286
deforestation xix, 1, **6**, 7, 10, 43, 48, 51, 54, 61–2, 64, 66, 70, 74, 76–8, 81–3,

324 Index

95, 139, 141–3, 152, 157, 164, 174, 184, 190–2, 194, 203–4, 206, 208–9, 213, 220–4, 226–7, 229–30, 254, 260–1, 274–6, 278–80, 282
degraded forests 7, 279–80
Democratic Republic of the Congo 211
deterritorialization 18, 305, 308–13; *see also* territorialization
disadvantaged communities 123, 178–9; *see also* marginalization
discourse 12, 27, 37, 40, 108–9, 124, 137, 141, 172, 174, 180, 213, 255
displacement 3, 12, 66, 79, 140, 142, 144, 305
dispossession 33, 40, 53–4, 83, 286, 310, 314
distributional equity *see* equity

eco-governmentality 313
ecological economics 46
ecological limits 46
economic efficiency 10, 44, 46–8, 52–3, 74, 120, 123, 211
economic externalities 15
economic growth 8, 26–7, 37, 40–1, 46, 107, 156, 174
Economics of Ecosystems and Biodiversity, The 28, 33, 38
ecosystem 3, 11, 28, 38, 43, 52, 66, 77, 159; metrics 30; restoration 276, 286–298
ecosystem services 5, 10, 16, 28–9, 37, 41, 45, 47, 50–2, 78, **81**, 83, 154, 159, 162, 208, 211, 222, 235, 296–7; *see also* Payment for Environmental Services
ecotourism 78, 138–40, 143, 277
EcoViva 291–3, 295, 297–8
Ecuador 25, 65, 286
EJAC *see* Environmental Justice Advisory Committee
El Llorón mangrove restoration project (El Salvador) 18, 293, 296–8
El Salvador 18, 286–298
emission reduction unit 92
emissions trading **6**, 84, 95–6, 119, 122; *see also* European Union Emissions Trading System
empowerment, empowering 174, 196, 262–3, 273–85
energy 41–3, 45, 50, 53, 64–5, 110–11, 113–4, 119, 126, 130, 138, 141–2, 144–5, 240, 276
environmental data 109, 111–15

environmental economics 13, 44
environmental enclosure 2, 70
environmental governance 1, 8, 115, 137–9, 179; and carbon 2, 4, **6**, 12–13, 15, 18, 38, 74–84, 298; centralized 3, 75, 294, 302; local 18, 75, 290, 292, 298; and markets 4, 18, 44–7, 119, 123, 125, 286–7; natural resource management 234; restoration 18, 291–2; women 292; *see also* carbon market; carbon offsets; climate governance; forest governance; governance; market environmentalism; market-based governance; markets
environmental impacts xx, 99, 138, 140, 172, 257, 279
environmentality 303, 310, 313, 314n3
environmental justice 15, 17, 83, 120–4, 128–9, 147, 173, **176**, 177–8, 239, 248; *see also* equity; justice
Environmental Justice Advisory Committee 123
environmental management 1–2, 109, 137–8, 172, 295
environmental narrative 16, 172–3, 175, 179–80
environmental policy xx, 1–2, 46, 55n2, 121
environmental racism 175
equality 14–15, 47–9, 67–8, 70, 74, 76, 94, 111, 138, 147, 163, **176**, 192; *see also* racial inequality
equity xxi, 15, 61–2, 66, 68–70, 74–84, 100, 119–30, 161, 163–5, 184–97, 208, 213, 220, 229–30, 232–5, 238–50, 309; contextual 75; distributional 75, 78, 80–2; procedural 75, 81, 180, 188; social 1–18, 171–80, 221, 287, 298, 302–14; spatial 125; *see also* justice
ERU *see* emission reduction unit
ethnography 107, 117, 138, 159, 239–40, 249, 287
EU *see* European Union
EU ETS *see* European Union Emissions Trading System
European Union 42, 63–4, 101, 125, 156, *see also* European Union Emissions Trading System
European Union Emissions Trading System xix, **6**, 55n13, 127, 130n2
exclusion 68, 82–3, 111, 153, 185, 194, 291, 304, 313
expulsion 18, 49, 303, 305, 308, 310–14

fairness *see* equity
fair treatment *see* equity
false promises 18, 303, 305, 308, 312–14
FCPF *see* Forest Carbon Partnership Facility
FIAES *see* Fund of the Initiative of the Americas
fictitious capital 27
financialization 2, 48, 122
Finland 191
FMLN *see* Frente Farabundo Martí para la Liberación Nacional
forest access 238
forest carbon: projects 3–4, 7–8, 11, 76, 155, 189, 223, 280, 284; sinks, definition of 60–1; *see also* carbon forestry
Forest Carbon Partnership Facility 39, 43, 96, 99, 154, 184, 187, 189, 214, 222, 263, 290; *see also* World Bank
forest-dependent people 3, 8–9
forest certification 92, 259
forest governance 1, 3, 18, 91, 98–9, 101, 188, 257, 261–2, 273, 277, 284, 286, 302, 303–4, 313
forest management xix–xx, **6**, 11–12, 16, 42–3, 52–4, 65, 74, 156–7, 184–5, 189–91, 196, 205, 255–6, 259, 261–4, 273–4, 278, 312
Forest Protection Committees 273
forest reserves 240, 306
forestry *see* carbon forestry; community forestry
Forest Stewardship Council 92, 205, 304
FPCs *see* Forest Protection Committees
FPIC *see* informed consent
Free, Prior, and Informed Consent *see* informed consent
Frente Farabundo Martí para la Liberación Nacional 287–92, 298
FSC *see* Forest Stewardship Council
FUNAI *see* National Indian Foundation
Fund of the Initiative of the Americas 290–3, 295–8

Germany 39, 97, 126, 151, 191
GFC *see* Guyana Forestry Commission
Ghana 188
GHGs *see* greenhouse gases
global North *see* climate change: North/South
Global Reporting Initiative 109
global South *see* climate change: North/South

Global Warming Solutions Act *see* California Global Warming Solutions Act
Gold Standard 108, 115–16, 247, 249, 280
governance 2–4, 5, 14–15, 17, 53, 96–8, 101–3, 122, 138, 180, 188, 190, 203, 212, 262–5, 277, 286, 290, 295, 303–4, 313; participation 3, 8, 11–12; *see also* climate governance; environmental governance; forest governance; market-based governance
governments 9, 13, 37, 39–43, 45–6, 49, 51–4, 60, 64, 69, 75–7, 79–80, 91–6, 98–103, 107–8, 120, 122, 138–9, 142, 151, 153–9, 161, 165, 171, 174, 188–9, 191, 195, 204–5, 209–11, 213–14, 220–4, 227–8, 231, 240–2, 244, 249, 255, 257, 259–61, 264, 273–80, 282, 284, 286–95, 297–8, 303, 308, 313; *see also* indigenous and rural peoples: governments
grassroots 277, 279, 283
green economy 13, 37–41, 44–6, 54, 120, 173
green grabbing 3, 12, 40, 49, 82, 146; *see also* land grabbing
greenhouse gases xi, 1–2, 5, **6**, 13, 18, 37, 39–45, 49–50, 54, 60–2, 65, 80, 109, 115, 119, 122–3, 125, 129, 140–1, 144, 171, 175, 179, 204, 220, 222, 225, 227, 231, 255, 284
greening capitalism 107
greenwashing 54
GRI *see* Global Reporting Initiative
GRIF *see* Guyana REDD+ Investment Fund
Guatemala 65, 288
Guyana Forestry Commission 93–4, 96–7, 101
Guyana REDD+ Investment Fund 93, 97, 99–102

Hague, The *see* United Nations Framework Convention on Climate Change
health 30, 123, 126–7, 130, 175, 178–9, 196
Honduras 79, 288, 298
human-animal conflict 238, 243–5
Hyderabad *see* United Nations Framework Convention on Climate Change
hydroelectric 15, 79, 138, 140–1, 144

ICDPs *see* Integrated Conservation and Development Projects
IFM *see* independent forest monitoring
independent forest monitoring 97, 101

Indigenous Amazonian REDD+ 83
indigenous and rural peoples 2, 4, 9, 12, 18, 32, 76, 94, 142, 152, 174, **176**, 177, 214, 222–3, 233, 264, 278, 280, 287–8, 298, 311, 313; communities 17, 94, 213, 221, 233, 256, 262, 274, 286; effects of CDM on 5, 14, 78–9, 84; effects of REDD+ on 3, 7, 17, 38, 40, 44, 48–9, 53–4, 80, 82, 158–9, 161, 178, 191, 220, 223–35, 249, 308–9; governments 17, 75–6, 82, 274, 276–80, 282; landscape management 12, 53, 223, 255–6, 258, 278–9; responses to REDD+ and forest carbon 3, 10, 40, 82–3, 142, 254; rights 3, 7–8, 11, 33, 82, 84, 98, 101, 144–5, 191, 211, 214, 223, 232, 257, 261–3; *see also* Reducing Emissions from Deforestation and Forest Degradation: indigenous REDD+
Indigenous Peoples Forum on Climate Change 64
Indonesia 40, 92, 188, 207, 211, 256, 273, 283–4
informed consent 3, 7, 11, 53, 82–3, 94, 100, 153, 155, 164, 189, 192, 196, 205–6, 210, 213–16, 234, 239, 265, 313
institutional reforms 14, 80
Integrated Conservation and Development Projects 14, 77, 139, 208
interests 2–3, 9, 11, 14, 38, 45, 50, 52, 68, 75–6, 137, 140, 142–4, 146, 152, 155, 160–1, 163–5, 180, 184, 209, 211, 223, 232, 234, 255, 261, 298, 310, 313
Inter-Ethnic Association for the Development of the Peruvian Amazon 82–3
Intergovernmental Panel on Climate Change **6**, 7, 60
Intergovernmental Platform on Biodiversity and Ecosystem Services 37–8
intermediary institutions 18, 259
international aid 231
International Emissions Trading Association 122
International Union for the Conservation of Nature 139, **307**
IPBES *see* Intergovernmental Platform on Biodiversity and Ecosystem Services
IPCC *see* Intergovernmental Panel on Climate Change
ISO 14001 109
IUCN *see* International Union for the Conservation of Nature

Japan 191
Japan International Cooperation Agency 165n4
JCN *see* Joint Concept Note
JICA *see* Japan International Cooperation Agency
Jiquilisco Bay (El Salvador) *289*, 290–3, 298
JNR *see* Reducing Emissions from Deforestation and Forest Degradation: jurisdictional
Joint Concept Note 92, 97–101, 103
Joint Implementation Mechanism 5
Jurisdictional and Nested REDD+ program *see* Reducing Emissions from Deforestation and Forest Degradation: jurisdictional
justice xxi, 1, 8–9, 11, 14–18, 69, 74, 78, 82–4, 120–4, 128–9, 147, 173, 175, **176**, 177–8, 209, 238–9, 248–9, 287; *see also* environmental justice; equity

Kayapo 228
Kenya 32, 142, 207, 210
Khasi Hills Project (India) 17–18, *275*, 276, 280–2, 284
Kyoto Protocol xix, 5, **6**, 13, 18, 37–8, 42, 49, 60, 63, 65, 67–8, 75, 92, 95, 141–2, 144, 171, 221, 303

labor 47, 53, 60, 110–11, 255–60, 263, 288–9, 296, 309
Lake District (England) 30
land grabbing 49, **81**; *see also* green grabbing
land rights 83, 163–4, 259; *see also* land tenure
landscape restoration 291
land tenure xx, 3, 7, 10–12, 15–17, 48, 53, 62, 75–6, 146, 153, 155, 191, 207–8, 210–11, 254–6, 260, 262–5, 277, 293, 299n3, 303, **307**, 310, 314n1, 315n10; common property 79; customary 12, 15, 151–65; forest estate 190, 303–5, 310–11; Incorporated Land Group (ILG) 153, 157–61, 165; kinship-based 151, 153; private 11, 191, 208, 240, 286, 303, **307**, 309–10, 313; state 10–12, 51, 94, 102, 153, 156, 185, 190–1, 208, 240–1, 244, 255–7, 263, 273, 291, 294, 303; *see also* land rights
Land Use, Land Use Change, and Forestry 60, 63

Latin America 39, 44, 46, 50, 53, 63, 65, 140, 191, 203, 225, 255–8, 286–8, 298
La Via Campesina 53
LCDS *see* Low Carbon Development Strategy
leakage 7, 51–2, 62, 79, 125, 127, 227, 276; *see also* carbon accounting
Lima *see* United Nations Framework Convention on Climate Change
livelihoods xxi, 3, 8, 10–12, 16, 18, 54, 75, 77, 102, 158, 179, 185, 187–91, 193–7, 207–9, 213, 232, 234, 238, 242, 244–5, 248–9, 273–5, 277, 286, 290, 292, 296–7, 303, 305; livelihood-positive benefits 16, 184
lobbying groups 121
logging 43, 48, 51, 54, 77–8, 94, 101–2, 146, 151–3, 156–7, 161, 163–4, 190, 194, 205, 208, 224, 228, 230, 240, 242, 259, 261, 280
logic of repair 29–30
Low Carbon Development Strategy 92, 96, 100
low-carbon economy 120
Lower Lempa region (El Salvador) 18, 287–93, 295, 297–8
low-income communities 175, 178–9
LULUCF *see* Land Use, Land Use Change, and Forestry

Malawi 17, 238–50
Mali 40
mangroves 5, 18, 66, 158–161, 286–98
marginalization 18, 83, 138, 147, 305, 308, 310–13; *see also* disadvantaged communities
market-based governance 2, 4, 8–15, 18, 38, 44, 54, 119; *see also* climate governance; environmental governance; forest governance; governance
market environmentalism 18, 33, 305, 313–14
markets 4, **6**, 7–8, 14–16, 27, 38–41, 44, 61–4, 66–7, 75, 81, 84, 92, 154, 174, **176**, 203, 224, 226, 231–2, 256–9, 265, 279, 281; in conservation and environmental policy xx, 1, 44–7, 50–2, 80, 107–8, 117, 146, 190, 221–2, 302–5; efficiency 44, 46–8, 52–3; equity 3, 8–12, 70, 82, 308–10; failures 8, 122, 50–2; logic 1, 46; market-based interventions 18, 303; mechanisms 37, 65, 76, 122, 147, 179, 185, 211, 304; *see also* carbon markets

Marrakesh *see* United Nations Framework Convention on Climate Change
Marrakesh Accords **6**, 64–5
Mesoamerica 290
meta-analyses 254, 258, 260–1, 265
Mexico 9, 16, 32, 40, 42, 46, 53–4, 78, 129, 171–2, 177, 257, 259, 262, 287, 290
Milan *see* United Nations Framework Convention on Climate Change
ministerial level 17
models 2, 5, 13, 26–31, 34, 38–9, 45, 54, 74, 92, 114, 120, 139, 142, 147, 156, 189, 191, 205, 211, 213, 222, 225, 241, 258, 265, 276, 280, 283–4, 310
monetary valuation 39, 44
monitoring, reporting, and verification 96–7, 156, 190, 238, 242, 245–8, 278, 284 *see also* accountability; additionality; leakage
Montreal Protocol 5, **6**
moral hazard 51
Mozambique 188
MRV *see* monitoring, reporting, and verification
Murik (Papua New Guinea) *152*, 159–61, 164

Namibia 30
National Indian Foundation 223, 226–7
Natura Carbon Neutral Initiative 226
natural resources 188, 193, 205, 211–12, 224; access xxi, 12, 48–9, 75, 79, **81**, 164, 238–9, 241, 243, 245, 247–9, 255, 261–4, 286, 291, 298, 309, 311, 314; rights 153, 261, 263, 308; use 65, 242
nature xx, 2, 5, 13, 37, 39, 44–5, 47, 52, 61, 76, 112, 122, 221, 286, 290, 292; for contemplation 26, 30–3; exchangeable 13, 25–6, 30–4; for speculation 26–33
neoclassical economics 26, 33, 44–5, 122–3
neoliberalism 1–2, 4, 5, 15, 18, 46, 60, 108, 173, 286–8, 290, 292, 298, 304, 310–11, 313–14; *see also* market-based governance
Nepal 95, 273, 276, 284
Ngäbe 79–80
NGOs *see* non-governmental organizations
Nicaragua 65, 75
non-governmental organizations 17, 28, 32, 40, 42–3, 51–4, 62, 64–5, 67, 75–6, 95, 110–11, 115, 120, 140, 143, 155–6, 162, 190, 195, 204, 214–15, 221–3, 225, 228–31, 240–2, 249, 254, 259–61, 264,

276–7, 279–83, 287, 290–2, 294, 297–8, 303, 305, **307**, 308, 313
non-timber forest products 240–1, 257, 259, 280–1
NORAD *see* Norwegian Agency for Development Cooperation
No REDD in Africa Network 142, **176**, 210
Norway 14, 39, 63, 65, 91–103, 143, 191, 222
Norwegian Agency for Development Cooperation 92–3, 95, **305**, **307**
NRAN *see* No REDD in Africa Network
NTFPs *see* non-timber forest products

OECD *see* Organisation for Economic Co-operation and Development
off-site mitigation 30
Ontario (Canada) 171
Organisation for Economic Co-operation and Development 108–9

PA *see* protected areas
Paiter-Suruí *see* Suruí
Panama 75–6, 79, 80, 82, 261, 292
Panama's National Coordinator of Indigenous Peoples 76, 82
Papua New Guinea 7, 15, 151–65, 221
Paris *see* United Nations Framework Convention on Climate Change
participation 5, 14, 17, 79, 82, 101, 145, 224, 228, 233, 249, 273, 276, 296, 298, 309, 312–13; in environmental narratives 172–3, 175–9; local 16, 188, 190–3, 207; multi-stakeholder 204; by women 292, 298; *see also* environmental governance; forest governance; governance; Reducing Emissions from Deforestation and Forest Degradation: participatory
participatory forest management 16, 185
participatory processes 16, 80–1, 153, 184–5, 187, 206, 215, 291, *see also* climate governance; environmental governance; forest governance; governance; market-based governance; participation
Payment for Environmental Services 9–10, 14–15, 18, 38–41, 44–6, 48–54, 74–5, 78, 142, 145, 147, 151–155, 158, 161, 212, 234, 258, 273, 276, 278, 281–7, 291, 297, 304, 314
pay to pollute 83, 120, 129
PDD *see* Project Design Document
performance-based contracting 211–12

performance measures 290
permanence 7, 51, 207
Peru 3, 64, 75, 82, 259, 262
perverse incentives 51
PES *see* Payment for Environmental Services
Philippines 32, 263, 273, 284
PIN *see* Project Identification Note
plantation forestry 141, 260, 304
Plan Vivo 10, 242, 276–7, 280–1, 283–4, **307**
PNG *see* Papua New Guinea
point source emitters 175, 179
political ecology 91, 138, 174
political expediency 120
polluter pays 124, 129
population growth 152, 279
postcolonial state 11, 151–4
poverty 8, 13, 40, 48, 50, 61, 77–8, 84, 94, 139, 145, 184, 193–4, 196, 216, 250, 291, 304
power relations 8, 47, 177, 180, 211, 249, 311
privatization 83, 124, 286
profit xx, 3, 9, 13, 25, 37–54, 67, 75, 95–6, 117, 129, 146–7, 161, 173, 204, 223, 225, 232–4, 239–41, 263, 281–2, 291, 293, 308
project design 16, 47, 51–2, 78, 158, 207, 230, 247, 261, 274, 276, 278, 280–5, 313
Project Design Document 276, 282, 284
project developers xxi, 5, 66, 79, 206, 208, 210, 274, 279–81, 283
Project Identification Note 276
project outcomes 17, 220–1, 229, 234, 239
project proponents 223, 232, 234
pro-poor projects 10, 48–9, 195–7
protected areas 12, 17, 28, 43, 139, 143, 190, 222–3, 226, 228, 238–47, 288, 290–1, 293–5, 297, 303, **305**, 306, **307**, 308–14; protected forest estate 303, 305, 310
protests *see* resistance

Quebec (Canada) 16, 42, 171

RA *see* Rainforest Alliance
racial inequality 74
rainforest **6**, 7, 32, 43, 83, 97–8, 102, 151–3, 157, 160, 222, 224, 292, **307**
Rainforest Alliance 97–101, **307**
rangelands 16
REDD *see* Reducing Emissions from Deforestation and Forest Degradation
REDD-Monitor 93, 96, 206

REDD+ *see* Reducing Emissions from Deforestation and Forest Degradation
Reducing Emissions from Deforestation and Forest Degradation 38, 74, 220, 242; architecture 17, 43, 77, 143, 184, 205; benefit distribution 3, 17, 39, 158, 163, 190, 194, 221, 229, 239, 241, 245, 250, 265; capacity building 7, 14, 80, 95, 189, 210, 225, 264, 277, 305; community-based 10, 15–18, 77, 162, 255, 263, 273–85, 293–4; community benefits 3–4, 5, 17, 38–40, 48, 53, 77, 161, 177–8, 184, 193–7, 221, 263, 296; and community forestry 17–18, 77, 191, 254–65, 273–4, 276, 284; conflict resolution *162*, 163, 165, 248, 264, 311–12; cost effectiveness 3, 184, 231, 280, 282; credits 1–2, 7–8, 12, 17, 40, 43, 49, 51–2, 77, 171–2, 175, 209, 223, 225, 232, 234, 275–7, 279, 283, **305**; critiques 8–10, 17, 40–1, 47–54, 80, **81**, 82, 91, 95–103, 174–5, 178, 185, 203–4, 206, 239, 248–9, 257, 259, 286, 297, 312–13; financing xix–xx, 2, 4, **6**, 8, 39–40, 43–4, 53–4, 78, 95, 185, 188, 190, 192–3, 220, 259, 277, 282–4, 293–4, 303, 310; governance 2–3, 7, 12, 14–15, 17, 44, 80, 91, 179, 188, 302; indigenous REDD+ 3, 82, 231; international policy, roles 7, 14, 39, 91, 103, 154, 174, 188, 204, 255, 283–4; jurisdictional 16, 52, 171–80, 214; national policy, roles 7, 8, 12, 15–16, 92–3, 154–5, 164, 190–1, 194, 204, 208–9, 273, 283; offsets 1–2, 7–8, 40, 129, 142, 171, 177, 180, 238; participatory 14, 16, 48, 165, 172, 175, 178–80, 185, 187–93, 195–7, 207, 209–10, 221, 223, 274, 291; pilot projects 8, 16, 40, 43, 48, 74, 77–8, 155–8, 184, 187, 191, 207–9, 223–35, 245, 274, 278, 295, **305**, **307**; policy guidance 177, 180, 215, 248–9, 260–5, 284–5, 298, 309; readiness 7, 14, 16, 75–7, 80, 82–3, 96, 142, 156, 184–5, 187–91, 193–4, 196, 204, 287, 290–1, 293, **305**; resistance 3–4, 40, 142; risk 4, 8, 11, 17, 192, 196, 203, 207, 210–12, 215, 231–3, 239, 248–50, 261, 264, 276, 279–83, 310; safeguards xx, 3, **6**, 7, 10–11, 16–17, 53, 82, 84, 94, 153, 158, 164, 174, 185, 187, 189, 191–2, 194–7, 203, 205–7, 210–12, 214–16, 233, 249, 265, 298, 313; social contracts 10, 16, 203, 208–16; social feasibility 203–9, 212, 215–16;

social soundness 17, 205–6, 254, 265; stakeholders 16, 40, 48, 54, 98–101, 165, 173, **176**, 177–8, 185, 192, 203–5, 207, 210–16, 220, 223, 230–5, 250, 262, 280, 290, 298, 310; state policy, roles 264; sustainability 10, 16, 38, 154, 204–7, 210, 212–15, 265, 293; timeline 5, **6**, 13, 44, 61, 70, 77–8, 92, 152, 154–6, 190–1, 205, 221–2, 284; *see also* indigenous and rural peoples: effects of REDD+ on; indigenous and rural peoples: responses to REDD+ and forest carbon
reforestation 66, **81**, 141, 145, 190, 224, 242, 255, 258, 260, 263, 277, 282, 286, 292, 303, **305**, 309; *see also* afforestation
regeneration 66, 141, 266n10, 279, 296–7; *see also* assisted natural regeneration
regulated entities/industries xix–xx, 38, 41–3, 74, 120, 122–9, 294
rent-seeking 51
reporting **6**, 7, 11, 61, 76, 82, 109, 111–13, 117, 187, 189, 195–6, 204, 279, 280, 282, 284; *see also* monitoring, reporting, and verification
resistance 4, 141, 144, 311; *see also* Reducing Emissions from Deforestation and Forest Degradation: resistance
resource shuffling 126–7
RIA *see* Indigenous Amazonian REDD+
rights: human 14, 53, 79–80, 84, 142, 144–5, 152, 174, 226, 257; negative 212; positive 213; *see also* carbon rights; informed consent; land rights
Rio+20 Earth Summit 40
rural and urban poor 9, 38, 48, 53, 288, 291, 308–9, 313
rural communities 5, 11, 53–4, 158–9, 256, 280, 286

SBSTA *see* Subsidiary Body on Scientific and Technical Advice
scarcity 13, 45, 46, 67, 70
scientific uncertainty 50, 67
security 75–6, **81**, 137, 142, 207, 210–11, 213, 255, 261, 288, 292
Sengwer 142, 210
Serengeti 31
Smithsonian Tropical Research Institute 80–1
social capital 261, 279–80
social safeguards *see* Reducing Emissions from Deforestation and Forest Degradation: safeguards

socio-economic performance indicators 276
South Asia 203
South Pacific 203
spiritual practices and spaces 2, 10, 238–9, 245–9, 274
standards *see* carbon standards; certification; forest certification
STRI *see* Smithsonian Tropical Research Institute
structural change 120, 188
subnational jurisdiction 16
Sub-Saharan Africa 203, 256
Subsidiary Body on Scientific and Technical Advice 189
subsidies 9, 42, 45, 51, 145, 259, 264
Suruí 16, 221, 223–6, 229–33
Suruí Carbon Fund 223
sustainable development 4, 13, 18, 38, 46, 61, 65, 92, 108–9, 140, 145, 174, 184, 223, 225, 308
swidden horticulture 152

Tanzania 78, 92, 188, 208, 296
technocratic solution 259
technology, investments in 37, 140
TEEB *see* The Economics of Ecosystems and Biodiversity
territorialization 304; *see also* deterritorialization
Thailand 284, 298
trade-offs 17, 46, 74, 232, 234
transaction costs 10, 44, 48, 67, 70, 210, 232, 281
transformative change 14, 80, 102
transnational market 5, 16
tree planting projects 5, 42, 190, 238, 242–5, 303–4, 308, 311

Uganda 18, 78, 302–14
UN-REDD Programme 222, 239
uncertainty *see* scientific uncertainty
UNDRIP *see* United Nations Declaration on the Rights of Indigenous Peoples
UNEP *see* United Nations Environment Programme
UNESCO *see* United Nations Educational, Scientific, and Cultural Organization
UNFCCC *see* United Nations Framework Convention on Climate Change
Union of Concerned Scientists 66, 96
United Nations Declaration on the Rights of Indigenous Peoples 11, 82, 84, 257, 261

United Nations Educational, Scientific, and Cultural Organization **6**
United Nations Environment Programme **6**, 38–9, 43–4, **307**
United Nations Framework Convention on Climate Change xix–xx, 1–3, 5, **6**, 7, 11, 13, 18, 37, 39, 41–2, 61–8, 70, 74, 76, 79, 84, 142, 154, 187–8, 190, 195–6, 204, 208, 284, 287, 291, 293, 295; COP-6 (The Hague) 63; COP-7 (Marrakesh) **6**, 65–6; COP-9 (Milan) 66; COP-11 (Hyderabad) **6**, 152; COP-13 (Bali) **6**, 154, 187, 208; COP-15 (Copenhagen) **6**, 225; COP-16 (Cancun) **6**, 7, 187, 189, 194–6, 221, 290–1, 296–7; COP-18 (Doha) **6**, 297; COP-20 (Lima) 3; COP-21 (Paris) xix, **6**; COP-22 (Marrakesh) **6**
United Nations Global Compact programme 109
United States 42, 52, 63, 65, 68, 93, 119–20, 125, 128, 171, 191, 241, 288, 290, 292
Uruguay 65
use value 27–9, 279

validation 17, 79–80, 98, 129, 205–6, 214, 238, 245, 247–9, 276, 281
valuation 39, 44, 68, 76, **81**, 155, 196
VCM *see* voluntary carbon market
VCS *see* Verified Carbon Standard
Venezuela 286
verifiable emissions reduction 212
verification 77, 79–80, 97–102, 205–6, 214–15, 245, 249; *see also* monitoring, reporting, and verification
Verified Carbon Standard 158, **176**, 225, 242, **307**
Vietnam 16, 78, 184–97, 284
village forest areas 242, 251n3
Viridor Waste Management Ltd. 227
voice, privilege of 16, 61, 115, 163, 172, 175, 177–80, 192
voluntary carbon market 5, 41, 43, 52, 108, 115–17, 208, 225, 229, 302–3, **305**, **307**, 308–13; accounting 107–8, 117, 188; offset (VCO) market 143, 238, 241; projects 209, 239, 308; standards 115, 117, 239
vulnerability 286–7, 298
vulnerable groups 9, 11, 15, 83, 177, 179–80, 195, 288, 310; *see also* disadvantaged communities; marginalization

WBCSD *see* World Business Council for Sustainable Development
WCI *see* Western Climate Initiative
well-being 101, 139, 205, 224, 250, 286, 314
Western Climate Initiative 125
WHO *see* World Health Organization
wildlife corridor 274
win-win solutions 3, 17
WMO *see* World Meteorological Organization
woodlots 242–4, 293
World Bank 28, 38–9, 43, 45–6, 48, 54, 76, 82, 95–6, 99, 142, 152, 154, 184, 187, 189, 204, 206, 214, 222, 224, 257, 259–60, 290, 293–4, **305**, **307**

World Business Council for Sustainable Development 108–9
World Conservation Congress 27–8
World Health Organization **6**
World Meteorological Organization **6**

Xingu 228–30
Xingu Indigenous Lands Ecosystem Services Project 17, 228–33
Xingu National Park 228
Xingu SocioEnvironmental Carbon Project 221, 228–33
XSEC *see* Xingu SocioEnvironmental Carbon Project